国家出版基金项目
NATIONAL PUBLICATION FOUNDATION

中国蜻蜓大图鉴

DRAGONFLIES AND DAMSELFLIES OF CHINA

— 张浩淼 著 —

HAOMIAO ZHANG

下册 Vol.2

重庆大学出版社

差翅亚目
SUBORDER ANISOPTERA

8 蜻科 Family Libellulidae

　　本科世界性分布，全球已知142属1000余种，是蜻蜓目最庞大的一个科。中国已经发现42属140余种。本科蜻蜓有些体型很小，如侏红小蜻，是全球最小型的蜻蜓之一，有些体型也很巨大，如彩虹蜻；但多数是中型身材。大部分种类体色非常艳丽，色彩丰富，体形千姿百态，是蜻蜓目中观赏性最高的一类。多数种类通过身体色彩即可以识别，少数较相似的种类可以通过肛附器、钩片及下生殖板构造来区分。

　　本科蜻蜓主要栖息于各种静水水域，在水草茂盛的湿地种类繁多。少数种类生活在溪流、河流等流水环境。雄性具有显著的领域行为，在晴朗的天气，通常停落在水面附近占据领地，有些种类具有长时间悬停飞行的本领。雌性不常见，仅在产卵时才会靠近水面。

This, the largest family of Odonata is distributed worldwide with over a thousand species in 142 genera. Over 140 species in 42 genera are currently known from China. Their body size is variable, some of them are among the smallest of anisopterous dragonflies, e. g. *Nannophya pygmaea*, others are large, e. g. *Zygonyx iris,* but most species are medium-sized. The majority of species possess brilliant colours and the shape of the body widely varies. They are among the most recognizable dragonflies of the order. Most species can be identified in the field by their unique color patterns, some species are similar and must be identified by morphology of the male anal appendages, hamulus and female vulvar lamina.

Libellulids frequently occur at standing water habitats including marshes, swamps and ponds with plenty of aquatic plants. A few species are found along streams and rivers. Males often exhibit territorial behavior by perching close to water's margin and darting out when confronting an intruder and returning to its perch, some species can hover for long periods. Females are less frequently seen and only approach the water to lay eggs.

晓褐蜻 雄
Trithemis aurora, male

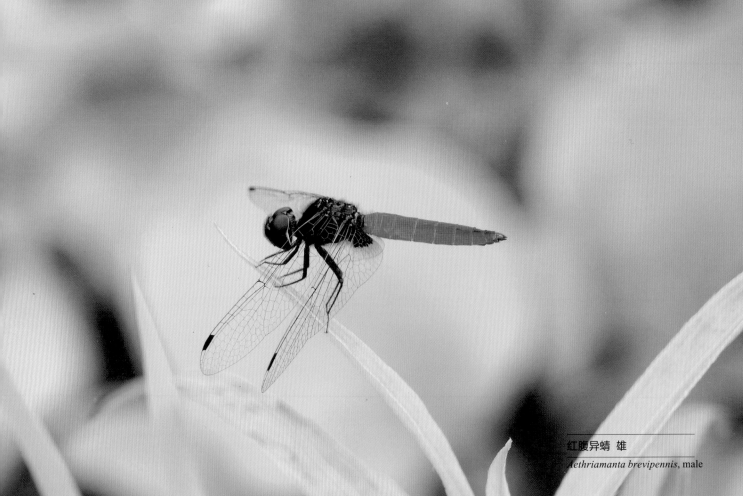

红腹异蜻 雄
Aethriamanta brevipennis, male

锥腹蜻属 Genus *Acisoma* Rambur, 1842

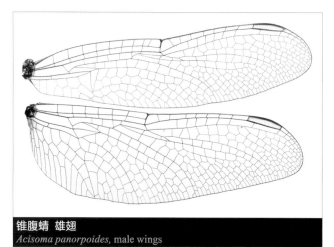

锥腹蜻 雄翅
Acisoma panorpoides, male wings

本属全球已知6种,广泛分布于亚洲和非洲。中国已知1种,分布广泛且常见。本属是体型细小的蜻科物种,腹部呈锥形;翅透明,翅痣较长但色彩较淡,弓脉位于第1条与第2条结前横脉之间,最末端的结前横脉不完整,盘区基方具1~2列翅室,基臀区具1条横脉,臀圈靴状,后翅三角室的基边与弓脉的位置相当。

本属蜻蜓栖息于水草茂盛的湿地、沼泽地等静水环境,经常隐藏在茂盛的植物丛中。

The genus contains six species, widespread in Asia and Africa. One species is recorded from China, common and widespread. Species of the genus are small-sized, abdomen is strongly inflated basally. Wings hyaline with pale and long pterostigma, arc between antenodals 1 and 2, the distal antenodal incomplete, discoidal field with 1-2 rows of cells basally, cubital space with one crossvein, anal loop boot-shaped, base of triangle in hind wing at the level of arc.

Acisoma species inhabit standing water habitats, including marshes and ponds, the adults usually perch in the grass.

锥腹蜻 雄
Acisoma panorpoides, male

锥腹蜻 *Acisoma panorpoides* Rambur, 1842

【形态特征】雄性复眼蓝色,面部蓝白色;胸部蓝白色,布满黑色细条纹,翅透明;腹部基方半部膨胀,前7节主要蓝白色并具黑色斑纹,后3节黑色。雌性复眼绿色,身体黄色具黑色斑纹。【长度】体长 25~28 mm,腹长 16~18 mm,后翅 19~20 mm。【栖息环境】海拔 2500 m 以下水草茂盛的沼泽、池塘和水稻田。【分布】广泛分布于中国南方;亚洲广布。【飞行期】全年可见。

[Identification] Male eyes blue, face bluish white. Thorax bluish white with fine black stripes, wings hyaline. Abdomen with basal half inflated, basal seven segments bluish white with black markings, distal three segments black. Female eyes green, body yellow with black markings. [Measurements] Total length 25-28 mm, abdomen 16-18 mm, hind wing 19-20 mm. [Habitat] Marshes, ponds and paddy fields below 2500 m elevation. [Distribution] Widespread in the south of China; Widespread in Asia. [Flight Season] Throughout the year.

锥腹蜻 雄,云南(德宏)
Acisoma panorpoides, male from Yunnan (Dehong)

锥腹蜻 雌,云南(德宏)
Acisoma panorpoides, female from Yunnan (Dehong)

锥腹蜻 雄,云南(德宏)
Acisoma panorpoides, male from Yunnan (Dehong)

异蜻属 Genus *Aethriamanta* Kirby, 1889

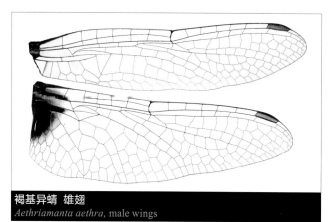

褐基异蜻 雄翅
Aethriamanta aethra, male wings

本属全球已知6种，主要分布在亚洲和非洲的热带地区。中国已知3种，仅发现于云南的低海拔湿地。本属均为体型较小的蜻科物种。本属雄性体色较艳丽，有些红色，有些覆盖蓝白色粉霜；雌性体色黄褐色，通常为黄褐色。翅大面积透明，基方具深色斑，翅脉稀疏，具较少的结前横脉，弓脉位于第1条与第2条结前横脉之间，盘区基方具1~2列翅室，基臀区具1条横脉，臀圈靴状。

本属蜻蜓喜欢水草茂盛的静水环境，尤其偏爱水葫芦滋生的池塘。雄性会停立在水葫芦的叶片顶端占据领地，并时而飞行，时而定点悬停。

The genus contains six species, mainly distributed in tropical Asia and Africa. Three species are recorded from China, only found in lowland wetlands of Yunnan. Species of the genus are small-sized. Males are colorful, red or pruinosed with bluish white whereas. Females are yellowish brown. Wings largely hyaline but hind wing bases with dark spots, venation sparse, antenodals in small number, arc between antenodals 1 and 2, discoidal field with 1-2 rows of cells basally, cubital space with one crossvein, anal loop boot-shaped.

Aethriamanta species inhabit well vegetated wetlands and are common inhabitants of ponds containing water hyacinth. Males usually perch on tips of emergent leaves of hyacinth, and sometimes leave their perch and hover.

霜蓝异蜻 雄
Aethriamanta gracilis, male

褐基异蜻 *Aethriamanta aethra* Ris, 1912

褐基异蜻 雄，云南（德宏）
Aethriamanta aethra, male from Yunnan (Dehong)

褐基异蜻 雌，云南（德宏）
Aethriamanta aethra, female from Yunnan (Dehong)

褐基异蜻 雄，云南（德宏）
Aethriamanta aethra, male from Yunnan (Dehong)

【形态特征】雄性全身覆盖蓝色粉霜；复眼黑褐色，面部蓝黑色，额具金属光泽；后翅基方具甚大的黑褐色斑；腹部第1~6节覆盖蓝色粉霜，第7~10节黑色。雌性主要褐色；合胸黑色具黄褐色条纹；腹部黑色具黄褐色斑。【长度】体长 26~31 mm，腹长 16~19 mm，后翅 22~24 mm。【栖息环境】海拔 500 m 以下水葫芦滋生的池塘。【分布】云南（德宏）；柬埔寨、印度尼西亚、马来半岛、新加坡、泰国、越南。【飞行期】3—12月。

[Identification] Male with blue pruinosity throughout. Eyes blackish brown, face metallic bluish black. Hind wing bases with large blackish brown markings. Abdomen with blue pruinosity on S1-S6, S7-S10 black. Female mainly brown. Thorax black with yellowish brown markings. Abdomen black with yellowish brown spots. [Measurements] Total length 26-31 mm, abdomen 16-19 mm, hind wing 22-24 mm. [Habitat] Ponds with water hyacinth below 500 m elevation. [Distribution] Yunnan (Dehong); Cambodia, Indonesia, Peninsular Malaysia, Singapore, Thailand, Vietnam. [Flight Season] March to December.

红腹异蜻 *Aethriamanta brevipennis* (Rambur, 1842)

【形态特征】雄性复眼黑褐色，面部蓝黑色，额具金属光泽；胸部背面黑色，侧面黄褐色，足黑色，后足腿节末端具1个甚小的红色斑，后翅基方具红褐色斑；腹部鲜红色，短而粗壮。雌性主要黄褐色；后足腿节末端具1个甚小的黄色斑；腹部黄色具黑色条纹。【长度】体长 27~29 mm，腹长 16~19 mm，后翅 24~25 mm。【栖息环境】海拔500 m以下水葫芦滋生的池塘。【分布】云南（德宏）、广东、香港；孟加拉国、印度、尼泊尔、斯里兰卡、马来半岛、印度尼西亚、新加坡、缅甸、泰国、柬埔寨。【飞行期】3—12月。

[Identification] Male eyes blackish brown, face metallic bluish black. Thorax black dorsally and yellowish brown laterally, legs black with a small red spot on tip of hind femora, hind wing bases with reddish brown markings. Abdomen red and depressed. Female largely yellowish brown. Legs black with a small yellow spot on tip of hind femora. Abdomen yellow with black stripes. [Measurements] Total length 27-29 mm, abdomen 16-19 mm, hind wing 24-25 mm.

红腹异蜻 雄，云南（德宏）
Aethriamanta brevipennis, male from Yunnan (Dehong)

[Habitat] Ponds with water hyacinth below 500 m elevation. [Distribution] Yunnan (Dehong), Guangdong, Hong Kong; Bangladesh, India, Nepal, Sri Lanka, Peninsular Malaysia, Indonesia, Singapore, Myanmar, Thailand, Cambodia. [Flight Season] March to December.

红腹异蜻 雄，云南（德宏）
Aethriamanta brevipennis, male from Yunnan (Dehong)

红腹异蜻 雌，香港 | 梁嘉景 摄
Aethriamanta brevipennis, female from Hong Kong | Photo by Kenneth Leung

霜蓝异蜻 *Aethriamanta gracilis* (Brauer, 1878)

【形态特征】雄性复眼褐色,面部黄褐色,额黑色具金属光泽;胸部覆盖蓝白色粉霜,翅透明,后翅基方染琥珀色;腹部第1~6节覆盖蓝白色粉霜,第7~10节深蓝色至黑色。雌性胸部蓝黑色具黄褐色条纹;腹部蓝黑色。【长度】体长 26~28 mm,腹长 16~18 mm,后翅 23~24 mm。【栖息环境】海拔 500 m以下挺水植物茂盛的池塘。【分布】云南(西双版纳、德宏);柬埔寨、印度尼西亚、马来西亚、新加坡、泰国、老挝、菲律宾。【飞行期】4—10月。

[Identification] Male eyes brown, face yellowish brown, frons metallic black. Thorax with bluish white pruinosity, wings hyaline, hind wing bases with amber tint. Abdomen with bluish white pruinosity on S1-S6, S7-S10 dark blue to black. Female thorax bluish black with yellowish brown markings. Abdomen bluish black. [Measurements] Total length 26-28 mm, abdomen 16-18 mm, hind wing 23-24 mm. [Habitat] Wetlands with copious amounts of emergent vegetation below 500 m elevation. [Distribution] Yunnan (Xishuangbanna, Dehong); Cambodia, Indonesia, Malaysia, Singapore, Thailand, Laos, Philippines. [Flight Season] April to October.

霜蓝异蜻 雌,云南(西双版纳)
Aethriamanta gracilis, female from Yunnan (Xishuangbanna)

霜蓝异蜻 雄,云南(西双版纳)
Aethriamanta gracilis, male from Yunnan (Xishuangbanna)

豹纹蜻属 Genus *Agrionoptera* Brauer, 1864

本属全球已知6种，主要分布于亚洲的热带地区和大洋洲。中国已知2亚种，分布于云南和台湾。本属蜻蜓体型小至中型；翅透明而狭长，后翅基方未显著加阔，翅痣很长，具有较多条结前横脉，弓脉位于第2条与第3条结前横脉之间，前翅三角室的上缘很短，基臀区具1条横脉，盘区基方具2~3列翅室，臀圈袋状。

本属蜻蜓栖息于茂盛森林中的池塘，在有林荫遮盖的池塘周边可以发现。雄性会停落在池塘边的植物上。本属中国分布的豹纹蜻与亚洲秘蜻相似，但可以通过胸部的斑纹区分。

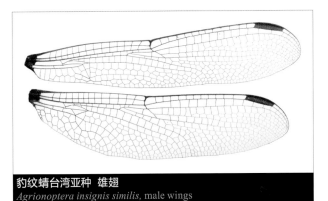

豹纹蜻台湾亚种 雄翅
Agrionoptera insignis similis, male wings

The genus contains six species distributed throughout tropical Asia and Oceania. Two subspecies are known from Yunnan and Taiwan, China. Species of the genus are small to medium sized dragonflies. Wings hyaline and narrow, but hind wing bases not broad, pterostigma long, antenodals abundant, arc between antenodals 2 and 3, the anterior margin of triangle short in fore wings, cubital space with one crossvein, discoidal field with 2-3 rows of cells, ananl loop sack-shaped.

Agrionoptera species inhabit ponds in dense forests. Territorial male perch on plants near water. *Agrionoptera insignis*, the only species in China, is similar to *Lathrecista asiatica*, but their thoracic color pattern is different.

豹纹蜻指名亚种 雄
Agrionoptera insignis insignis, male

豹纹蜻指名亚种 *Agrionoptera insignis insignis* (Rambur, 1842)

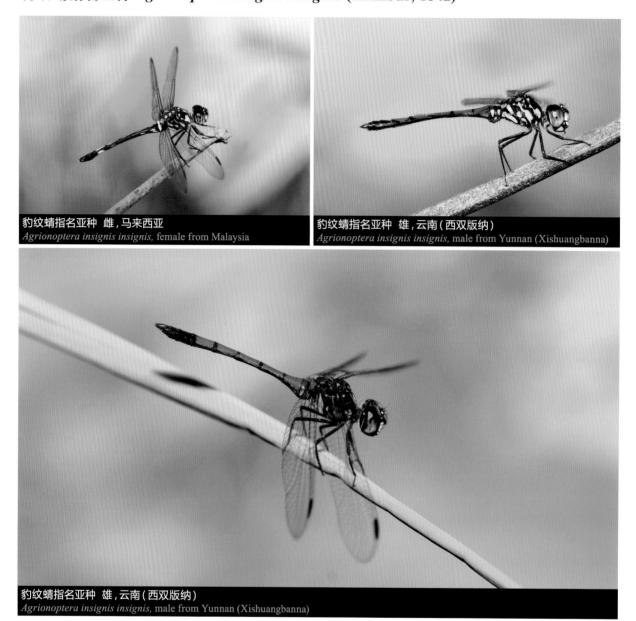

豹纹蜻指名亚种 雌, 马来西亚
Agrionoptera insignis insignis, female from Malaysia

豹纹蜻指名亚种 雄, 云南 (西双版纳)
Agrionoptera insignis insignis, male from Yunnan (Xishuangbanna)

豹纹蜻指名亚种 雄, 云南 (西双版纳)
Agrionoptera insignis insignis, male from Yunnan (Xishuangbanna)

【形态特征】雄性复眼褐色和绿色, 面部白色, 额具1个金属蓝黑色斑; 胸部黑色具不规则的黄色条纹和斑点, 足黑色, 翅透明; 腹部大面积红色, 第3~7节末端具黑色环纹, 第8~10节和肛附器黑色。【长度】雄性体长 37 mm, 腹长 25 mm, 后翅 28 mm。【栖息环境】海拔 1000 m 以下森林中的池塘。【分布】云南 (德宏、西双版纳); 从东南亚至大洋洲广布。【飞行期】9—12月。

[Identification] Male eyes brown and green, face white, frons with a metallic bluish black spot. Thorax black with irregular yellow markings, legs black, wings hyaline. Abdomen largely red, S3-S7 with distal black rings, S8-S10 and anal appendages black. [Measurements] Male total length 37 mm, abdomen 25 mm, hind wing 28 mm. [Habitat] Ponds in forest below 1000 m elevation. [Distribution] Yunnan (Dehong, Xishuangbanna); Widespread from Southeast Asia to Oceania. [Flight Season] September to December.

豹纹蜻台湾亚种 *Agrionoptera insignis similis* **Selys, 1879**

【形态特征】雄性复眼褐色和绿色，面部白色，额具1个金属蓝黑色斑；胸部黑色具不规则的黄色条纹和斑点，足黑色，翅透明；腹部大面积红色，第1~7节红色，第8~10节和肛附器黑色。本亚种与指名亚种合胸的黄色条纹形状有差异，而且腹部第3~7节末端无黑色环纹。【长度】雄性体长 42 mm，腹长 23 mm，后翅 31 mm。【栖息环境】海拔 500 m以下森林中的池塘。【分布】中国台湾；菲律宾至所罗门的热带岛屿。【飞行期】5—10月。

[Identification] Male eyes brown and green, face white, frons with a metallic bluish black spot. Thorax black with irregular yellow markings, legs black, wings hyaline. Abdomen largely red, S1-S7 red, S8-S10 and anal appendages black. The subspecies differs from the nominate subspecies by the shape of thoracic yellow markings and S3-S7 without black rings. [Measurements] Male total length 42 mm, abdomen 23 mm, hind wing 31 mm. [Habitat] Ponds in forest below 500 m elevation. [Distribution] Taiwan of China; Tropical islands from Philippines to Solomon islands. [Flight Season] May to October.

豹纹蜻台湾亚种 雄，台湾
Agrionoptera insignis similis, male from Taiwan

安蜻属 Genus *Amphithemis* **Selys, 1891**

长腹安蜻，雄
Amphithemis vacillans, male

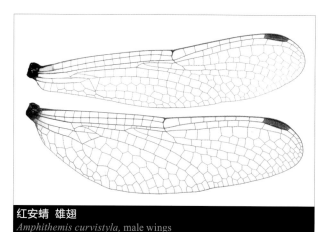

红安蜻 雄翅
Amphithemis curvistyla, male wings

本属全球已知3种，分布于亚洲的热带地区。中国已知2种，发现于云南西部和南部。本属蜻蜓体型较小；翅透明而狭长，后翅基方未显著加阔，翅痣很长，具有较多条结前横脉，弓脉位于第2条与第3条结前横脉之间，前翅三角室的上缘很短，基臀区具1条横脉，盘区基方具2列翅室，臀圈袋状。

本属蜻蜓栖息于水草茂盛的池塘，雄性停落在池塘的杂草丛中占据领地。

The genus contains three species distributed in tropical Asia. Two species are known from China, recorded from the west and south of Yunnan. Species of the genus are small-sized dragonflies. Wings hyaline and narrow, hind wing bases not clearly broad, pterostigma long, antenodals abundant, arc between antenodals 2 and 3, the anterior margin of triangle short in fore wings, cubital space with one crossvein, discoidal field with two rows of cells, anal loop sack-shaped.

Amphithemis species inhabit well vegetated ponds in forests. Territorial males perch on plants near water.

红安蜻 *Amphithemis curvistyla* Selys, 1891

【形态特征】雄性复眼黑褐色，面部白色，额具1个金属蓝黑色斑；胸部黄褐色，肩条纹甚阔，黑色，翅透明；腹部第1~8节红色，第3~8节末端具黑色斑，第9~10节和肛附器黑色。雌性多型，黄色型的雌性腹部黄色，具宽阔的黑色条纹；红色型雌性和雄性的色彩相似。【长度】雄性体长 31~32 mm，腹长 21~22 mm，后翅 24~25 mm。【栖息环境】海拔 1500 m 以下挺水植物茂盛的湿地。【分布】云南（西双版纳）；缅甸、泰国、越南。【飞行期】9—12月。

红安蜻 雌，红色型，云南（西双版纳）
Amphithemis curvistyla, female, the red morph from Yunnan (Xishuangbanna)

[Identification] Male eyes blackish brown, face white, frons with a metallic bluish black spot. Thorax yellowish brown, antehumeral stripes broad and black, wings hyaline. S1-S8 red, S3-S8 with distal black spots, S9-S10 and anal appendages black. Female polymorphic, the yellow morph with yellow abdomen and broad black stripes. The red morph similar to male. [Measurements] Male total length 31-32 mm, abdomen 21-22 mm, hind wing 24-25 mm. [Habitat] Wetlands with plenty emerging plants below 1500 m elevation. [Distribution] Yunnan (Xishuangbanna); Myanmar, Thailand, Vietnam. [Flight Season] September to December.

红安蜻 雄,云南(西双版纳)|莫善濂 摄
Amphithemis curvistyla, male from Yunnan (Xishuangbanna) | Photo by Shanlian Mo

红安蜻 雄,云南(西双版纳)|莫善濂 摄
Amphithemis curvistyla, male from Yunnan (Xishuangbanna) | Photo by Shanlian Mo

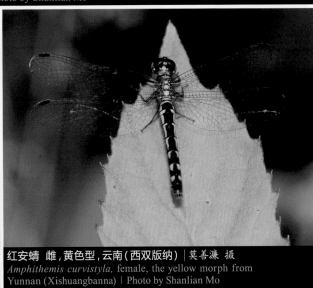

红安蜻 雌,黄色型,云南(西双版纳)|莫善濂 摄
Amphithemis curvistyla, female, the yellow morph from Yunnan (Xishuangbanna) | Photo by Shanlian Mo

长腹安蜻 *Amphithemis vacillans* Selys, 1891

【形态特征】雄性复眼黑褐色，面部白色，额具金属蓝黑色斑；胸部完全黑色，翅透明；腹部大面积黑色，仅第3节覆盖白色粉霜，上肛附器甚长。【长度】雄性体长 37 mm，腹长 26.5 mm，后翅 23.5 mm。【栖息环境】海拔 500 m 以下茂盛森林中的小型水潭。【分布】云南（德宏）；印度、缅甸。【飞行期】9—11月。

[Identification] Male eyes blackish brown, face white, frons with a metallic bluish black spot. Thorax entirely black, wings hyaline. Abdomen largely black, S3 whitishly pruinosed, superior appendages quite long. [Measurements] Male total length 37 mm, abdomen 26.5 mm, hind wing 23.5 mm. [Habitat] Small pools in dense forests below 500 m elevation. [Distribution] Yunnan (Dehong); India, Myanmar. [Flight Season] September to November.

长腹安蜻 雄，云南（德宏）
Amphithemis vacillans, male from Yunnan (Dehong)

黑斑蜻属 Genus *Atratothemis* Wilson, 2005

本属全球仅1种，分布于中国南部、泰国、越南和老挝。本属蜻蜓体中型；翅大面积透明，基方和端方具黑色斑，最末端的结前横脉完整，弓脉位于第1条与第2条结前横脉之间，盘区具3~4列翅室，基臀区具2条横脉，臀圈靴状。

本属蜻蜓栖息于茂盛森林中较阴暗的池塘，飞行快速。在一天的较炎热时段可以发现它们停落在树荫中。有集群捕食的习性，通常在空旷地较高处飞行。

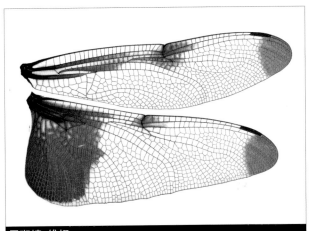

黑斑蜻 雄翅
Atratothemis reelsi, male wings

The genus contains a single species found in Southern China, Thailand, Vietnam and Laos. It is a medium-sized species. Wings hyaline with basal and apical black spots, arc between antenodals 1 and 2, discoidal field with 3-4 rows of cells, cubital space with two crossveins, anal loop boot-shaped.

Atratothemis species inhabits shady ponds in dense forests. Individuals hang pendently among dense tree foliage during the hot hours of the day. They often forage in small group flying high in the exposed areas.

黑斑蜻 雄
Atratothemis reelsi, male

黑斑蜻 *Atratothemis reelsi* Wilson, 2005

黑斑蜻 雄，广西
Atratothemis reelsi, male from Guangxi

【形态特征】通体黑色，面部具金属光泽；前翅的亚前缘脉至肘脉之间具黑褐色带，直达翅结处，翅端具黑褐色斑，后翅基方具甚大的黑褐色斑，端部具黑褐色斑；腹部较短。【长度】体长 40～41 mm，腹长 27～29 mm，后翅 40～43 mm。【栖息环境】海拔 1000 m 以下森林中树荫遮蔽的池塘。【分布】贵州、广西、海南、云南（红河）；泰国、越南、老挝。【飞行期】4—7月。

黑斑蜻 雌，云南（红河）
Atratothemis reelsi, female from Yunnan (Honghe)

[Identification] Black throughout, face metallic. Fore wings with blackish brown band between Sc and Cup, the band extending to nodus, wings tips with blackish brown spots, hind wing bases with large blackish brown markings, tips with blackish brown spots. Abdomen short. [Measurements] Total length 40-41 mm, abdomen 27-29 mm, hind wing 40-43 mm. [Habitat] Shady ponds in forest below 1000 m elevation. [Distribution] Guizhou, Guangxi, Hainan, Yunnan (Honghe); Thailand, Vietnam, Laos. [Flight Season] April to July.

疏脉蜻属 Genus *Brachydiplax* Brauer, 1868

本属全球已知8种，主要分布于亚洲的热带和亚热带区域。中国已知5种，分布于中部至南部。本属为小至中型的蜻科物种；雄性具蓝灰色粉霜，面部具较大的金属色斑，翅大面积透明，仅在基方具色斑，弓脉位于第1条与第2条结前横脉之间，臀圈靴状，基臀区仅有1条横脉，后翅的三角室位于弓脉以外。

本属蜻蜓多栖息于水草茂盛的池塘，雄性会停落在水草的端部占据领地，并经常展开争斗。有些种类在城市的荷花池中也较容易遇见。

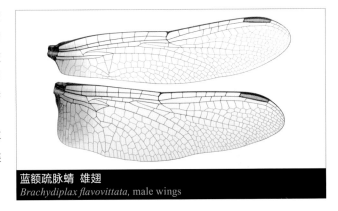

蓝额疏脉蜻 雄翅
Brachydiplax flavovittata, male wings

The genus contains eight species and confined to tropical and subtropical Asia. Five species are known from Southern and Central China. They are small to medium sized species. Males usually bluish grey pruinosed, face with big metallic spots, wings largely hyaline with bases tinted with color spots, arc between antenodals 1 and 2, anal loop boot-shaped, cubital space with one crossvein, triangle in hind wing distal to arc.

Brachydiplax species inhabit well vegetated ponds. Territorial males usually perch on tops of plants and often fight. Some species can be seen at lotus ponds within city limits.

霜白疏脉蜻 雄
Brachydiplax farinosa, male

褐胸疏脉蜻 *Brachydiplax chalybea* Brauer, 1868

【形态特征】雄性复眼下方灰色，上方褐色，面部白色，额具1个金属蓝黑色斑；合胸背面覆盖蓝白色粉霜，侧面黄褐色，翅透明；腹部第1～6节具蓝白色粉霜，第1～3节侧面具黄褐色斑，第7～10节和肛附器黑色。【长度】体长 33～37 mm，腹长 21～23 mm，后翅 24～27 mm。【栖息环境】海拔 500 m以下挺水植物茂盛的池塘。【分布】云南（德宏、西双版纳）；南亚、东南亚广布。【飞行期】3—12月。

[Identification] Male eyes grey below and brown above, face white, frons with a metallic bluish black spot. Dorsal part of thorax with bluish white pruinosity, lateral part yellowish brown, wings hyaline. Abdomen with bluish white pruinosity on S1-S6, S1-S3 with lateral yellowish brown markings, S7-S10 and anal appendages black. [Measurements] Total length 33-37 mm, abdomen 21-23 mm, hind wing 24-27 mm. [Habitat] Wetlands with plenty of emergent plants below 500 m elevation. [Distribution] Yunnan (Dehong, Xishuangbanna); Widespread in South and Southeast Asia. [Flight Season] March to December.

褐胸疏脉蜻 雄，云南（德宏）
Brachydiplax chalybea, male from Yunnan (Dehong)

褐胸疏脉蜻 雄，云南（德宏）
Brachydiplax chalybea, male from Yunnan (Dehong)

霜白疏脉蜻 *Brachydiplax farinosa* Krüger, 1902

霜白疏脉蜻 雌，黄色型，云南（德宏）
Brachydiplax farinosa, female, the yellow morph from Yunnan (Dehong)

霜白疏脉蜻 雌，白色型，云南（德宏）
Brachydiplax farinosa, female, the white morph from Yunnan (Dehong)

霜白疏脉蜻 雄，云南（德宏）
Brachydiplax farinosa, male from Yunnan (Dehong)

【形态特征】雄性复眼黑褐色，面部白色，额具1个金属蓝黑色斑；合胸背面稍微覆盖蓝白色粉霜，侧面蓝黑色，翅透明；腹部基方至第6节中部具蓝白色粉霜，第7~10节和肛附器黑色。雌性多型，白色型个体与雄性相似，但色彩稍淡；黄色型个体身体棕黄色，合胸和腹部具黑色条纹。【长度】体长 23~25 mm，腹长 14~15 mm，后翅 17~18 mm。【栖息环境】海拔 1000 m以下挺水植物茂盛的池塘。【分布】云南（西双版纳、临沧、德宏）；孟加拉国、文莱、印度、缅甸、泰国、老挝、越南、马来西亚、印度尼西亚。【飞行期】全年可见。

[Identification] Male eyes blackish brown, face white, frons with a metallic bluish black spot. Dorsal part of thorax slightly bluish white pruinosed, laterally bluish black, wings hyaline. Abdomen with bluish white pruinosity on S1-S6, S7-S10 and anal appendages black. Female polymorphic, the white morph is similar to male but paler. The yellow morph body largely brownish yellow with black thoracic and abdominal markings. [Measurements] Total length 23-25 mm, abdomen 14-15 mm, hind wing 17-18 mm. [Habitat] Wetlands with plenty of emergent plants below 1000 m elevation. [Distribution] Yunnan (Xishuangbanna, Lincang, Dehong); Bangladesh, Brunei Darussalam, India, Myanmar, Thailand, Laos, Vietnam, Malaysia, Indonesia. [Flight Season] Throughout the year.

蓝额疏脉蜻 *Brachydiplax flavovittata* Ris, 1911

蓝额疏脉蜻 雌,广东 | 宋睿斌 摄
Brachydiplax flavovittata, female from Guangdong | Photo by Ruibin Song

蓝额疏脉蜻 雄,广东
Brachydiplax flavovittata, male from Guangdong

　　【形态特征】雄性复眼上方褐色,下方绿色,面部白色,额具1个金属蓝黑色斑;合胸背面覆盖蓝白色粉霜,侧面黑色具2条宽阔的黄条纹,翅透明,基方具琥珀色斑;腹部第1~6节中部具蓝白色粉霜,第7~10节和肛附器黑色。雌性主要黑褐色具黄色条纹。本种一直被作为褐胸疏脉蜻的亚种,此处将其提升至种。【长度】体长 34~40 mm,腹长 22~25 mm,后翅 27~29 mm。【栖息环境】海拔 1500 m以下挺水植物茂盛的池塘。【分布】广泛分布于中国南方;日本、越南。【飞行期】3—11月。

[Identification] Male eyes brown above and green below, face white, frons with a metallic bluish black spot. Dorsal part of thorax with bluish white pruinosity, laterally black with two broad yellow stripes, wings hyaline with basal amber tints. Abdomen with bluish white pruinosity on S1-S6, S7-S10 and anal appendages black. Female mainly blackish brown with yellow markings. The species was regarded as a subspecies of *Brachydiplax chalybea* but it is considered a valid species here. [Measurements] Total length 34-40 mm, abdomen 22-25 mm, hind wing 27-29 mm. [Habitat] Wetlands with plenty of emergent plants below 1500 m elevation. [Distribution] Widespread in the south of China; Japan, Vietnam. [Flight Season] March to November.

暗色疏脉蜻 *Brachydiplax sobrina* (Rambur, 1842)

　　【形态特征】雄性复眼黑褐色，面部白色，额具1个金属蓝黑色斑；合胸背面覆盖蓝白色粉霜，侧面黑褐色，翅透明；腹部第1~6节具白色粉霜，第7~10节和肛附器黑色，第7节具1对黄色背斑。本种与霜白疏脉蜻外观相似，但后者胸部和腹部无黄色斑点，而本种即使老熟仍然可见黄色斑点。【长度】雄性体长 30~35 mm，腹长 19~23 mm，后翅 22~23 mm。【栖息环境】海拔 1000 m以下挺水植物茂盛的池塘。【分布】云南（德宏）；孟加拉国、印度、斯里兰卡、尼泊尔、缅甸、泰国、柬埔寨、越南。【飞行期】4—10月。

　　[Identification] Male eyes blackish brown, face white, frons with a metallic bluish black spot. Dorsal part of thorax with bluish white pruinosity, laterally blackish brown, wings hyaline. Abdomen with white pruinosity on S1-S6, S7-S10 and anal appendages black, S7 with a pair of yellow spots dorsally. The species is similar to *B. farinosa* in general habitus, the thoracic and abdominal yellow spots absent in *B. farinosa* but present even in aged individuals of this species. [Measurements] Male total length 30-35 mm, abdomen 19-23 mm, hind wing 22-23 mm. [Habitat] Wetlands with plenty of emergent plants below 1000 m elevation. [Distribution] Yunnan (Dehong); Bangladesh, India, Sri Lanka, Nepal, Myanmar, Thailand, Cambodia, Vietnam. [Flight Season] April to October.

暗色疏脉蜻 雄，云南（德宏）
Brachydiplax sobrina, male from Yunnan (Dehong)

疏脉蜻属待定种 *Brachydiplax* sp.

疏脉蜻属待定种 雄，云南（德宏）
Brachydiplax sp., male from Yunnan (Dehong)

【形态特征】雄性复眼黑褐色，面部白色，额具1个金属蓝黑色斑；合胸背面覆盖蓝白色粉霜，侧面具3条黄色条纹，随年纪增长色彩加深，翅透明；腹部第1~6节具白色粉霜，第7~10节和肛附器黑色。【长度】雄性体长 33 mm，腹长 22 mm，后翅 27 mm。【栖息环境】海拔 500 m以下植物茂盛的池塘。【分布】云南（德宏）。【飞行期】4—10月。

[Identification] Male eyes blackish brown, face white, frons with a metallic bluish black spot. Dorsal part of thorax with bluish white pruinosity, laterally with three yellow stripes but darkened with age, wings hyaline. Abdomen with white pruinosity on S1-S6, S7-S10 and anal appendages black. [Measurements] Male total length 33 mm, abdomen 22 mm, hind wing 27 mm. [Habitat] Wetlands with plenty emerging plants below 500 m elevation. [Distribution] Yunnan (Dehong). [Flight Season] April to October.

黄翅蜻属 Genus *Brachythemis* Brauer, 1868

本属全球已知6种，主要分布于亚洲和非洲。中国已知1种，分布广泛且常见。本属是一类小型蜻蜓；翅痣较长，弓脉位于第1条与第2条结前横脉之间，前翅最末端的结前横脉不完整，盘区基方具2~3列翅室，基臀区具1条横脉，臀圈靴状，后翅三角室基边与弓脉的位置相当。

本属蜻蜓主要栖息于各种静水环境，包括一些大型水库的边缘，在流速缓慢且具有水草的溪流和河流也可以见到。雄性在水边的植物上占据领地，白天通常是长时间停落，很少飞行，黄昏时非常活跃，雌性在黄昏时产卵。

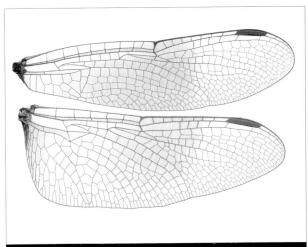

黄翅蜻，雄翅
Brachythemis contaminata, male wings

The genus contains six species, confined to Asia and Africa. One species is recorded from China, common and widespread. Species of the genus are small-sized. Pterostigma long, arc between antenodals 1 and 2, distal antenodal in fore wings incomplete, discoidal field with 2-3 rows of cells basally, cubital space with one crossvein, anal loop boot-shaped, base of triangle in hind wing at the level of arc.

Brachythemis species mainly inhabit large bodies of standing water and are often seen at the margins of reservoirs, also found at edges of slow flowing streams and large rivers. Males perch on the plants close to water during the daytime and become active at twilight, females may also be seen laying eggs during this time.

黄翅蜻 雄
Brachythemis contaminata, male

黄翅蜻 *Brachythemis contaminata* (Fabricius, 1793)

【形态特征】雄性复眼褐色，面部黄褐色；胸部褐色具不清晰的深褐色条纹，翅具甚大的橙红色斑，仅端部透明；腹部橙红色具细小的褐色斑。雌性黄褐色，胸部和腹部具甚细的褐色斑纹。【长度】体长 27~31 mm，腹长 17~19 mm，后翅 21~23 mm。【栖息环境】海拔 1500 m以下的池塘、湖泊、水库等静水环境。【分布】广泛分布于中国南方；孟加拉国、印度、斯里兰卡、尼泊尔、缅甸、泰国、老挝、柬埔寨、越南、马来西亚、印度尼西亚、菲律宾、新加坡、日本。【飞行期】全年可见。

[Identification] Male eyes brown, face yellowish brown. Thorax brown with unclear dark brown markings, wings with large areas of orange red with only the wing tip hyaline. Abdomen orange red with small brown spots. Female generally yellowish brown with brown thoracic and abdominal markings. [Measurements] Total length 27-31 mm, abdomen 17-19 mm, hind wing 21-23 mm. [Habitat] Standing water, including ponds, lakes and reservoirs below 1500 m elevation. [Distribution] Widespread in the south of China; Bangladesh, India, Sri Lanka, Nepal, Myanmar, Thailand, Laos, Cambodia, Vietnam, Malaysia, Indonesia, Philippines, Singapore, Japan. [Flight Season] Throughout the year.

黄翅蜻 雄，云南（德宏）
Brachythemis contaminata, male from Yunnan (Dehong)

黄翅蜻 雌，湖北
Brachythemis contaminata, female from Hubei

黄翅蜻 雄，湖北
Brachythemis contaminata, male from Hubei

岩蜻属 Genus *Bradinopyga* Kirby, 1893

本属全球已知3种，分布于亚洲和非洲。中国已知1种，仅在云南西部的中缅边境发现。本属蜻蜓身体暗色具斑驳的斑纹，与它们经常停落的岩石色彩一致；翅透明，弓脉位于第1条与第2条结前横脉之间，前翅最末端的结前横脉不完整，盘区基方具2~3列翅室，基臀区具1条横脉，臀圈靴状。

赭岩蜻 雄
Bradinopyga geminata, male

The genus contains three species found in Asia and Africa. One species is recorded from China but only found in the Sino-Burmese border of western Yunnan. Species of the genus are dark with a variegated colored body that matches the rocky substrate upon which they frequently rest. Wings hyaline, arc between antenodals 1 and 2, distal antenodal in fore wings incomplete, discoidal field with 2-3 rows of cells basally, cubital space with one crossvein, anal loop boot-shaped.

赭岩蜻 *Bradinopyga geminata* (Rambur, 1842)

【形态特征】雄性复眼黑褐色，面部灰色；胸部和腹部深灰色，具灰白色和黑色斑纹，肛附器白色。【长度】雄性体长 41 mm，腹长 26 mm，后翅 32 mm。【栖息环境】不详。【分布】云南（德宏）；印度、斯里兰卡、泰国。【飞行期】不详。

[Identification] Male eyes blackish brown, face grey. Thorax and abdomen dark grey with greyish white and black markings, anal appendages white. [Measurements] Male total length 41 mm, abdomen 26 mm, hind wing 32 mm. [Habitat] Unknown. [Distribution] Yunnan (Dehong); India, Sri Lanka, Thailand. [Flight Season] Unknown.

赭岩蜻 雄, 云南 (德宏)
Bradinopyga geminate, male from Yunnan (Dehong)

巨蜻属 Genus *Camacinia* Kirby, 1889

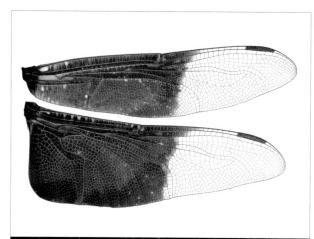

亚洲巨蜻 雄翅
Camacinia gigantea, male wings

本属全球已知3种, 分布于亚洲的热带地区和大洋洲。中国已知2种, 主要分布在华南和西南的热带区域。本属是一类体大型且粗壮的蜻科物种; 翅具大块色斑, 翅痣较长, 后翅基方较阔, 弓脉位于第1条与第2条结前横脉之间, IR3 呈明显的波状, 盘区基方具多列翅室, 基臀区具多条横脉, 臀圈靴状。

本属蜻蜓主要栖息于森林中水草茂盛的湿地, 比较偏爱水葫芦滋生的池塘。雄性在池塘中来回飞行占据领地, 有时悬停, 有时停歇在树枝上。雄性之间常展开追逐争斗。交尾在空中进行, 时间较短。

The genus contains three species distributed in tropical Asia and Oceania. Two species are recorded from China, mainly distributed in the tropical area of South and Southwest regions. Species of the genus are large and robust libellulids. Wings colourful, pterostigma long, hind wing bases broad, arc between antenodals 1 and 2, IR3 clearly waved, discoidal field with many rows of cells basally, cubital space with many crossveins, anal loop boot-shaped.

Camacinia species inhabit well vegetated wetlands preferring ponds with plenty water hyacinth. Males fly above water, sometimes hovering and sometimes perching on trees. Males often fight for territory. Mating takes place in flight, but lasts for only a few seconds.

亚洲巨蜻 雄
Camacinia gigantea, male

亚洲巨蜻 *Camacinia gigantea* (Brauer, 1867)

【形态特征】雄性复眼和面部红褐色；胸部深红色，翅具大块的红褐色斑，端部1/3透明；腹部红色。雌性多型，黄色透翅型身体黄色，翅大面积透明；黄色斑翅型身体黄色，翅基方1/2具色斑；红色斑翅型与雄性色彩相似，但色彩稍淡。【长度】体长 50~58 mm，腹长 30~37 mm，后翅 44~52 mm。【栖息环境】海拔 500 m以下水葫芦滋生的池塘。【分布】云南（西双版纳、临沧）、西藏；印度、缅甸、泰国、老挝、柬埔寨、越南、马来西亚、印度尼西亚、巴布亚新几内亚、菲律宾、新加坡。【飞行期】全年可见。

[Identification] Male eyes and face reddish brown. Thorax dark red, wings with large reddish brown bands at basal two thirds and apical one third hyaline. Abdomen red. Female polymorphic, the yellow hyaline wing morph with body yellow and largely hyaline wings. The yellow spotted wing morph with body yellow and basal half of wings spotted.

亚洲巨蜻 雄，云南（西双版纳）
Camacinia gigantea, male from Yunnan (Xishuangbanna)

The red spotted wing morph similar to male but paler. [Measurements] Total length 50-58 mm, abdomen 30-37 mm, hind wing 44-52 mm. [Habitat] Ponds with plenty water hyacinth below 500 m elevation. [Distribution] Yunnan (Xishuangbanna, Lincang), Tibet; India, Thailand, Laos, Cambodia, Myanmar, Vietnam, Malaysia, Indonesia, Papua New Guinea, Philippines, Singapore. [Flight Season] Throughout the year.

亚洲巨蜻 雌，黄色透翅型，云南（西双版纳）
Camacinia gigantea, female, the yellow hyaline wing morph from Yunnan (Xishuangbanna)

亚洲巨蜻 雌，黄色斑翅型，云南（西双版纳）
Camacinia gigantea, female, the yellow spotted wing morph from Yunnan (Xishuangbanna)

亚洲巨蜻 雄，云南（西双版纳）
Camacinia gigantea, male from Yunnan (Xishuangbanna)

亚洲巨蜻 雌，红色斑翅型，云南（西双版纳）
Camacinia gigantea, female, the red spotted wing morph from Yunnan (Xishuangbanna)

森林巨蜻 *Camacinia harterti* Karsch, 1890

【形态特征】雄性复眼和面部棕褐色；胸部棕色，翅大面积透明，后翅基部具黑褐色和黄色斑纹，端部具小褐斑；腹部红色。雌性大面积黄色。【长度】雄性体长 59 mm，腹长 39 mm，后翅 47 mm。【栖息环境】海拔 500~1500 m森林中的池塘。【分布】广东、云南（西双版纳、普洱）；泰国、老挝、越南、马来西亚、印度尼西亚。【飞行期】3—10月。

[Identification] Male eyes and face brown. Thorax brown, wings largely hyaline, hind wings with basal blackish brown and yellow markings, wing tips with small brown spots. Abdomen red. Female largely yellow. [Measurements] Male total length 59 mm, abdomen 39 mm, hind wing 47 mm. [Habitat] Ponds in forest at 500-1500 m elevation. [Distribution] Guangdong, Yunnan (Xishuangbanna, Pu'er); Thailand, Laos, Vietnam, Malaysia, Indonesia. [Flight Season] March to October.

森林巨蜻 雄，云南（西双版纳）
Camacinia harterti, male from Yunnan (Xishuangbanna)

林蜻属 Genus *Cratilla* Kirby, 1900

本属全球仅2种，主要分布于亚洲的热带地区。中国已知2亚种，主要分布在云南、广西、海南和台湾。本属是一类中型蜻蜓；面部具金属色斑；翅透明而狭长，翅痣很长，具有较多条结前横脉，弓脉位于第2条与第3条结前横脉之间，前翅三角室的上缘很短，基臀区具1条横脉，盘区基方具2～3列翅室，臀圈靴状。

本属蜻蜓栖息于森林中的小型池塘，包括林道上的季节性和暂时性水潭。雄性在池塘边缘的树枝上停落等待雌性，有时在水面上快速来回飞行。

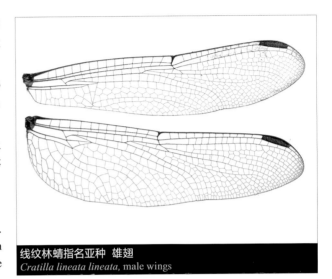

线纹林蜻指名亚种 雄翅
Cratilla lineata lineata, male wings

The genus contains two tropical Asian species. Two subspecies are known from China, recorded from Yunnan, Guangxi, Hainan and Taiwan. Species of the genus are medium-sized. Face with metallic spots. Wings hyaline and narrow, pterostigma long, antenodals abundant, arc between antenodals 2 and 3, the anterior margin of triangle short in fore wings, cubital space with one crossvein, discoidal field with 2-3 rows of cells, anal loop boot-shaped.

Cratilla species inhabit small ponds in forests, including seasonal and temporary pools. Territorial males perch on plants near water and sometimes rapidly fly above water.

线纹林蜻指名亚种 雄
Cratilla lineata lineata, male

线纹林蜻台湾亚种 *Cratilla lineata assidua* Lieftinck, 1953

线纹林蜻台湾亚种 雄，台湾
Cratilla lineata assidua, male from Taiwan

【形态特征】雄性复眼上方褐色，下方黄绿色，面部白色，额具1个金属蓝黑色斑；胸部深褐色具数条并行排列的黄色条纹，翅透明；腹部黑褐色，具短而细的黄色条纹。本亚种与指名亚种未见显著区别。【长度】体长 44~50 mm，腹长 29~33 mm，后翅 34~40 mm。【栖息环境】海拔 1500 m 以下森林中林道上的小积水潭和季节性水塘。【分布】中国台湾；菲律宾、印度尼西亚。【飞行期】4—10月。

[Identification] Male eyes brown above and yellowish green below, face white, frons with a metallic bluish black spot. Thorax dark brown with some parallel yellow stripes, wings hyaline. Abdomen blackish brown with short and fine yellow stripes. This subspecies shows no clear difference from the nominate subspecies. [Measurements] Total length 44-50 mm, abdomen 29-33 mm, hind wing 34-40 mm. [Habitat] Small pools on the paths in forest and seasonal pools below 1500 m elevation. [Distribution] Taiwan of China; Philippines, Indonesia. [Flight Season] April to October.

线纹林蜻指名亚种 *Cratilla lineata lineata* (Brauer, 1878)

【形态特征】雄性复眼上方褐色,下方黄绿色,面部白色,额具1个金属蓝黑色斑;胸部深褐色具数条并行排列的黄色条纹,翅透明;腹部黑褐色,具短而细的黄色条纹。【长度】体长 45~50 mm,腹长 30~33 mm,后翅 34~41 mm。【栖息环境】海拔 2000 m以下森林中林道上的小积水潭和季节性水塘。【分布】云南、广西、海南;斯里兰卡、缅甸、泰国、越南、马来西亚、印度尼西亚、菲律宾、新加坡。【飞行期】全年可见。

[Identification] Male eyes brown above and yellowish green below, face white, frons with a metallic bluish black spot. Thorax dark brown with some parallel yellow stripes, wings hyaline. Abdomen blackish brown with short and fine yellow stripes. [Measurements] Total length 45-50 mm, abdomen 30-33 mm, hind wing 34-41 mm. [Habitat] Small pools on the paths in forest and seasonal pools below 2000 m elevation. [Distribution] Yunnan, Guangxi, Hainan; Sri Lanka, Myanmar, Thailand, Vietnam, Malaysia, Indonesia, Philippines, Singapore. [Flight Season] Throughout the year.

线纹林蜻指名亚种 雌,云南(西双版纳)
Cratilla lineata lineata, female from Yunnan (Xishuangbanna)

线纹林蜻指名亚种 雄,云南(西双版纳)
Cratilla lineata lineata, male from Yunnan (Xishuangbanna)

红蜻属 Genus *Crocothemis* Brauer, 1868

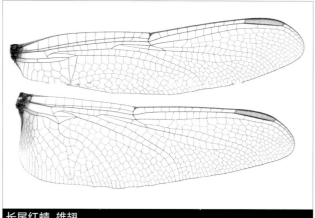

长尾红蜻 雄翅
Crocothemis erythraea, male wings

本属全球已知约10种,世界性分布。中国已知3种及亚种,全国广布且常见。本属蜻蜓体中型;翅透明,翅痣较长,弓脉位于第1条与第2条结前横脉之间,前翅最末端的结前横脉不完整,盘区基方具2~3列翅室,基臀区具1条横脉,臀圈靴状。

本属蜻蜓主要栖息于各类静水环境和流速缓慢的溪流。雄性经常出现在水面的植物上,时而飞行,时而停落。中国已知种外观十分近似,雄性鲜红色,雌性红色或黄色,可以通过雄性钩片的形状和雌性的下生殖板长度来区分。

The genus contains about ten species, distributed worldwide. Three species and subspecies have been recorded from China, common and widespread. Species of the genus are medium-sized libellulids. Wings hyaline, pterostigma long, arc between antenodals 1 and 2, distal antenodal in fore wings incomplete, discoidal field with 2-3 rows of cells basally, cubital space with one crossvein, anal loop boot-shaped.

Crocothemis species inhabit standing water habitats and slow flowing streams. Males perch on the plants above water and fly frequently. The Chinese species are similar in appearance, males red, females red or yellow, they can be separated by male hamulus and length of female ovipositor.

长尾红蜻
Crocothemis erythraea

红蜻
Crocothemis servilia

红蜻属 雌性下生殖板
Genus *Crocothemis*, female vulvar lamina

长尾红蜻
Crocothemis erythraea

红蜻
Crocothemis servilia

红蜻属 雄性钩片
Genus *Crocothemis*, male hamulus

长尾红蜻 雄
Crocothemis erythraea, male

长尾红蜻 *Crocothemis erythraea* (Brullé, 1832)

长尾红蜻 雄,云南(德宏)
Crocothemis erythraea, male from Yunnan (Dehong)

长尾红蜻 雌,红色型,云南(西双版纳)
Crocothemis erythraea, female, the red morph
from Yunnan (Xishuangbanna)

长尾红蜻 雌,黄色型,云南(西双版纳)
Crocothemis erythraea, female, the yellow morph
from Yunnan (Xishuangbanna)

【形态特征】雄性通体红色;翅透明,通常后翅基方具橙色斑。雌性多型,分为黄色型和红色型,下生殖板较长。本种与红蜻非常相似,但雄性钩片末端通常具有2个齿,而红蜻仅有1个齿。本种雌性的下生殖板明显长于红蜻。【长度】体长 39~42 mm,腹长 25~28 mm,后翅 30~32 mm。【栖息环境】海拔 2000 m以下水草茂盛的湿地、池塘和水稻田。【分布】云南、贵州、广东、广西;亚洲、欧洲、非洲广布。【飞行期】全年可见。

[Identification] Male red throughout. Wings hyaline, hind wings with basal orange spots. Female polymorphic, color yellow or red, vulvar lamella projected and long. The species is similar to *C. servilia* but the distal tip of hamulus usually (but not always) with two teeth, only one in *C. servilia*. Female vulvar lamina longer than *C. servilia*. [Measurements] Total length 39-42 mm, abdomen 25-28 mm, hind wing 30-32 mm. [Habitat] Wetlands with plenty of emergent plants, ponds and paddy fields below 2000 m elevation. [Distribution] Yunnan, Guizhou, Guangdong, Guangxi; Widespread in Asia, Europe, Africa. [Flight Season] Throughout the year.

红蜻古北亚种 *Crocothemis servilia mariannae* Kiauta, 1983

【形态特征】雄性通体红色；翅透明，基方具橙色斑。雌性多型，分为黄色型和红色型。本亚种分布于中国的古北界区域。本亚种的雄性比指名亚种的雄性颜色更深，翅基方的色斑面积更大，色彩更深，且雌性多型。并未在指名亚种中发现红色型雌性。【长度】体长 44~47 mm，腹长 28~31 mm，后翅 34~35 mm。【栖息环境】海拔 1000 m 以下水草茂盛的湿地。【分布】广泛分布于中国北方；朝鲜半岛、日本。【飞行期】5—9月。

[Identification] Male red throughout. Wings hyaline, hind wings with basal orange spots. Female polymorphic, color yellow or red. This subspecies is confined to the Palearctic China whereas the nominated subspecies is confined to the Oriental region. Males of this subspecies has darker ground color, the spots at bases of wings are broader and darker, female is polymorphic. No red morph female has been found in nominate subspecies. [Measurements] Total length 44-47 mm, abdomen 28-31 mm, hind wing 34-35 mm. [Habitat] Wetlands with plenty of emergent plants below 1000 m elevation. [Distribution] Widespread in the north of China; Korean peninsula, Japan. [Flight Season] May to September.

红蜻古北亚种 雌，黄色型，吉林 | 金洪光 摄
Crocothemis servilia mariannae, female, the yellow morph from Jilin | Photo by Hongguang Jin

红蜻古北亚种 雌，红色型，吉林 | 金洪光 摄
Crocothemis servilia mariannae, female, the red morph from Jilin | Photo by Hongguang Jin

红蜻古北亚种 雄，吉林 | 金洪光 摄
Crocothemis servilia mariannae, male from Jilin | Photo by Hongguang Jin

红蜻指名亚种 *Crocothemis servilia servilia* (Drury, 1773)

红蜻指名亚种 雄,云南(红河)
Crocothemis servilia servilia, male from Yunnan (Honghe)

红蜻指名亚种 雌,贵州
Crocothemis servilia servilia, female from Guizhou

【形态特征】雄性通体红色;翅透明,基方具橙色斑;有时腹部有黑色的背中线。雌性棕黄色。【长度】体长 40~44 mm,腹长 26~29 mm,后翅 32~33 mm。【栖息环境】海拔 2500 m以下水草茂盛的湿地、池塘和水稻田。【分布】广泛分布于中国南方;广泛分布于亚洲的热带和亚热带区域、中东、美国(引入)、牙买加、波多黎各、古巴。【飞行期】全年可见。

[Identification] Male red throughout. Wings hyaline, hind wings with basal orange spots. Abdomen with or without black stripes along carina. Female brownish yellow. [Measurements] Total length 40-44 mm, abdomen 26-29 mm, hind wing 32-33 mm. [Habitat] Wetlands with plenty of emergent plants, ponds and paddy fields below 2500 m elevation. [Distribution] Widespread in the south of China; Widespread in tropical and subtropical Asia, Middle East, introduced to United States, Jamaica, Puerto Rico, Cuba. [Flight Season] Throughout the year.

多纹蜻属 Genus *Deielia* Kirby, 1889

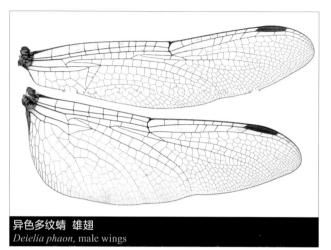

异色多纹蜻 雄翅
Deielia phaon, male wings

本属全球仅知1种，是体中型的蜻科物种，在亚洲的温带和亚热带地区广布，包括日本、朝鲜半岛和中国。雄性完全成熟后身体覆盖蓝白色粉霜，雌性多型。翅大面积透明，翅痣较长，弓脉位于第1条与第2条结前横脉之间，前翅最末端的结前横脉完整，盘区基方具2~3列翅室，基臀区具1条横脉，臀圈靴状。

本属蜻蜓栖息于大型池塘和水库，喜欢较大型的水体。雄性通常停落在水体边缘等待配偶。

The genus contains a single medium-sized species, widespread in temperate and subtropical Asia, including Japan, Korean peninsula and China. Males pruinosed when fully mature, females polymorphic. Wings largely hyaline, pterostigma long, arc between antenodals 1 and 2, distal antenodal in fore wings complete, discoidal field with 2-3 rows of cells basally, cubital space with one crossvein, anal loop boot-shaped.

Deielia phaon inhabits large ponds and reserviors. Males perch at the edge of water waiting for females.

异色多纹蜻 雌
Deielia phaon, female

异色多纹蜻 *Deielia phaon* (Selys, 1883)

【形态特征】雄性成熟以后胸部和腹部覆盖蓝白色粉霜；翅透明。雌性多型；蓝色型与雄性相似；橙色型身体为黄色并具黑色条纹，翅橙色，近翅端常具有褐色条纹。【长度】体长 40~42 mm，腹长 28~30 mm，后翅 32~36 mm。【栖息环境】海拔 2000 m以下的水库周边和中型池塘。【分布】东北至西南广布；朝鲜半岛、日本、俄罗斯远东。【飞行期】4—9月。

[Identification] Thorax and abdomen with bluish white pruinosity in fully mature male. Wings hyaline. Female polymorphic. The blue morph similar to male. The orange morph generally yellow with black markings, wings orange with brown bands near the tips. [Measurements] Total length 40-42 mm, abdomen 28-30 mm, hind wing 32-36 mm. [Habitat] Margin of reservoirs and medium-sized ponds below 2000 m elevation. [Distribution] Widespread from Northeast to Southwest China; Korean peninsula, Japan, Russian Far East. [Flight Season] April to September.

异色多纹蜻 雄,吉林│金洪光 摄
Deielia phaon, male from Jilin │ Photo by Hongguang Jin

异色多纹蜻 雄,黑龙江│莫善濂 摄
Deielia phaon, male from Heilongjiang │ Photo by Shanlian Mo

异色多纹蜻 雌,橙色型,湖北│吴宏道 摄
Deielia phaon, female, the orange morph from Hubei │ Photo by Hongdao Wu

异色多纹蜻 雌,蓝色型,黑龙江│莫善濂 摄
Deielia phaon, female, the blue morph from Heilongjiang │ Photo by Shanlian Mo

蓝小蜻属 Genus *Diplacodes* Kirby, 1889

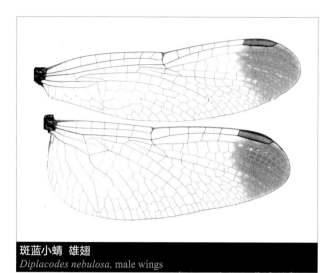

斑蓝小蜻 雄翅
Diplacodes nebulosa, male wings

本属全球已知11种,分布于亚洲、欧洲、大洋洲和非洲的热带区域。中国已知2种。本属蜻蜓体型很小;翅大面积透明,有时具色斑,翅痣较长,弓脉位于第1条与第2条结前横脉之间,前翅最末端的结前横脉不完整,盘区基方具1~2列翅室,基臀区具1条横脉,臀圆靴状,后翅三角室基边与弓脉的位置相当。

本属蜻蜓主要栖息于浅水池塘,包括季节性池塘和水稻田。它们喜欢停落在路面上,但飞行速度快,有时突然消失,不容易追踪。

The genus contains 11 species distributed in tropical Asia, Europe, Ocearia and Africa. Two species are known from China. Species of the genus are small-sized libellulids. Wings hyaline, color spots on wing tips present in some species, pterostigma long, arc between antenodals 1 and 2, distal antenodal in fore wings incomplete, discoidal field with 1-2 rows of cells basally, cubital space with one crossvein, anal loop boot-shaped, base of triangle in hind wings at the level of arc.

Diplacodes species inhabit standing water habitats, preferring shallow ponds, seasonal pools and paddy fields. They usually perch on ground and can fly very fast when disturbed.

纹蓝小蜻 雄
Diplacodes trivialis, male

斑蓝小蜻 *Diplacodes nebulosa* (Fabricius, 1793)

【形态特征】雄性完全成熟后身体覆盖深蓝色粉霜；翅端具褐色斑。雌性身体黄色具较丰富的黑色条纹；翅透明。【长度】体长 21~22 mm，腹长 14~15 mm，后翅 16~17 mm。【栖息环境】海拔 500 m以下水草茂盛的池塘。【分布】广东、广西、海南、香港、云南；南亚、东南亚、澳大利亚、新几内亚。【飞行期】3—12月。

斑蓝小蜻 雄,云南(德宏)
Diplacodes nebulosa, male from Yunnan (Dehong)

斑蓝小蜻 雌,广东│宋睿斌 摄
Diplacodes nebulosa, female from Guangdong │ Photo by Ruibin Song

[Identification] Male body covered by dark blue pruinosity. Wings with apical brown spots. Female body yellow with black markings. Wings hyaline. [Measurements] Total length 21-22 mm, abdomen 14-15 mm, hind wing 16-17 mm. [Habitat] Well vegetated ponds below 500 m elevation. [Distribution] Guangdong, Guangxi, Hainan, Hong Kong, Yunnan; South and Southeast Asia, Australia, New Guinea. [Flight Season] March to December.

斑蓝小蜻 雄,云南(德宏)
Diplacodes nebulosa, male from Yunnan (Dehong)

纹蓝小蜻 *Diplacodes trivialis* (Rambur, 1842)

【形态特征】雄性完全成熟后身体覆盖蓝色粉霜；翅透明。雌性身体黄色具黑色条纹；翅透明。【长度】体长 29~32 mm，腹长 19~22 mm，后翅 21~25 mm。【栖息环境】海拔 2000 m 以下的沼泽、池塘和暂时性水塘。【分布】广泛分布于华南和西南地区；亚洲、大洋洲广布。【飞行期】全年可见。

[Identification] Fully mature male body covered by blue pruinosity. Wings hyaline. Female body yellow with black markings. Wings hyaline. [Measurements] Total length 29-32 mm, abdomen 19-22 mm, hind wing 21-25 mm. [Habitat] Marshes, ponds and temporary pools below 2000 m elevation. [Distribution] Widespread in South and Southwest China; Widespread in Asia, Oceania. [Flight Season] Throughout the year.

纹蓝小蜻 雌，云南（德宏）
Diplacodes trivialis, female from Yunnan (Dehong)

纹蓝小蜻 雄，云南（德宏）
Diplacodes trivialis, male from Yunnan (Dehong)

楔翅蜻属 Genus *Hydrobasileus* Kirby, 1889

本属全球已知3种,分布于亚洲和大洋洲。中国已知1种。本属蜻蜓体型较大;翅具色彩,后翅基方甚阔,弓脉位于第1条与第2条结前横脉之间,盘区基方具3~4列翅室,基臀区具1条横脉,臀圈靴状。

本属蜻蜓主要栖息于水草茂盛的池塘。雄性会在池塘上具有一定高度的区域滑行,巡视领地,飞行速度较慢。雌雄连结产卵。

The genus contains three species distributed throughout Asia and Oceania. One species is recorded from China. Species of the genus are moderately large. Wings colourful, hind wing bases broad, arc between antenodals 1 and 2, discoidal field with 3-4 rows of cells basally, cubital space with one crossvein, anal loop boot-shaped.

Hydrobasileus species inhabit well vegetated ponds. Males patrol by slowly flying continually above water. They lay eggs in tandem.

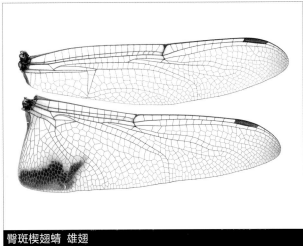

臀斑楔翅蜻 雄翅
Hydrobasileus croceus, male wings

臀斑楔翅蜻 雄 | 严少华 摄
Hydrobasileus croceus, male | Photo by Shaohua Yan

臀斑楔翅蜻 *Hydrobasileus croceus* (Brauer, 1867)

臀斑楔翅蜻 雄, 云南 (西双版纳)
Hydrobasileus croceus, male from Yunnan (Xishuangbanna)

臀斑楔翅蜻 雌, 云南 (西双版纳)
Hydrobasileus croceus, female from Yunnan (Xishuangbanna)

臀斑楔翅蜻 雄, 云南 (西双版纳)
Hydrobasileus croceus, male from Yunnan (Xishuangbanna)

　　【形态特征】通体橙色。雄性腹部具黑黄相间的斑点; 翅橙色, 后翅臀区具甚大的褐色斑。雌性腹部无显著斑点。【长度】体长 47~54 mm, 腹长 31~35 mm, 后翅 41~50 mm。【栖息环境】海拔 1000 m 以下水草茂盛的湿地。【分布】广泛分布于中国南方; 日本、文莱、孟加拉国、印度、斯里兰卡、缅甸、泰国、老挝、柬埔寨、越南、马来西亚、印度尼西亚、菲律宾、新加坡。【飞行期】3—12月。

　　[Identification] Body orange throughout. Male abdomen with black and yellow spots. Wings orange, hind wings with large brown spots at anal field. Female abdomen without clear markings. [Measurements] Total length 47-54 mm, abdomen 31-35 mm, hind wing 41-50 mm. [Habitat] Wetlands with plenty of emergent vegatation below 1000 m elevation. [Distribution] Widespread in the south of China; Japan, Brunei, Bangladesh, India, Sri Lanka, Myanmar, Thailand, Laos, Cambodia, Vietnam, Malaysia, Indonesia, Philippines, Singapore. [Flight Season] March to December.

沼蜻属 Genus *Hylaeothemis* Ris, 1909

本属全球已知4种,分布于亚洲。中国已知1种,分布于云南和海南的热带雨林。本属蜻蜓体型较小;体黑色具黄色斑纹;翅狭长,结前横脉完整,弓脉位于第2条和第3条结前横脉之间,盘区基方仅有1列翅室,无臀圈。

本属蜻蜓栖息于森林中的小型池塘和沼泽地。雄性会停落在水边的植物上占据领地,并时而巡逻。

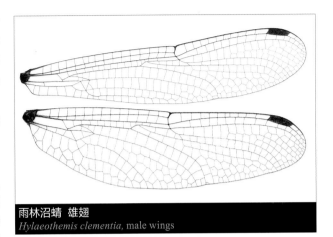

雨林沼蜻 雄翅
Hylaeothemis clementia, male wings

The genus contains four species distributed in Asia. One is recorded from China confined to the tropical forests in Yunnan and Hainan. Species of the genus are small-sized. Body black with yellow markings. Wings narrow and long, distal antenodal complete, arc between antenodals 2 and 3, discoidal field with only one row of cells, anal loop absent.

Hylaeothemis species prefer small pools and marshes in forest. Territorial males perch on plants near water, sometimes they patrol and hover.

雨林沼蜻 雄
Hylaeothemis clementia, male

雨林沼蜻 *Hylaeothemis clementia* Ris, 1909

【形态特征】复眼蓝绿色，面部白色，额蓝黑色具金属光泽；胸部和腹部主要黑色，合胸脊具黄白色条纹，胸部侧面具2条甚阔的黄白色条纹；腹部基方4节具斑点，第7节具1对黄白色斑点。【长度】体长 34~35 mm，腹长 22~23 mm，后翅 28~30 mm。【栖息环境】海拔 1000 m以下森林中的小型沼泽地和渗流地。【分布】云南、海南；文莱、泰国、老挝、越南、马来西亚、印度尼西亚。【飞行期】4—7月。

[Identification] Eyes bluish green, face white, frons metallic bluish black. Thorax and abdomen mainly black, thoracic dorsal carina with yellowish white stripes, sides with two broad yellowish white stripes. Basal four abdominal segments spotted, S7 with a pair of yellowish white spots. [Measurements] Total length 34-35 mm, abdomen 22-23 mm, hind wing 28-30 mm. [Habitat] Small marshes and seepages in forest below 1000 m elevation. [Distribution] Yunnan, Hainan; Brunei Darussalam, Thailand, Laos, Vietnam, Malaysia, Indonesia. [Flight Season] April to July.

雨林沼蜻 雄，海南
Hylaeothemis clementia. male from Hainan

雨林沼蜻 雌，海南
Hylaeothemis clementia, female from Hainan

雨林沼蜻 雄，云南（西双版纳）
Hylaeothemis clementia, male from Yunnan (Xishuangbanna)

印蜻属 Genus *Indothemis* Ris, 1909

本属全球已知仅2种，主要分布于亚洲的热带气候区。在中国它们分布于广东、广西和云南。雄性身体被深蓝色粉霜覆盖，雌性黄色；翅透明而狭长，翅痣较长，弓脉位于第1条与第2条结前横脉之间，前翅最末端的结前横脉不完整，盘区基方具1~3列翅室，基臀区具1条横脉，臀圈靴状，后翅三角室基边与弓脉的位置相当。

本属蜻蜓栖息于具有宽阔水面而水草匮乏的池塘。雄性会停落在水面附近的枝条和植物上等待配偶。交尾在空中完成，交尾后雌雄连结产卵。

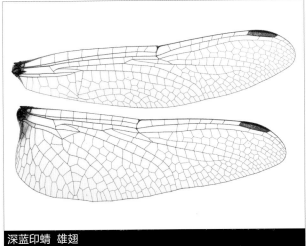

深蓝印蜻 雄翅
Indothemis carnatica, male wings

The genus contains only two species, confined to tropical Asia. Both are recorded from China, found in Guangdong, Guangxi and Yunnan. Males heavily pruinosed by dark blue, female yellow fundamentally. Wings hyaline, pterostigma long, arc between antenodals 1 and 2, distal antenodal in fore wings incomplete, discoidal field with 1-3 rows of cells basally, cubital space with one crossvein, anal loop boot-shaped, base of triangle in hind wing at the level of arc.

Indothemis species inhabit ponds with exposed water body. Males perch on the branches or plants. Mating takes place in flight, after which the pair lay eggs in tandem.

深蓝印蜻 交尾
Indothemis carnatica, mating pair

深蓝印蜻 *Indothemis carnatica* (Fabricius, 1798)

【形态特征】雄性复眼和面部黑褐色；胸部和腹部覆盖深蓝色粉霜；翅透明，后翅基方具1个较小的褐色斑。雌性身体黄色具黑色条纹；翅透明。【长度】体长 33~39 mm，腹长 22~25 mm，后翅 28~30 mm。【栖息环境】海拔 1500 m以下的池塘。【分布】云南、广西、广东；印度、斯里兰卡、泰国、柬埔寨、越南。【飞行期】4—11月。

[Identification] Male eyes and face blackish brown. Thorax and abdomen covered by dark blue pruinosity. Wings hyaline, hind wing bases with a small brown spot. Female mostly yellow with black markings. Wings hyaline. [Measurements] Total length 33-39 mm, abdomen 22-25 mm, hind wing 28-30 mm. [Habitat] Ponds below 1500 m elevation. [Distribution] Yunnan, Guangxi, Guangdong; India, Sri Lanka, Thailand, Cambodia, Vietnam. [Flight Season] April to November.

深蓝印蜻 雄,广东 | 宋睿斌 摄
Indothemis carnatica, male from Guangdong | Photo by Ruibin Song

深蓝印蜻 雌,云南(红河)
Indothemis carnatica, female from Yunnan (Honghe)

蓝黑印蜻 *Indothemis limbata* (Selys, 1891)

蓝黑印蜻 雌，云南（德宏）
Indothemis limbata, female from Yunnan (Dehong)

蓝黑印蜻 雄，云南（德宏）
Indothemis limbata, male from Yunnan (Dehong)

【形态特征】雄性复眼黑褐色，面部大面积黑色，额灰色；胸部黑色，翅大面积透明，末端边缘染有褐色，后翅基方具较大的黑褐色斑；腹部完全成熟以后覆盖深蓝色粉霜，未熟时具黄色条纹。雌性身体主要黄色具黑色条纹；翅大面积透明，末端边缘染有淡褐色，基方染有橙色。【长度】体长 30~37 mm，腹长 20~25 mm，后翅 25~28 mm。【栖息环境】海拔 500 m以下的池塘。【分布】云南（德宏）；印度、斯里兰卡、缅甸、泰国、老挝、越南、马来西亚、新加坡。【飞行期】不详，仅有迁移个体的记录。

[Identification] Male eyes blackish brown, face largely black, frons grey. Thorax black, wings largely hyaline, the tips tinted with brown, hind wing bases with big blackish brown spots. Abdomen covered by dark blue pruinosity when fully mature, abdominal yellow stripes present in immatures. Female mostly yellow with black markings. Wings largely hyaline, tips tinted with light brown, bases tinted with orange. [Measurements] Total length 30-37 mm, abdomen 20-25 mm, hind wing 25-28 mm. [Habitat] Ponds below 500 m elevation. [Distribution] Yunnan (Dehong); India, Sri Lanka, Myanmar, Thailand, Laos, Vietnam, Malaysia, Singapore, [Flight Season] Unknown, only migrations recorded.

秘蜻属 Genus *Lathrecista* Kirby, 1889

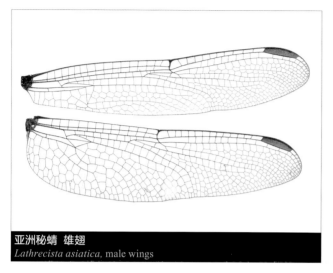

亚洲秘蜻 雄翅
Lathrecista asiatica, male wings

本属全球仅1种,广泛分布于亚洲和大洋洲,在中国分布于云南、海南和台湾等地的热带区域。本属蜻蜓体中型;翅透明而狭长,翅痣很长,具有较多条结前横脉,弓脉位于第2条与第3条结前横脉之间,前翅三角室的上缘很短,基臀区具1条横脉,盘区基方具2～3列翅室,臀圈靴状。

本属蜻蜓主要栖息于茂盛森林中具有树荫的池塘。雄性停落在水边的枝条上。

The genus contains a single species widely distributed in tropical Asia and Oceania. In China, it is recorded from Yunnan, Hainan and Taiwan. Species of the genus is medium-sized. Wings hyaline and narrow, pterostigma long, antenodals abundant, arc between antenodals 2 and 3, the anterior margin of triangle shot in fore wings, cubital space with one crossvein, discoidal field with 2-3 rows of cells, anal loop boot-shaped.

Lathrecista species inhabits shady ponds in forests. Territorial males perch for exterded periods on branches near water.

亚洲秘蜻 雄
Lathrecista asiatica, male

亚洲秘蜻 *Lathrecista asiatica* (Fabricius, 1798)

【形态特征】雄性复眼上方褐色下方蓝色，面部乳白色，额具1个甚大的褐色斑；胸部褐色，老熟以后覆盖淡淡的灰白色粉霜，翅透明；腹部鲜红色，末端2节和肛附器黑色。雌性体色暗淡；胸部褐色，侧面具发达的黄色条纹；腹部褐色具黄色条纹。【长度】体长 43~47 mm，腹长 29~32 mm，后翅 34~35 mm。【栖息环境】海拔 500 m 以下的小型林荫池塘。【分布】云南、海南、台湾；亚洲的热带和亚热带区域、大洋洲。【飞行期】全年可见。

[Identification] Male eyes brown above and blue below, face milk white, frons with a large brown spot. Thorax brown, slightly greyish white pruinosed when aged, wings hyaline. Abdomen red, distal two segments and anal appendages black. Female paler. Thorax brown with lateral yellow stripes. Abdomen brown with yellow stripes. [Measurements] Total length 43-47 mm, abdomen 29-32 mm, hind wing 34-35 mm. [Habitat] Small shady pools below 500 m elevation. [Distribution] Yunnan, Hainan, Taiwan; Tropical and subtropical Asia, Oceania. [Flight Season] Throughout the year.

亚洲秘蜻 雄，海南 | 莫善濂 摄
Lathrecista asiatica, male from Hainan | Photo by Shanlian Mo

亚洲秘蜻 雌，云南 (西双版纳)
Lathrecista asiatica, female from Yunnan (Xishuangbanna)

亚洲秘蜻 雄，海南
Lathrecista asiatica, male from Hainan

白颜蜻属 Genus *Leucorrhinia* Brittinger, 1850

本属全球已知15种，主要分布于欧洲、北美洲和亚洲气候较冷的区域。中国已知2种，分布于黑龙江和吉林。本属蜻蜓体型较小，体色较暗；翅大面积透明，基方具黑褐色斑，弓脉位于第1条与第2条结前横脉之间，臀圈靴状，盘区基方具有3列或者更多的翅室，后翅三角室基方与弓脉位置相当。

本属蜻蜓栖息于水草茂盛的湿地。雄性停落在水面的植物上等待配偶，时而飞行巡逻。

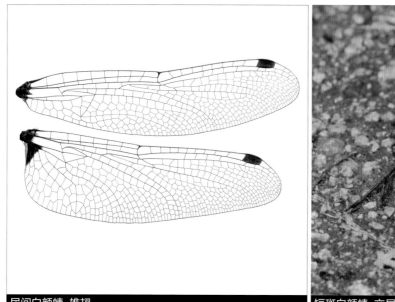

居间白颜蜻 雄翅
Leucorrhinia intermedia, male wings

短斑白颜蜻 交尾 | 莫善濂 摄
Leucorrhinia dubia, mating pair | Photo by Shanlian Mo

The genus contains 15 species confined to cooler climatic regions in Europe, North America and Asia. Two species are recored from China, found in Heilongjiang and Jilin. Species of the genus are small and generally dark dragonflies. Wings largely hyaline with basal blackish brown spots, arc between antenodals 1 and 2, anal loope boot-shaped, discoidal field with 3 rows of cells or more, the triangle in hind wing at the level of arc.

Leucorrhinia species inhabit well vegetated ponds. Territorial males perch for extended periods on plants near water, sometimes patroling for short distances.

短斑白颜蜻 *Leucorrhinia dubia* (Vander Linden, 1825)

【形态特征】雄性复眼灰绿色，面部白色；胸部黑褐色，具金属光泽，背板红色，翅透明；腹部主要黑色，基方具1个甚大红斑，第6~7节具黄斑。雌性身体通常黑褐色具黄色斑纹。【长度】体长 35~37 mm，腹长 24~25 mm，后翅 26~29 mm。【栖息环境】海拔 500 m以下挺水植物茂盛的湿地。【分布】黑龙江；朝鲜半岛、日本、西伯利亚、欧洲。【飞行期】5—8月。

[Identification] Male eyes greyish green, face white. Thorax blackish brown with metallic reflections, dorsum red, wings hyaline. Abdomen mainly black, base with a large red spot, S6-S7 with yellow spots. Female generally blackish brown with yellow markings. [Measurements] Total length 35-37 mm, abdomen 24-25 mm, hind wing 26-29 mm. [Habitat] Wetlands with plenty of emergent vegetation below 500 m elevation. [Distribution] Heilongjiang; Korean peninsula, Japan, Siberia, Europe. [Flight Season] May to August.

短斑白颜蜻 交尾，黑龙江｜莫善濂 摄
Leucorrhinia dubia, mating pair from Heilongjiang｜Photo by Shanlian Mo

短斑白颜蜻 雄，黑龙江｜莫善濂 摄
Leucorrhinia dubia, male from Heilongjiang｜Photo by Shanlian Mo

居间白颜蜻 *Leucorrhinia intermedia* Bartenev, 1912

【形态特征】雄性复眼灰绿色,面部白色;胸部黑褐色具褐色条纹,背板红色,翅透明;腹部主要黑色,第2~7节具红色和黄色斑。本种与短斑白颜蜻近似,但后者腹部第4~5节无黄色斑点。【长度】体长 33~37 mm,腹长 22~26 mm,后翅 26~28 mm。【栖息环境】海拔 500 m以下挺水植物茂盛的湿地。【分布】黑龙江、吉林;朝鲜半岛、日本、西伯利亚。【飞行期】5—8月。

居间白颜蜻 雄,黑龙江 | 莫善濂 摄
Leucorrhinia intermedia, male from Heilongjiang | Photo by Shanlian Mo

[Identification] Male eyes greyish green, face white. Thorax blackish brown with brown markings, dorsum red, wings hyaline. Abdomen mainly black, S2-S7 with red and yellow spots. The species is similar to *Leucorrhinia dubia*, but yellow spots on S4-S5 absent in *L. dubia*. [Measurements] Total length 33-37 mm, abdomen 22-26 mm, hind wing 26-28 mm. [Habitat] Wetlands with plenty of emergent vegetation below 500 m elevation. [Distribution] Heilongjiang, Jilin; Korean peninsula, Japan, Siberia. [Flight Season] May to August.

蜻属 Genus *Libellula* Linnaeus, 1758

本属全球已知约30种,主要分布在美洲,少数分布于欧洲和亚洲。中国已知5种,全国广布。本属蜻蜓较粗壮,身体色彩多样;翅透明而狭长,常具色斑,翅痣很长;具有较多条结前横脉,弓脉位于第2条与第3条结前横脉之间,前翅三角室的上缘很短,IR3呈明显的波浪状,基臀区具1条横脉,盘区具3~4列翅室,臀圈靴状。

本属蜻蜓主要栖息于水草茂盛的池塘。雄性经常在水面四处巡逻,时而悬停,时而停歇。交尾通常在空中进行,时间较短。雄性护卫产卵。

The genus contains about 30 species, most of which are found in Americas, a few others found in Europe and Asia. Five species are recorded from China, widespread throughout the country. Species of the genus are rather robust often with colourful body patterns. Wings hyaline and long with color spots, pterostigma long. Antenodals abundant, arc between antenodals 2 and 3, the anterior margin of triangle short in fore wings, IR3 clearly waved, cubital space with one crossvein, discoidal field with 3-4 rows of cells, anal loop boot-shaped.

Libellula species inhabit well vegetated ponds where males can be seen flying, sometimes hovering or perching. Mating usually takes place when flying but lasts for only a few seconds, after which the female oviposits alone with the hovering male guarding nearby.

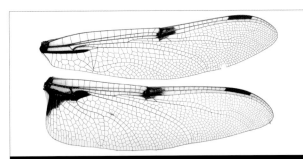

高斑蜻 雄翅
Libellula basilinea, male wings

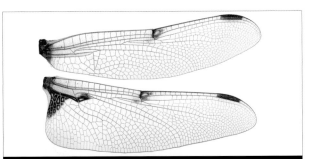

小斑蜻 雄翅
Libellula quadrimaculata, male wings

低斑蜻 雄
Libellula angelina, male

低斑蜻 *Libellula angelina* Selys, 1883

【形态特征】成熟的雄性身体黑褐色；翅透明具三角形黑色斑。雌性身体通常黄褐色；翅透明具褐色斑；腹部背面中央具1条黑色宽条纹。未熟的雄性与雌性色彩相似。【长度】体长 38~43 mm，腹长 25~27 mm，后翅 30~32 mm。【栖息环境】海拔 500 m 以下挺水植物茂盛的湿地。【分布】北京、河北、山西、江苏、安徽、湖北；朝鲜半岛、日本。【飞行期】3—5月。

[Identification] Fully mature male body blackish brown. Wings hyaline with triangular black spots. Female mostly yellowish brown. Wings hyaline with brown spots. Abdomen with a broad mid-dorsal black stripe. Immature male similar to female. [Measurements] Total length 38-43 mm, abdomen 25-27 mm, hind wing 30-32 mm. [Habitat] Wetlands with plenty of emergent vegetation below 500 m elevation. [Distribution] Beijing, Hebei, Shanxi, Jiangsu, Anhui, Hubei; Korean peninsula, Japan. [Flight Season] March to May.

低斑蜻 雌,北京 | 安起迪 摄
Libellula angelina, female from Beijing | Photo by Qidi An

低斑蜻 雄,安徽
Libellula angelina, male from Anhui

低斑蜻 雄,河北 | 陈炜 摄
Libellula angelina, male from Hebei | Photo by Wei Chen

高斑蜻 *Libellula basilinea* McLachlan, 1894

高斑蜻 雌，湖北
Libellula basilinea, female from Hubei

高斑蜻 雄，湖北
Libellula basilinea, male from Hubei

【形态特征】头部复眼褐色，面部黄色；胸部背面黄褐色，侧面鲜黄色，翅透明，前缘金黄色，翅结处和基方具黑色斑；腹部基方5节黄褐色，后方5节主要黑色，第2~9节侧面具黄色斑点。【长度】体长 42~52 mm，腹长 28~34 mm，后翅 38~40 mm。【栖息环境】海拔 1500~2500 m 的高山湿地。【分布】中国特有，分布于湖北、四川、重庆、云南。【飞行期】5—8月。

[Identification] Eyes brown, face yellow. Thorax dorsally yellowish brown, laterally yellow, wings hyaline, the anterior margin golden, bases and nodus with black spots. Abdomen with basal five segments yellowish brown, the remainder black, S2-S9 with lateral yellow spots. [Measurements] Total length 42-52 mm, abdomen 28-34 mm, hind wing 38-40 mm. [Habitat] Wetlands on high mountains at 1500-2500 m elevation. [Distribution] Endemic to China, recorded from Hubei, Sichuan, Chongqing, Yunnan. [Flight Season] May to August.

基斑蜻 *Libellula depressa* Linnaeus, 1758

【形态特征】雄性头部和胸部褐色，合胸具白色的肩前条纹；翅透明，基方具甚大的黑褐色斑；腹部甚阔，覆盖蓝白色粉霜，第3～4节侧面具黄色斑点。雌性通常黄褐色，头部和胸部与雄性相似；腹部黄褐色具黄色斑点。本种与米尔蜻相似，但米尔蜻雄性腹部仅第3～7节覆盖粉霜，而本种几乎全部覆盖；本种雄性具白色的肩前条纹，而米尔蜻无肩前条纹。【长度】体长 39～48 mm，腹长 22～31 mm，后翅 32～38 mm。【栖息环境】挺水植物茂盛的湿地。【分布】新疆；欧洲和亚洲中部广布。【飞行期】5—7月。

[Identification] Male eyes and thorax brown, synthorax with white antehumeral stripes. Wings hyaline with large basal blackish brown markings. Abdomen broad and covered by bluish white pruinosity, S3-S4 with lateral yellow spots. Female mostly yellowish brown, head and thorax similar to male but abdomen yellowish brown with yellow spots. The species is similar to *Libellula melli* but male abdomen is pruinosed in almost all segments, only S3-S7 pruinosed in *L. melli*. Male of the species possesses white antehumeral stripes but absent in *L. melli*. [Measurements] Total length 39-48 mm, abdomen 22-31 mm, hind wing 32-38 mm. [Habitat] Wetlands with plenty of emergent plants. [Distribution] Xinjiang; Widespread in Europe and Central Asia. [Flight Season] May to July.

基斑蜻 雄,芬兰 | Sami Karjalainen 摄
Libellula depressa, male from Finland | Photo by Sami Karjalainen

基斑蜻 雄,芬兰 | Matti Hämäläinen 摄
Libellula depressa, male from Finland | Photo by Matti Hämäläinen

基斑蜻 雌,芬兰 | Sami Karjalainen 摄
Libellula depressa, female from
Finland | Photo by Sami Karjalainen

米尔蜻 *Libellula melli* Schmidt, 1948

【形态特征】雄性复眼和胸部褐色；翅透明，基方具甚大的黑褐色斑；腹部甚阔，第3~7节覆盖蓝白色粉霜，其余各节深褐色。雌性主要黄褐色。【长度】体长 43~46 mm，腹长 26~30 mm，后翅 38~40 mm。【栖息环境】海拔 500~2000 m森林中的水塘。【分布】四川、贵州、湖北、湖南、安徽、浙江、福建、广东；越南。【飞行期】4—8月。

[Identification] Male eyes and thorax brown. Wings hyaline with large basal blackish brown markings. Abdomen broad, S3-S7 covered by bluish white pruinosity, remaining segments dark brown. Female mostly yellowish brown.

米尔蜻 雄,安徽 | 秦彧 摄
Libellula melli, male from Anhui | Photo by Yu Qin

米尔蜻 雄,安徽 | 陈尽 摄
Libellula melli, male from Anhui | Photo by Jin Chen

米尔蜻 雌,湖北
Libellula melli, female from Hubei

[Measurements] Total length 43-46 mm, abdomen 26-30 mm, hind wing 38-40 mm. [Habitat] Ponds in forest at 500-2000 m elevation. [Distribution] Sichuan, Guizhou, Hubei, Hunan, Anhui, Zhejiang, Fujian, Guangdong; Vietnam. [Flight Season] April to August.

小斑蜻 *Libellula quadrimaculata* Linnaeus, 1758

【形态特征】头部复眼褐色,面部黄色;胸部黄褐色,翅透明,前缘稍染金黄色,翅结处和基方具形状多变的褐色斑;腹部基方6节褐色,后方4节主要黑色,第2～9节侧面具黄色斑点。【长度】体长 42～47 mm,腹长 27～30 mm,后翅 34～36 mm。【栖息环境】海拔 500 m 以下水草茂盛的池塘。【分布】黑龙江、吉林、辽宁、内蒙古;朝鲜半岛、日本、西伯利亚、欧洲、北美洲。【飞行期】5—8月。

[Identification] Eyes brown, face yellow. Thorax yellowish brown, wings hyaline, costal margin with golden tint, bases and nodus with variably sized brown spots. Abdomen with basal six segments brown, the remainder black, S2-

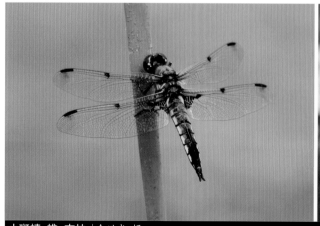

小斑蜻 雄,吉林 | 金洪光 摄
Libellula quadrimaculata, male from Jilin | Photo by Hongguang Jin

S9 with lateral yellow spots. **[Measurements]** Total length 42-47 mm, abdomen 27-30 mm, hind wing 34-36 mm. **[Habitat]** Wetlands with plenty of emergent vegetation below 500 m elevation. **[Distribution]** Heilongjiang, Jilin, Liaoning, Inner Mongolia; Korean peninsula, Japan, Siberia, Europe, North America. **[Flight Season]** May to August.

小斑蜻 雌，黑龙江 | 莫善濂 摄
Libellula quadrimaculata, female from Heilongjiang | Photo by Shanlian Mo

宽腹蜻属 Genus *Lyriothemis* Brauer, 1868

本属全球已知16种，仅分布于亚洲和新几内亚。中国已知5种，分布广泛。本属蜻蜓体小至中型，腹部扁平而宽阔；翅透明而狭长，翅痣很长，具有较多条结前横脉，弓脉位于第2条与第3条结前横脉之间，前翅三角室的上缘很短，后翅基臀区具2条横脉，盘区基方具2～3列翅室，臀圈靴状。

本属蜻蜓主要栖息于静水环境，常见于沼泽地、森林中的小型水潭等环境，有些可以在热带雨林的树洞中繁殖。雄性会在水面附近的枝条和植物上停落等待配偶。

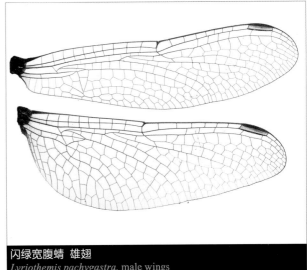

闪绿宽腹蜻 雄翅
Lyriothemis pachygastra, male wings

The genus contains 16 species only distributed in Asia and New Guinea. Five species are recorded from China, widespread. Species of the genus are small to medium sized, abdomen is broadly flattened. Wings hyaline, pterostigma long, antenodals abundant, arc between antenodals 2 and 3, the anterior margin of triangle is short in fore wings, cubital space with two crossveins in hind wings, discoidal field with 2-3 rows of cells, anal loop boot-shaped.

Lyriothemis species inhabit standing water, preferring small ponds, marshes in forest, some species breed in tree holes in tropical rain forests. Territorial males usually perch on the plants.

卡米宽腹蜻 雄
Lyriothemis kameliyae, male

双纹宽腹蜻 *Lyriothemis bivittata* (Rambur, 1842)

【形态特征】雄性复眼黑褐色，面部黄色，额具1个甚大的深蓝色金属斑；合胸背面红褐色，侧面黑色具2条宽阔的黄色条纹，翅透明，基方具甚细的褐色条纹；腹部红色，背面中央具1条甚细的褐色线纹。雌性色彩稍淡，腹部第8节侧面具甚小的片状突起。【长度】体长 50~55 mm，腹长 33~36 mm，后翅 35~38 mm。【栖息环境】海拔 500 m 以下热带雨林中的小型水潭和树洞中的积水潭。【分布】云南（红河、西双版纳）；印度、尼泊尔、缅甸、泰国、老挝、越南。【飞行期】4—6月。

[Identification] Male eyes blackish brown, face yellow, frons with a large metallic dark blue spot. Thorax dorsally reddish brown, sides black with two broad yellow stripes, wings hyaline with narrow brown basal stripes. Abdomen red

双纹宽腹蜻 雄，云南（红河）
Lyriothemis bivittata , male from Yunnan (Honghe)

双纹宽腹蜻 雌，云南（红河）｜莫善濂 摄
Lyriothemis bivittata, female from Yunnan (Honghe) ｜ Photo by Shanlian Mo

with a central narrow brown band along carina. Female paler, S8 with small lateral leaf-like flaps. [Measurements] Total length 50-55 mm, abdomen 33-36 mm, hind wing 35-38 mm. [Habitat] Small pools and small water in tree holes in dense forest below 500 m elevation. [Distribution] Yunnan (Honghe, Xishuangbanna); India, Nepal, Myanmar, Thailand, Laos, Vietnam. [Flight Season] April to June.

双纹宽腹蜻 雄，云南（红河）
Lyriothemis bivittata , male from Yunnan (Honghe)

华丽宽腹蜻 *Lyriothemis elegantissima* Selys, 1883

【形态特征】雄性复眼绿褐色，面部白色，额具1个甚大的深蓝色金属斑；胸部黄褐色具黑色条纹，翅透明；腹部鲜红色，背面中央具甚细的黑色线纹，末端2节黑色。雌性黄褐色并具黑色条纹，第8节侧面有较小的片状突起。【长度】体长36～41 mm，腹长23～26 mm，后翅30～35 mm。【栖息环境】海拔 1000 m以下的林荫池塘。【分布】福建、广东、广西、海南、香港、台湾；日本、泰国、柬埔寨、越南。【飞行期】5—10月。

[Identification] Male eyes greenish brown, face white, frons with a large metallic dark blue spot. Thorax yellowish brown with black stripes, wings hyaline. Abdomen red with narrow black stripes along dorsal carina, distal two segments black. Female yellowish brown with black markings, S8 with small lateral leaf-like flaps. [Measurements] Total length 36-41 mm, abdomen 23-26 mm, hind wing 30-35 mm. [Habitat] Shady ponds below 1000 m elevation. [Distribution] Fujian, Guangdong, Guangxi, Hainan, Hong Kong, Taiwan; Japan, Thailand, Cambodia, Vietnam. [Flight Season] May to October.

华丽宽腹蜻 雌,广东 │莫善濂 摄
Lyriothemis elegantissima, female from Guangdong │ Photo by Shanlian Mo

华丽宽腹蜻 雄,广东 │宋睿斌 摄
Lyriothemis elegantissima, male from Guangdong │ Photo by Ruibin Song

华丽宽腹蜻 雄,广东 │莫善濂 摄
Lyriothemis elegantissima, male from Guangdong │ Photo by Shanlian Mo

金黄宽腹蜻 *Lyriothemis flava* Oguma, 1915

【形态特征】雄性复眼黑褐色，面部白色，额具1个甚大的深蓝色金属斑；胸部黑褐色，具甚阔的肩前条纹，胸部侧面具2条宽阔的黄条纹，翅透明；腹部甚阔，橙黄色，沿背面中央具1条黑色线纹，末端黑褐色。雌性与雄性相似，腹部第8~10节主要黑褐色。【长度】体长 46~47 mm，腹长 30~31 mm，后翅 34~38 mm。【栖息环境】海拔1000 m以下茂盛森林中的小型水潭和树洞中的积水潭。【分布】贵州、福建、广东、广西、海南、台湾；日本、孟加拉国、印度、缅甸、越南。【飞行期】3—9月。

[Identification] Male eyes blackish brown, face white, frons with a large metallic dark blue spot. Thorax blackish brown with broad antehumeral stripes, sides with two broad yellow stripes, wings hyaline. Abdomen broad and orange

金黄宽腹蜻 雄，广东 | 宋睿斌 摄
Lyriothemis flava, male from Guangdong | Photo by Ruibin Song

yellow with a central black band along the carina, the apex blackish brown. Female similar to male, S8-S10 mainly blackish brown. [Measurements] Total length 46-47 mm, abdomen 30-31 mm, hind wing 34-38 mm. [Habitat] Small pools and small water in tree wholes in dense forest below 1000 m elevation. [Distribution] Guizhou, Fujian, Guangdong, Guangxi, Hainan, Taiwan; Bangladesh, India, Myanmar, Vietnam. [Flight Season] March to September.

金黄宽腹蜻 雌，广西
Lyriothemis tricolor, female from Guangxi

卡米宽腹蜻 *Lyriothemis kameliyae* Kompier, 2017

【形态特征】雄性复眼黑褐色，面部白色，额具1个甚大的蓝黑色斑；胸部黑褐色，侧面具2条黄色条纹，翅透明，基方具甚小的黑斑；腹部橙红色，末端黑色。雌性多型，橙色型与雄性相似，但色彩稍淡；黄色型黄褐色具黑色条纹；腹部第8节侧面有较小的片状突起。【长度】体长 36~46 mm，腹长 24~31 mm，后翅 33~39 mm。【栖息环境】海拔 1000 m 以下茂盛森林中的小型水潭和树洞中的积水潭。【分布】云南、贵州、广西、广东；越南。【飞行期】4—8月。

[Identification] Male eyes blackish brown, face white, frons with a large bluish black spot. Thorax blackish brown, sides with two broad yellow stripes, wings hyaline with very small black basal spots. Abdomen orange red with

卡米宽腹蜻 雌，广西
Lyriothemis kameliyae, female from Guangxi

卡米宽腹蜻 雄,广西
Lyriothemis kameliyae, male from Guangxi

black tip. Female polymorphic, the orange morph similar to male but paler. The yellow morph yellowish brown with black markings. S8 with small lateral leaf-like flaps. [Measurements] Total length 36-46 mm, abdomen 24-31 mm, hind wing 33-39 mm. [Habitat] Small pools and small water in tree wholes in dense forest below 1000 m elevation. [Distribution] Yunnan, Guizhou, Guangxi, Guangdong; Vietnam. [Flight Season] April to August.

闪绿宽腹蜻 *Lyriothemis pachygastra* (Selys, 1878)

【形态特征】雄性复眼黑褐色,面部白色,额具1个甚大的深蓝色金属斑;胸部黑褐色,年老后稍染蓝色粉霜,翅透明;腹部覆盖深蓝色粉霜,沿背面中央具1条黑色线纹。雌性黄褐色并具黑色条纹,腹部第8节侧面具甚小的片状突起。云南高山地区的雄性与其他省份明显不同,胸部侧面具发达的黄色斑。【长度】体长 32~35 mm,腹长 21~24 mm,后翅 24~26 mm。【栖息环境】海拔 2500 m以下的沼泽地。【分布】中国除华南地区全国广布;朝鲜半岛、日本、俄罗斯远东、泰国。【飞行期】5—9月。

[Identification] Male eyes blackish brown, face white, frons with a large metallic dark blue spot. Thorax blackish brown and covered by blue pruinosity when aged, wings hyaline. Abdomen covered by dark blue pruinosity with a black band along the carina. Female yellowish brown with black markings, S8 with small lateral leaf-like flaps. Males from Yunnan are clearly different from those from other provinces by the presence of yellow thoracic spots laterally. [Measurements] Total length 32-35 mm, abdomen 21-24 mm, hind wing 24-26 mm. [Habitat] Marshes below 2500 m elevation. [Distribution] Widespread throughout China except the South region; Korean peninsula, Japan, Russian Far East, Thailand. [Flight Season] May to September.

闪绿宽腹蜻 雌,湖北
Lyriothemis pachygastra, female from Hubei

闪绿宽腹蜻 雄,湖北
Lyriothemis pachygastra, male from Hubei

闪绿宽腹蜻 雄,安徽 | 秦彧 摄
Lyriothemis pachygastra, male from Anhui | Photo by Yu Qin

闪绿宽腹蜻 雄，云南（大理）
Lyriothemis pachygastra, male from Yunnan (Dali)

漭蜻属 Genus *Macrodiplax* Brauer, 1868

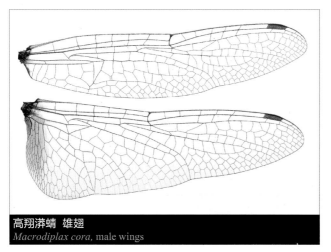

高翔漭蜻 雄翅
Macrodiplax cora, male wings

本属全球仅2种，一种广泛分布于亚洲、大洋洲和非洲，另一种分布于美洲。中国已知1种，分布于华南沿海地区。本属蜻蜓体中型，腹部扁平宽阔；翅大面积透明，基方具深色斑，翅脉稀疏，具较少的结前横脉，弓脉位于第1条与第2条结前横脉之间，盘区基方具2列翅室，基臀区具1条横脉，臀圈靴状。

本属蜻蜓可以忍耐咸水环境，在沿海的湿地和池塘较常见。

The genus contains two species, one widespread in Asia, Oceania and Africa, the other confined to Americas. One species is recorded from China and seen only along the southern coast. Species of the genus are medium-sized, abdomen flattened and broad. Wings largely hyaline but hind wing bases have dark color spots, venation sparse, antenodals few in number, arc between antenodals 1 and 2, discoidal field with 2 rows of cells basally, cubital space with one crossvein, anal loop boot-shaped.

Macrodiplax spcies are known for their to tolerance lerance of salty lagoons and estuaries.

高翔漭蜻 雄｜梁嘉景 摄
Macrodiplax cora, male｜Photo by Kenneth Leung

高翔湴蜻 *Macrodiplax cora* (Brauer, 1867)

高翔湴蜻 雌,香港 | 梁嘉景 摄
Macrodiplax cora, female from Hong Kong | Photo by Kenneth Leung

高翔湴蜻 雌,广东 | 莫善濂 摄
Macrodiplax cora, female from Guangdong | Photo by Shanlian Mo

高翔湴蜻 雄,广东 | 吴宏道 摄
Macrodiplax cora, male from Guangdong | Photo by Hongdao Wu

【形态特征】雄性头甚阔,复眼红褐色,面部红色;胸部红褐色,侧面具黑色条纹,翅透明,基方具红褐色斑;腹部红色,沿背面中央具1条黑色宽条纹。雌性黄褐色具褐色条纹。【长度】体长 40~45 mm,腹长 27~29 mm,后翅 31~34 mm。【栖息环境】河口和沿海附近的池塘。【分布】广东、海南、香港、台湾;广泛分布于亚洲、非洲、大洋洲。【飞行期】3—11月。

[Identification] Male head broad, eyes reddish brown, face red. Thorax reddish brown with lateral black stripes, wings hyaline, bases with reddish brown markings. Abdomen red with a central black stripe. Female yellowish brown with brown markings. [Measurements] Total length 40-45 mm, abdomen 27-29 mm, hind wing 31-34 mm. [Habitat] Estuaries and coastal ponds. [Distribution] Guangdong, Hainan, Hong Kong, Taiwan; Widespread in Asia, Africa and Oceania. [Flight Season] March to November.

红小蜻属 Genus *Nannophya* Rambur, 1842

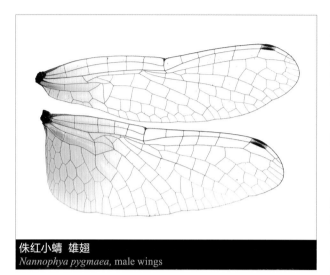

侏红小蜻 雄翅
Nannophya pygmaea, male wings

本属全球已知5种，中国仅知1种，侏红小蜻，是全球体型最小的蜻蜓之一。本属蜻蜓是一类十分小型的蜻科物种；翅大面积透明，仅在基方具色斑，翅脉稀疏，弓脉位于第1条与第2条结前横脉之间，基臀区仅具1条横脉。

本属蜻蜓栖息于水草茂盛的池塘。雄性常停落在植物顶部。交尾时停落在水草上。雄性护卫产卵。

The genus contains five species but only *Nannophya pygmaea* is recorded from China, a species among the smallest dragonflies. Species of the genus are tiny. Wings largely hyaline with colored spots at bases, venation sparse, arc between antenodals 1 and 2, cubital space with one crossvein.

Nannophya species inhabit well vegetated ponds. Males usually perch on top of plants. Mating pairs also perch. Female lays eggs with the hovering male guarding nearby.

侏红小蜻 雄 | 宋睿斌 摄
Nannophya pygmaea, male | Photo by Ruibin Song

侏红小蜻 *Nannophya pygmaea* Rambur, 1842

侏红小蜻 雄,广东 | 宋睿斌 摄
Nannophya pygmaea, male from Guangdong | Photo by Ruibin Song

【形态特征】雄性通体红色；后翅基部具大块红褐色斑。雌性黑褐色具浅黄色斑点。【长度】体长 17~19 mm，腹部 9~11 mm，后翅 12~15 mm。【栖息环境】海拔 1500 m 以下水草茂盛的池塘和水稻田。【分布】江苏、浙江、安徽、福建、湖南、江西、广东、广西、海南、香港、台湾；日本、东南亚、美拉尼西亚。【飞行期】4—9月。

[Identification] Male red throughout. Hind wing bases with large reddish brown spots. Female blackish brown with pale yellow markings. [Measurements] Total length 17-19 mm, abdomen 9-11 mm, hind wing 12-15 mm. [Habitat] Wetlands with plenty of emergent vegetation and paddy fields below 1500 m elevation. [Distribution] Jiangsu, Zhejiang, Anhui, Fujian, Hunan, Jiangxi, Guangdong, Guangxi, Hainan, Hong Kong, Taiwan; Japan, Southeast Asia, Melanesia. [Flight Season] April to September.

侏红小蜻 雌,广东 | 宋睿斌 摄
Nannophya pygmaea, female from Guangdong | Photo by Ruibin Song

斑小蜻属 Genus *Nannophyopsis* Lieftinck, 1935

本属全球已知2种，中国已知1种。本属是十分小型的黑色蜻科物种；翅短而阔，后翅的形状较特别，翅脉稀疏，弓脉位于第1条与第2条结前横脉之间，基臀区仅有1条横脉，臀圈靴状。

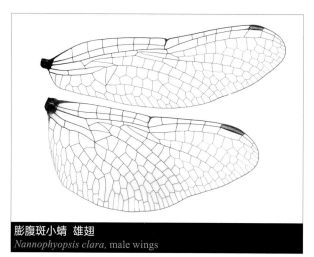

膨腹斑小蜻 雄翅
Nannophyopsis clara, male wings

本属蜻蜓栖息于水草茂盛的池塘。雄性在水面附近的枝条和植物上停落占据领地。飞行时腹部弯曲，形似蜂。

The genus contains two species and one of them is recorded from China. Species of the genus are small black species. Wings short and broad, the shape of hind wing special, venation sparse, arc between antenodals 1 and 2, cubital space with one crossvein, anal loop boot-shaped.

Nannophyopsis species inhabit well vegetated ponds. Territorial males usually perch on plants. The abdomen is curved when flying, looking like a wasp.

膨腹斑小蜻 雄
Nannophyopsis clara, male

膨腹斑小蜻 *Nannophyopsis clara* (Needham, 1930)

【形态特征】身体黑绿色具金属光泽，复眼深绿色。雄性后翅具1个甚大的琥珀色斑；腹部末端膨胀。雌性腹部末端未膨大。【长度】体长 22~24 mm，腹部 13~15 mm，后翅 14~16 mm。【栖息环境】海拔 1000 m 以下的池塘。【分布】浙江、福建、广东、海南、香港、台湾；泰国、越南。【飞行期】2—9月。

[Identification] Body blackish green with metallic reflection, eyes dark green. Male hind wings with a large amber spot. Abdominal tip exponded. Female abdominal tip not expanded. [Measurements] Total length 22-24 mm, abdomen 13-15 mm, hind wing 14-16 mm. [Habitat] Ponds below 1000 m elevation. [Distribution] Zhejiang, Fujian, Guangdong, Hainan, Hong Kong, Taiwan; Thailand, Vietnam. [Flight Season] February to September.

膨腹斑小蜻 雌,广东 | 莫善濂 摄
Nannophyopsis clara, female from Guangdong | Photo by Shanlian Mo

膨腹斑小蜻 交尾,广东 | 莫善濂 摄
Nannophyopsis clara, mating pair from Guangdong | Photo by Shanlian Mo

膨腹斑小蜻 雄,广东
Nannophyopsis clara, male from Guangdong

脉蜻属 Genus *Neurothemis* Brauer, 1867

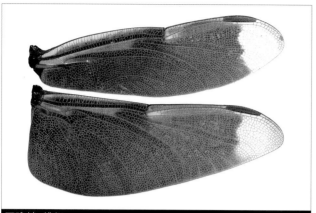

网脉蜻 雄翅
Neurothemis fulvia, male wings

本属全球已知10余种，分布于亚洲的热带区域、澳大利亚和美拉尼西亚。中国已知5种，主要分布于华南和西南地区。本属为小至中型蜻蜓；翅具色彩，翅痣较长，翅脉非常密集，盘区基方具多列翅室，基臀区具较多数量的横脉，臀圈靴状。本属蜻蜓可以通过翅的色彩和身体条纹区分。

本属蜻蜓栖息于水草茂盛的浅水池塘、沼泽地和水稻田。雄性经常在水面附近的枝条和水草上停落，时而飞行，一些小型种类经常在池塘附近的灌木丛中游荡。交尾时间较短，雄性护卫产卵。

The genus contains over ten species, confined to tropical Asia, Australia and Melanesia. Five species are recorded from China, often seen in the South and Southwest. Species of the genus are small to medium sized libellulids. Wings colorfully patterned, pterostigma long, venation dense, discoidal field with many rows of cells basally, cubital space with many crossveins, anal loop boot-shaped. Color pattern of body and wings is diagnostic character.

Neurothemis species inhabit well vegetated ponds, marshes and paddy fields. Males perch on the branches or plants, and sometimes fly above water, some small-sized species are often found amongst short bushes. Mating duration is short and females lay eggs with male guarding nearby.

网脉蜻 雄
Neurothemis fulvia, male

月斑脉蜻 *Neurothemis fluctuans* (Fabricius, 1793)

【形态特征】雄性通体红色；翅大面积红色，仅端部透明；腹部具黑褐色斑。雌性通体黄褐色；翅稍染棕黄色；腹部具褐色条纹。【长度】体长 28~34 mm，腹部 19~26 mm，后翅 24~29 mm。【栖息环境】海拔 500 m以下森林中的池塘。【分布】云南（德宏）；广泛分布于亚洲的热带和亚热带区域。【飞行期】10—12月。

[Identification] Male red throughout. Wings largely red with tips hyaline. Abdomen with blackish brown markings. Female yellowish brown throughout. Wings slightly tinted with brownish yellow. Abdomen with brown stripes. [Measurements] Total length 28-34 mm, abdomen 19-26 mm, hind wing 24-29 mm. [Habitat] Ponds in forest below 500 m elevation. [Distribution] Yunnan (Dehong); Widespread in tropical and subtropical Asia. [Flight Season] October to December.

月斑脉蜻 雄，马来西亚
Neurothemis fluctuans, male from Malaysia

月斑脉蜻 雌，云南（德宏）
Neurothemis fluctuans, female from Yunnan (Dehong)

月斑脉蜻 雄，云南（德宏）
Neurothemis fluctuans, male from Yunnan (Dehong)

网脉蜻 *Neurothemis fulvia* (Drury, 1773)

【形态特征】雄性通体红色；翅大面积红色，仅端部透明。雌性通体黄褐色。【长度】体长 35～40 mm，腹部 20～26 mm，后翅 26～32 mm。【栖息环境】海拔 2000 m以下的湿地和水稻田。【分布】云南、福建、广东、广西、海南、香港、台湾；南亚、东南亚广布。【飞行期】全年可见。

[Identification] Male red throughout. Wings largely red with tips hyaline. Female yellowish brown throughout. [Measurements] Total length 35-40 mm, abdomen 20-26 mm, hind wing 26-32 mm. [Habitat] Wetlands and paddy fields below 2000 m elevation. [Distribution] Yunnan, Fujian, Guangdong, Guangxi, Hainan, Hong Kong, Taiwan; Widespread in South and Southeast Asia. [Flight Season] Throughout the year.

网脉蜻 雄，广东
Neurothemis fulvia, male from Guangdong

网脉蜻 雄, 云南 (德宏)
Neurothemis fulvia, male from Yunnan (Dehong)

网脉蜻 雌, 云南 (西双版纳)
Neurothemis fulvia, female from Yunnan (Xishuangbanna)

褐基脉蜻 *Neurothemis intermedia* (Rambur, 1842)

【形态特征】雄性黄褐色；翅基部具较大的琥珀色斑；腹部橙红色。雌性黄色具褐色斑纹；翅透明。【长度】体长 29~32 mm，腹长 19~21 mm，后翅 23~24 mm。【栖息环境】海拔 1500 m 以下的湿地。【分布】云南 (德宏)、海南；广泛分布于亚洲的热带和亚热带区域。【飞行期】全年可见。

[Identification] Male yellowish brown. Wing base with large amber spots. Abdomen orange red. Female yellow

褐基脉蜻 雄, 泰国 | Oleg E. Kosterin 摄
Neurothemis intermedia, male from Thailand | Photo by Oleg E. Kosterin

褐基脉蜻 雌, 云南 (德宏)
Neurothemis intermedia, female from Yunnan (Dehong)

with brown markings. Wings hyaline. [Measurements] Total length 29-32 mm, abdomen 19-21 mm, hind wing 23-24 mm. [Habitat] Wetlands below 1500 m elevation. [Distribution] Yunnan (Dehong), Hainan; Widespread in tropical and subtropical Asia. [Flight Season] Throughout the year.

褐基脉蜻 雄，云南（德宏）
Neurothemis intermedia, male from Yunnan (Dehong)

台湾脉蜻 *Neurothemis taiwanensis* Seehausen & Dow, 2016

【形态特征】雄性通体红色；翅大面积红色，仅端部透明；腹部具黑色斑纹。雌性黄褐色；翅色彩稍淡；腹部具黑斑。【长度】体长 34～42 mm，腹部 22～25 mm，后翅 27～30 mm。【栖息环境】海拔 1500 m以下的湿地和水稻田。【分布】中国台湾；日本。【飞行期】全年可见。

[Identification] Male red throughout. Wings largely red with tips hyaline. Abdomen with black markings. Female yellowish brown. Wings paler. Abdomen with black markings. [Measurements] Total length 34-42 mm, abdomen 22-25 mm, hind wing 27-30 mm. [Habitat] Wetlands and paddy fields below 1500 m elevation. [Distribution] Taiwan of China; Japan. [Flight Season] Throughout the year.

台湾脉蜻 雄, 台湾 | 嘎嘎 摄
Neurothemis taiwanensis, male from Taiwan | Photo by Gaga

台湾脉蜻 雌, 台湾 | 嘎嘎 摄
Neurothemis taiwanensis, female from Taiwan | Photo by Gaga

截斑脉蜻 *Neurothemis tullia* (Drury, 1773)

【形态特征】雄性身体黑褐色；胸部背脊黄色，翅基方2/3处黑色，近翅端或具乳白色斑，或缺如；腹部黑色，背面中央具1条黄色条纹。雌性主要黄色；翅基部和端部具琥珀色斑；腹部背面具黑色条纹。【长度】体长 25～30 mm，腹部 16～20 mm，后翅 19～23 mm。【栖息环境】海拔 1500 m 以下的湿地和水稻田。【分布】福建、广东、广西、海南、香港、台湾；南亚、东南亚广布。【飞行期】3—12月。

[Identification] Male body blackish brown. Thorax with dorsal carina yellow, wings with basal two thirds black, milk white subapical spots sometimes present. Abdomen black with a central yellow stripe. Female mainly yellow. Wing bases and tips with amber markings. Abdomen with black markings dorsally. [Measurements] Total length 25-30 mm, abdomen 16-20 mm, hind wing 19-23 mm. [Habitat] Wetlands and paddy fields below 1500 m elevation. [Distribution] Fujian, Guangdong, Guangxi, Hainan, Hong Kong, Taiwan; Widespread in South and Southeast Asia. [Flight Season] March to December.

截斑脉蜻 雌，广东
Neurothemis tullia, female from Guangdong

截斑脉蜻 雄，广东
Neurothemis tullia, male from Guangdong

截斑脉蜻 雄，广东
Neurothemis tullia, male from Guangdong

脉蜻属待定种 *Neurothemis* sp.

脉蜻属待定种　雄，云南（德宏）
Neurothemis sp., male from Yunnan (Dehong)

脉蜻属待定种　雌，云南（德宏）
Neurothemis sp., female from Yunnan (Dehong)

【形态特征】雄性蓝黑色；翅基方2/3处黑色，有时近翅端具乳白色斑；腹部染有蓝灰色粉霜。雌性多型；黑色型与雄性相似，但腹部背面具黄色斑纹；黄色型主要黄色具褐色条纹，翅基部染有浅褐色，端部和中央具深褐色斑纹。【长度】体长 28～34 mm，腹长 19～22 mm，后翅 20～26 mm。【栖息环境】海拔 500 m 以下的小型池塘。【分布】云南（德宏）。【飞行期】全年可见。

[Identification] Male bluish black. Wings with basal two thirds black, milk white subapical spots sometimes present. Abdomen covered by bluish grey pruinosity. Female polymorphic. The black morph similar to male but abdomen with yellow markings dorsally. The yellow morph mainly yellow with brown markings, wing base pale brown, mid point and tip with dark brown markings. [Measurements] Total length 28-34 mm, abdomen 19-22 mm, hind wing 20-26 mm. [Habitat] Small ponds below 500 m elevation. [Distribution] Yunnan (Dehong). [Flight Season] Throughout the year.

爪蜻属 Genus *Onychothemis* Brauer, 1868

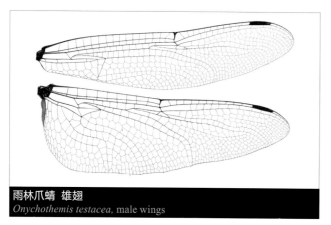

雨林爪蜻 雄翅
Onychothemis testacea, male wings

本属全球已知6种，仅分布于东洋界。中国已知3种。爪蜻是较大型的蜻科种类；翅透明，结前横脉数量较多，弓脉位于第1条与第2条结前横脉之间，R3和IR3在近翅端强烈弯曲，盘区具2~3列翅室，基臀区具1~2条横脉，臀圈靴状。本属蜻蜓的一个显著特征是爪上缺少第2趾。

本属蜻蜓栖息于溪流和河流。雄性在河岸带植被茂盛的区域活动，长时间飞行巡逻，有时多只雄性在同一处聚集。交尾时停落在树枝上，雌性将卵产在水面的漂浮物上。本属蜻蜓有捕食蝴蝶的习性。

The genus contains six species confined to the Oriental realm. Three species are recorded from China. Species of the genus are large-sized libellulids. Wings hyaline, antenodals adundant, arc between antenodals 1 and 2, R3 and IR3 curved distally, discoidal field with 2-3 rows of cells basally, cubital space with 1-2 crossveins, anal loop boot-shaped. Another diagnostic character for the genus is the absence of supplementary hook on the tarsal claws.

Onychothemis species inhabit exposed streams and rivers. Males usually occur along densely vegetated river margins where perched males frequently fly within a confined area. mating pairs perch on trees when mating, and females lay eggs on the floatings. Species of the genus are voracious predators and are often seen on butterflies.

雨林爪蜻 雄
Onychothemis testacea, male

红腹爪蜻 *Onychothemis culminicola* **Förster, 1904**

【形态特征】雄性复眼深绿色，面部褐色；胸部褐色，合胸脊黄色，侧面具2条黄色细条纹，翅透明；腹部红色，具较小的黑色斑。雌性较粗壮，腹部色彩较暗。【长度】体长 47~49 mm，腹长 30~32 mm，后翅 36~41 mm。【栖息环境】海拔 500 m以下的山区开阔小溪。【分布】云南（西双版纳、普洱、临沧）；文莱、缅甸、泰国、老挝、越南、马来西亚、印度尼西亚。【飞行期】4—11月。

[Identification] Male eyes dark green, face brown. Throax brown, dorsal carina yellow, laterally with two yellow narrow stripes, wings hyaline. Abdomen red with small black spots. Female stouter, abdomen darker. [Measurements] Total length 47-49 mm, abdomen 30-32 mm, hind wing 36-41 mm. [Habitat] Exposed montane streams below 500 m elevation. [Distribution] Yunnan (Xishuangbanna, Pu'er, Lincang); Brunei Darussalam, Myanmar, Thailand, Laos, Vietnam, Malaysia, Indonesia. [Flight Season] April to November.

红腹爪蜻 雌，云南（西双版纳）
Onychothemis culminicola, female from Yunnan (Xishuangbanna)

红腹爪蜻 雄，云南（普洱）
Onychothemis culminicola, male from Yunnan (Pu'er)

红腹爪蜻 雄，云南（西双版纳）
Onychothemis culminicola, male from Yunnan (Xishuangbanna)

雨林爪蜻 *Onychothemis testacea* **Laidlaw, 1902**

【形态特征】雄性复眼深绿色，面部黄色，额具蓝黑色斑；胸部主要黑色，合胸脊黄色，侧面具3条黄色细条纹，翅透明；腹部黑色具丰富的黄色斑点。雌性与雄性相似但较粗壮。【长度】体长 51~54 mm，腹长 35~36 mm，后翅 41~45 mm。【栖息环境】海拔 500 m 以下的山区开阔小溪。【分布】云南（西双版纳、普洱）；缅甸、泰国、老挝、马来西亚、新加坡。【飞行期】3—12月。

[Identification] Male eyes dark green, face yellow, frons with bluish black spot. Throax mainly black, dorsal carina yellow, laterally with three narrow yellow stripes, wings hyaline. Abdomen black with numerous yellow spots. Female similar to male but stouter. **[Measurements]** Total length 51-54 mm, abdomen 35-36 mm, hind wing 41-45 mm. **[Habitat]** Exposed montane streams below 500 m elevation. **[Distribution]** Yunnan (Xishuangbanna, Pu'er); Myanmar, Thailand, Laos, Malaysia, Singapore. **[Flight Season]** March to December.

雨林爪蜻 雌,云南(西双版纳)
Onychothemis testacea, female from Yunnan (Xishuangbanna)

雨林爪蜻 交尾,云南(西双版纳)
Onychothemis testacea, mating pair from Yunnan (Xishuangbanna)

雨林爪蜻 雄,云南(西双版纳)
Onychothemis testacea, male from Yunnan (Xishuangbanna)

海湾爪蜻 *Onychothemis tonkinensis* Martin, 1904

【形态特征】雄性复眼深绿色，面部黄色，额具蓝黑色斑；胸部主要黑色，合胸脊黄色，侧面具3条黄色细条纹，翅透明；腹部黑色具丰富的黄色斑点。雌性与雄性相似但更粗壮。本种与雨林爪蜻近似但腹部斑纹不同。【长度】体长 54～58 mm，腹部 34～37 mm，后翅 44～49 mm。【栖息环境】海拔 1000 m 以下的山区开阔小溪。【分布】云南（红河）、广东、广西、海南、香港、台湾；越南。【飞行期】3—10月。

[Identification] Male eyes dark green, face yellow, frons with bluish black spot. Throax mainly black, dorsal carina yellow, laterally with three narrow yellow stripes, wings hyaline. Abdomen black with numerons yellow spots. Female similar to male but stouter. The species is similar to *O. testacea* but the abdominal markings are different. [Measurements] Total length 54-58 mm, abdomen 34-37 mm, hind wing 44-49 mm. [Habitat] Exposed montane streams below 1000 m elevation. [Distribution] Yunnan (Honghe), Guangdong, Guangxi, Hainan, Hong Kong, Taiwan; Vietnam. [Flight Season] March to October.

海湾爪蜻 雄，海南
Onychothemis tonkinensis, male from Hainan

海湾爪蜻 雌，广东
Onychothemis tonkinensis, female from Guangdong

海湾爪蜻 雄，广东 | 宋睿斌 摄
Onychothemis tonkinensis, male from Guangdong | Photo by Ruibin Song

灰蜻属 Genus *Orthetrum* Newman, 1833

　　本属全球已知超过60种，是本科中最庞大的一个属。中国已知10余种，全国广布。本属多为中型蜻蜓，体色多变，许多种类具有粉霜；翅透明而狭长，有时基方具色斑，翅痣很长，具有较多条结前横脉，弓脉位于第2条与第3条结前横脉之间，前翅三角室的上缘很短，IR3呈明显的波浪状，基臀区具1条横脉，盘区基方具2～3列翅室，臀圈靴状。

　　本属蜻蜓栖息于各类静水环境和流速缓慢的溪流。雄性经常在水面上来回飞行，并时而停落在地面上或者植物上。交尾前先连结飞行片刻而后停落，交尾时间较长，雄性护卫产卵。

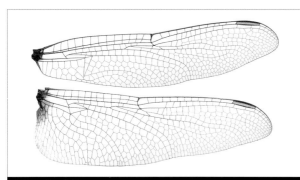

黑尾灰蜻 雄翅
Orthetrum glaucum, male wings

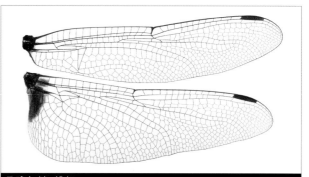

鼎脉灰蜻 雄翅
Orthetrum triangulare, male wings

　　The genus contains over 60 species, the most speciose genus of the family. Over ten species are recorded from China, widespread. Species of the genus are medium-sized and variably colored dragonflies, many species are heavily pruinosed. Wings hyaline and narrow, bases tinted with color spots sometimes, pterostigma long, antenodals abundant, arc between antenodals 2 and 3, the anterior margin of triangle short in fore wings, IR3 clearly waved, cubital space with one crossvein, discoidal field with 2-3 rows of cells, anal loop boot-shaped.

　　Orthetrum species inhabit both standing water and running water. Males often fly above water, sometimes perching on the ground or plants. Before mating, male and female will remain in tandem while flying, upon copulating they perch and remain in wheel for several minutes and male guards the egg laying female.

赤褐灰蜻中印亚种 雄
Orthetrum pruinosum neglectum, male

白尾灰蜻 *Orthetrum albistylum* Selys, 1848

【形态特征】雄性复眼深绿色，面部白色；胸部褐色，侧面具2条白色条纹，翅透明；腹部1～6节覆盖蓝白色粉霜，其余各节黑色。雌性多型，蓝色型与雄性相似；黄色型主要上黄色。【长度】体长 50～56 mm，腹长 35～38 mm，后翅 37～42 mm。【栖息环境】海拔 2000 m以下的湿地、水稻田、沟渠和流速缓慢的溪流。【分布】全国广布；朝鲜半岛、日本、西伯利亚、中亚、欧洲。【飞行期】3—10月。

[Identification] Male eyes dark green, face white. Thorax brown, laterally with two white stripes, wings hyaline. Abdomen with bluish white pruinosity on S1-S6, remaining segments black. Female polymorphic, the blue morph

白尾灰蜻 雄，湖北
Orthetrum albistylum, male from Hubei

similar to male. The yellow morph mainly khaki yellow. [Measurements] Total length 50-56 mm, abdomen 35-38 mm, hind wing 37-42 mm. [Habitat] Wetlands, paddy fields, ditches and slow flowing streams below 2000 m elevation. [Distribution] Widespread throughout the country; Korean peninsula, Japan, Siberia, Central Asia, Europe. [Flight Season] March to October.

白尾灰蜻 交尾, 湖北
Orthetrum albistylum, mating pair from Hubei

白尾灰蜻 雄, 湖北
Orthetrum albistylum, male from Hubei

白尾灰蜻 雌, 贵州 | 宋睿斌 摄
Orthetrum albistylum, female from Guizhou | Photo by Ruibin Song

白尾灰蜻 雌, 湖北
Orthetrum albistylum, female from Hubei

天蓝灰蜻 *Orthetrum brunneum* (Fonscolombe, 1837)

【形态特征】雄性复眼深绿色，面部白色；胸部和腹部覆盖天蓝色粉霜，翅透明。雌性主要黄色。【长度】体长41～50 mm，腹部25～32 mm，后翅33～37 mm。【栖息环境】狭窄的小溪、沟渠和渗流地。【分布】内蒙古、河北、北京；欧洲、阿富汗、阿尔及利亚。【飞行期】4—9月。

[Identification] Male eyes dark green, face white. Throax and abdomen with sky-blue pruinosity, wings hyaline. Female mainly yellow. [Measurements] Total length 41-50 mm, abdomen 25-32 mm, hind wing 33-37 mm. [Habitat] Narrow streams, ditches and seepages. [Distribution] Inner Mongolia, Hebei, Beijing; Europe, Afghanistan, Algeria. [Flight Season] April to September.

天蓝灰蜻 雌,比利时 | Sami Karjalainen 摄
Orthetrum brunneum, female from Bulgria | Photo by Sami Karjalainen

天蓝灰蜻 雄,比利时 | Sami Karjalainen 摄
Orthetrum brunneum, male from Bulgria | Photo by Sami Karjalainen

粗灰蜻 *Orthetrum cancellatum* (Linnaeus, 1758)

粗灰蜻 雄,芬兰 | Sami Karjalainen 摄
Orthetrum cancellatum, male from Finland | Photo by Sami Karjalainen

【形态特征】雄性复眼深绿色；胸部褐色，侧面具2条白色条纹，翅透明；腹部1~7节覆盖蓝白色粉霜，末端3节黑色，年长的雄性整个腹部覆盖粉霜；雌性黄色具褐色条纹；腹部较粗壮。【长度】体长44～52 mm，腹部28～35 mm，后翅35～40 mm。【栖息环境】较大型的池塘、水库和湖泊周边。【分布】吉林、内蒙古、北京；中亚、欧洲、西伯利亚、印度。【飞行期】4—9月。

[Identification] Male eyes dark green. Thorax brown, laterally with two white stripes, wings hyaline. Abdomen with bluish white pruinosity on S1-S7, remaining segments black, aged male with entire abdomen pruinosed. Female yellow with brown markings. Abdomen robust. [Measurements] Total length 44-

52 mm, abdomen 28-35 mm, hind wing 35-40 mm. **[Habitat]** Large ponds, reservoirs and lakes. **[Distribution]** Jilin, Inner Mongolia, Beijing; Central Asia, Europe, Siberia, India. **[Flight Season]** April to September.

粗灰蜻 雄,吉林 |金洪光 摄
Orthetrum cancellatum, male from Jilin | Photo by Hongguang Jin

粗灰蜻 雌,吉林 |金洪光 摄
Orthetrum cancellatum, female from Jilin | Photo by Hongguang Jin

华丽灰蜻 *Orthetrum chrysis* (Selys, 1891)

　　【形态特征】雄性复眼褐色,面部红褐色;胸部褐色,翅透明,后翅基方染有琥珀色斑;腹部鲜红色。雌性黄褐色,第8节侧面具片状突起。【长度】体长 42~51 mm,腹长 28~34 mm,后翅 33~38 mm。【栖息环境】海拔 1000 m 以下的湿地。【分布】云南、广西、广东、海南、香港;广泛分布于亚洲的热带和亚热带区域。【飞行期】全年可见。

华丽灰蜻 雄,云南(西双版纳)
Orthetrum chrysis, male from Yunnan (Xishuangbanna)

[Identification] Male eyes brown, face reddish brown. Thorax brown, wings hyaline, hind wing bases with amber spots. Abdomen red. Female yellowish brown, S8 with lateral leaf-like flaps. [Measurements] Total length 42-51 mm, abdomen 28-34 mm, hind wing 33-38 mm. [Habitat] Wetlands below 1000 m elevation. [Distribution] Yunnan, Guangxi, Guangdong, Hainan, Hong Kong; Widespread in tropical and subtropical Asia. [Flight Season] Throughout the year.

华丽灰蜻 交尾, 广东 | 莫善濂 摄
Orthetrum chrysis, mating pair from Guangdong | Photo by Shanlian Mo

华丽灰蜻 雌, 云南 (德宏)
Orthetrum chrysis, female from Yunnan (Dehong)

黑尾灰蜻 *Orthetrum glaucum* (Brauer, 1865)

【形态特征】雄性复眼深绿色, 面部黑褐色; 完全成熟后胸部和腹部覆盖蓝色粉霜, 翅透明, 后翅基方具甚小的琥珀色斑。雌性年轻时黄色具褐色条纹, 年老以后腹部覆盖灰色粉霜, 第8节侧面具不发达的片状突起。【长度】体长 42~51 mm, 腹长 27~31 mm, 后翅 32~35 mm。【栖息环境】海拔 2000 m 以下的湿地、沟渠和流速缓慢的溪流。【分布】中国中部至南部广布; 广泛分布于亚洲的热带和亚热带区域。【飞行期】全年可见。

[Identification] Male eyes dark green, face blackish brown. Thorax and abdomen covered by blue pruinosity when fully mature, wings hyaline, hind wing bases with small amber spots. Young female yellow with brown markings,

黑尾灰蜻 交尾, 云南 (德宏)
Orthetrum glaucum, mating pair from Yunnan (Dehong)

黑尾灰蜻 雌, 广东 | 宋睿斌 摄
Orthetrum glaucum, female from Guangdong | Photo by Ruibin Song

and gradually bluish pruinosed with age, S8 with small lateral leaf-like flaps. **[Measurements]** Total length 42-51 mm, abdomen 27-31 mm, hind wing 32-35 mm. **[Habitat]** Wetlands, ditches and slow flowing streams below 2000 m elevation. **[Distribution]** Widespread from central to the south of China; Widespread in tropical and subtropical Asia. **[Flight Season]** Throughout the year.

黑尾灰蜻 雄,云南(西双版纳)
Orthetrum glaucum, male from Yunnan (Xishuangbanna)

褐肩灰蜻 *Orthetrum internum* McLachlan, 1894

【形态特征】雄性复眼蓝绿色,面部黄白色;合胸背面覆盖白色粉霜,侧面具2条黄色宽条纹,翅透明;腹部较宽阔,覆盖白色粉霜。雌性黄色具丰富的黑色条纹,第8节侧面具不发达的片状突起。【长度】体长 41~44 mm,腹长 26~29 mm,后翅 32~34 mm。【栖息环境】海拔 2000 m以下的湿地和水稻田。【分布】中国中部至南部广布;朝鲜半岛、日本。【飞行期】3—9月。

[Identification] Male eyes bluish green, face yellowish white. Thorax with dorsal part covered by white pruinosity, laterally with two broad yellow stripes, wings hyaline. Abdomen broad and covered by white pruinosity. Female yellow with abundant black markings, S8 with small lateral leaf-like flaps. **[Measurements]** Total length 41-44 mm, abdomen 26-29 mm, hind wing 32-34 mm. **[Habitat]** Wetlands and paddy fields below 2000 m elevation. **[Distribution]** Widespread from central to the south of China; Korean peninsula, Japan. **[Flight Season]** March to September.

褐肩灰蜻 雄,湖北
Orthetrum internum, male from Hubei

褐肩灰蜻 雄,广东 | 吴宏道 摄
Orthetrum internum, male from Guangdong | Photo by Hongdao Wu

褐肩灰蜻 雌,广东 | 吴宏道 摄
Orthetrum internum, female from Guangdong | Photo by Hongdao Wu

线痣灰蜻 *Orthetrum lineostigma* (Selys, 1886)

【形态特征】雄性复眼蓝绿色，面部蓝白色；完全成熟后胸部和腹部覆盖蓝色粉霜，翅透明，翅端具褐色斑。雌性大面积黄褐色具丰富的黑色条纹；翅稍染褐色，翅端具褐斑；腹部第8节侧面具不发达的片状突起。【长度】体长 41~45 mm，腹长 27~30 mm，后翅 32~35 mm。【栖息环境】海拔 1000 m以下的狭窄小溪、宽阔河流中水草茂盛的浅水部分和水草茂盛的池塘。【分布】吉林、辽宁、北京、河北、河南、山西、陕西、山东、江苏；朝鲜半岛。【飞行期】4—9月。

[Identification] Male eyes bluish green, face bluish white. Thorax and abdomen with bluish white pruinosity when fully mature, wings hyaline with apical brown spots. Female largely yellowish brown with numerous black markings. Wings slightly tinted with brown tips. S8 with small lateral leaf-like flaps. [Measurements] Total length 41-45 mm, abdomen 27-30 mm, hind wing 32-35 mm. [Habitat] Narrow streams, shallow river banks and well vegetated ponds below 1000 m elevation. [Distribution] Jilin, Liaoning, Beijing, Hebei, Henan, Shanxi, Shaanxi, Shandong, Jiangsu; Korean peninsula. [Flight Season] April to September.

线痣灰蜻 雌，北京
Orthetrum lineostigma, female from Beijing

线痣灰蜻 交尾，北京
Orthetrum lineostigma, mating pair from Beijing

线痣灰蜻 雄，北京
Orthetrum lineostigma, male from Beijing

吕宋灰蜻 *Orthetrum luzonicum* (Brauer, 1868)

　　【形态特征】雄性复眼蓝绿色，面部蓝白色；完全成熟后胸部和腹部覆盖蓝色粉霜，翅透明。雌性大面积黄褐色具丰富的黑色条纹；翅稍染褐色；腹部第8节侧面具不发达的片状突起。本种与线痣灰蜻相似但翅痣色彩不同。【长度】体长 38~45 mm，腹长 27~32 mm，后翅 28~32 mm。【栖息环境】海拔 2500 m以下的狭窄小溪、宽阔河流中水草茂盛的浅水部分和水草茂盛的池塘。【分布】中国中部至南部广布；广泛分布于亚洲的热带和亚热带区域。【飞行期】全年可见。

　　[Identification] Male eyes bluish green, face bluish white. Thorax and abdomen with bluish white pruinosity when fully mature, wings hyaline. Female mainly yellowish brown with numerous black markings, wing slightly tinted with brown. S8 with small lateral leaf-like flaps. The species is similar to *Orthetrum lineostigma* but the color of pterostigma is different. [Measurements] Total length 38-45 mm, abdomen 27-32 mm, hind wing 28-32 mm. [Habitat] Narrow streams, shallow river banks and well vegetated ponds below 2500 m elevation. [Distribution] Widespread from central to the south of China; Widespread in tropical and subtropical Asia. [Flight Season] Throughout the year.

吕宋灰蜻 雌，广东
Orthetrum luzonicum, female from Guangdong

吕宋灰蜻 雄，贵州
Orthetrum luzonicum, male from Guizhou

吕宋灰蜻 雄，云南（德宏）
Orthetrum luzonicum, male from Yunnan (Dehong)

异色灰蜻 *Orthetrum melania melania* (Selys, 1883)

异色灰蜻 雌,安徽 | 秦彧 摄
Orthetrum melania melania, female from Anhui | Photo by Yu Qin

异色灰蜻 雄,贵州
Orthetrum melania melania, male from Guizhou

异色灰蜻 雄,贵州
Orthetrum melania melania, male from Guizhou

【形态特征】雄性全身覆盖蓝色粉霜;头部黑褐色;翅透明,翅端稍染褐色,后翅基方具黑褐色斑;腹部末端黑色。雌性主要黄色具大量黑色条纹;腹部第8节侧面具片状突起。【长度】体长 51～55 mm,腹长 33～35 mm,后翅 40～43 mm。【栖息环境】海拔 2000 m 以下的湿地和沟渠。【分布】中国华北、华南和西南广布;朝鲜半岛、日本、俄罗斯。【飞行期】全年可见。

[Identification] Male pruinosed throughout. Head blackish brown. Wings hyaline, tips with brown tint, hind wing bases with blackish brown markings. Abdomen tip black. Female mainly yellow with numerons black markings. S8 with small lateral leaf-like flaps. [Measurements] Total length 51-55 mm, abdomen 33-35 mm, hind wing 40-43 mm. [Habitat] Wetlands and ditches below 2000 m elevation. [Distribution] Widespread in North, South and Southwest China; Korean peninsula, Japan, Russia. [Flight Season] Throughout the year.

斑灰蜻 *Orthetrum poecilops* Ris, 1919

【形态特征】雄性复眼蓝绿色，面部白色具黑色条纹；胸部背面覆盖白色粉霜，侧面具黄白色条纹，翅透明；腹部覆盖蓝白色粉霜，第1~3节膨大。雌性主要黄色具黑色条纹，老熟后腹部覆盖粉霜；腹部第8节侧面具不发达的片状突起。【长度】体长 45~54 mm，腹部 30~35 mm，后翅 34~39 mm。【栖息环境】红树林中的小溪。【分布】福建、海南、广东、香港；日本。【飞行期】5—8月。

斑灰蜻 雄, 香港 | 梁嘉景 摄
Orthetrum poecilops, male from Hong Kong | Photo by Kenneth Leung

[Identification] Male eyes bluish green, face white with black markings. Thorax dorsally covered by white pruinosity, laterally with yellowish white stripes, wings hyaline. Abdomen covered by bluish white pruinosity, S1-S3 expanded. Female mainly yellow with black markings, abdomen pruinosed when aged. S8 with small lateral leaf-like flaps. [Measurements] Total length 45-54 mm, abdomen 30-35 mm, hind wing 34-39 mm. [Habitat] Streams in tidal mangroves. [Distribution] Fujian, Hainan, Guangdong, Hong Kong; Japan. [Flight Season] May to August.

斑灰蜻 雌, 香港 | 祁麟峰 摄
Orthetrum poecilops, female from Hong Kong | Photo by Mahler Ka

赤褐灰蜻西里亚种 *Orthetrum pruinosum clelia* (Selys, 1878)

【形态特征】雄性复眼灰褐色,面部红褐色;胸部褐色,翅透明,后翅基方具褐斑;腹部粉红色,基方2节和肛附器蓝灰色。雌性黄褐色;腹部第8节侧面具片状突起。【长度】体长 44~45 mm,腹长 30~31 mm,后翅 34~35 mm。【栖息环境】海拔 500 m以下的湿地和水稻田。【分布】中国台湾;菲律宾、苏拉威西岛、摩鹿加群岛。【飞行期】2—11月。

[Identification] Male eyes greyish brown, face reddish brown. Thorax brown, wings hyaline, hind wing bases with brown spots. Abdomen pink red with basal two segments and anal appendages bluish grey. Female yellowish brown, S8 with lateral leaf-like flaps. [Measurements] Total length 44-45 mm, abdomen 30-31 mm, hind wing 34-35 mm. [Habitat] Wetlands and paddy fields below 500 m elevation. [Distribution] Taiwan of China; Philippines, Sulawesi, Moluccas. [Flight Season] February to November.

赤褐灰蜻西里亚种 雄,台湾 | 嘎嘎 摄
Orthetrum pruinosum clelia, male from Taiwan | Photo by Gaga

赤褐灰蜻中印亚种 *Orthetrum pruinosum neglectum* (Rambur, 1842)

【形态特征】雄性复眼灰褐色,面部红褐色;胸部褐色,翅透明,后翅基方具褐斑;腹部粉红色。雌性黄褐色,第8节侧面具片状突起。本亚种与西里亚种腹部基方的色彩不同。【长度】体长 46~50 mm,腹长 31~33 mm,后翅 35~38 mm。【栖息环境】海拔 2500 m以下的各类池塘、水库、沟渠、水稻田和流速缓慢的溪流。【分布】中国南部广布;南亚、东南亚广布。【飞行期】全年可见。

[Identification] Male eyes greyish brown, face reddish brown. Thorax brown, wings hyaline, hind wing bases with brown spots. Abdomen pink red. Female yellowish brown, S8 with lateral leaf-like flaps. The subspecies is different from *O. pruinosum clelia* by the color of basal abdominal segments. [Measurements] Total length 46-50 mm, abdomen 31-33 mm, hind wing 35-38 mm. [Habitat] Ponds, reservoirs, ditches, paddy fields and slow flowing streams below 2500 m elevation. [Distribution] Widespread in the south of China; Widespread in South and Southeast Asia. [Flight Season] Throughout the year.

赤褐灰蜻中印亚种 雌, 云南 (西双版纳)
Orthetrum pruinosum neglectum, female from Yunnan (Xishuangbanna)

赤褐灰蜻中印亚种 交尾, 云南 (德宏)
Orthetrum pruinosum neglectum, mating pair from Yunnan (Dehong)

赤褐灰蜻中印亚种 雄, 云南 (西双版纳)
Orthetrum pruinosum neglectum, male from Yunnan (Xishuangbanna)

狭腹灰蜻 *Orthetrum sabina* (Drury, 1773)

狭腹灰蜻 雌，云南（德宏）
Orthetrum sabina, female from Yunnan (Dehong)

狭腹灰蜻 交尾，云南（德宏）
Orthetrum sabina, mating pair from Yunnan (Dehong)

狭腹灰蜻 雄，云南（德宏）
Orthetrum sabina, male from Yunnan (Dehong)

【形态特征】雄性复眼绿色，面部黄色；胸部黄色具黑色细纹，翅透明；腹部黑色具黄色和白色条纹，第1~3节膨大显著，第7~9节稍微膨大。雌性与雄性较相似但腹部较粗。【长度】体长 47~51 mm，腹长 34~37 mm，后翅33~35 mm。【栖息环境】海拔 2500 m以下的各类池塘、水库、沟渠、水稻田和流速缓慢的溪流。【分布】中国中部至南部广布；从地中海东部经亚洲南部至大洋洲广布。【飞行期】全年可见。

[Identification] Male eyes green, face yellow. Thorax yellow with fine black stripes, wings hyaline. Abdomen black with yellow and white markings, S1-S3 strongly expanded laterally, S7-S9 slightly expanded. Female similar to male but abdomen broader. [Measurements] Total length 47-51 mm, abdomen 34-37 mm, hind wing 33-35 mm. [Habitat] Ponds, reservoirs, ditches, paddy fields and slow flowing below 2500 m elevation. [Distribution] Widespread from central to the south of China; Widespread from the eastern Mediterranean throughout southern Asia to Oceania. [Flight Season] Throughout the year.

鼎脉灰蜻 *Orthetrum triangulare* (Selys, 1878)

【形态特征】雄性复眼深绿色,面部黑色;胸部黑褐色,翅透明,后翅基方具黑褐色斑;腹部黑色,通常第1~7节具蓝白色粉霜。雌性大面积黄色具褐色条纹,年老以后腹部覆盖蓝灰色粉霜,腹部第8节侧面具片状突起。【长度】体长 45~50 mm,腹长 29~33 mm,后翅 39~41 mm。【栖息环境】海拔 2500 m以下的各类池塘、水库、沟渠、水稻田和流速缓慢的溪流。【分布】中国中部至南部广布;广泛分布于亚洲的热带和亚热带区域。【飞行期】全年可见。

[Identification] Male eyes dark green, face black. Thorax blackish brown, wings hyaline, hind wing bases with blackish brown markings. Abdomen black with bluish white pruinosity usually on S1-S7. Female largely yellow with brown markings, abdomen gradually bluish grey pruinosed with age, S8 with small lateral leaf-like flaps. [Measurements] Total length 45-50 mm, abdomen 29-33 mm, hind wing 39-41 mm. [Habitat] Ponds, reservoirs, ditches, paddy fields and slow flowing streams below 2500 m elevation. [Distribution] Widespread from central to the south of China; Widespread in tropical and subtropical Asia. [Flight Season] Throughout the year.

鼎脉灰蜻 雌,广东|宋睿斌 摄
Orthetrum triangulare, female from Guangdong | Photo by Ruibin Song

鼎脉灰蜻 交尾,云南(德宏)
Orthetrum triangulare, mating pair from Yunnan (Dehong)

鼎脉灰蜻 雄,云南(西双版纳)
Orthetrum triangulare, male from Yunnan (Xishuangbanna)

灰蜻属待定种1 *Orthetrum* sp. 1

【形态特征】雄性复眼深绿色,面部黄色;胸部黄褐色,侧面具2条黄色条纹,翅透明;腹部主要黑褐色,第1~8节具黄色斑点,第3~4节基方覆盖蓝白色粉霜。雌性主要黄色,腹部具黑色条纹;腹部第8节侧面具不发达的片状突起。【长度】体长 48~51 mm,腹长 33~35 mm,后翅 39~42 mm。【栖息环境】海拔 1000~2500 m的沼泽、溪流和水库。【分布】湖北、贵州、重庆、云南。【飞行期】5—9月。

[Identification] Male eyes dark green, face yellow. Thorax yellowish brown, laterally with two yellow stripes, wings hyaline. Abdomen mainly blackish brown, S1-S8 with yellow spots, S3 and basal S4 with bluish white pruinosity. Female mainly yellow, abdomen with black stripes. S8 with small lateral leaf-like flaps. [Measurements] Total length 48-51 mm, abdomen 33-35 mm, hind wing 39-42 mm. [Habitat] Marshes, streams and reservoirs at 1000-2500 m elevation. [Distribution] Hubei, Guizhou, Chongqing, Yunnan. [Flight Season] May to September.

灰蜻属待定种1 雌,湖北
Orthetrum sp. 1, female from Hubei

灰蜻属待定种1 雄,贵州 | 莫善濂 摄
Orthetrum sp. 1, male from Guizhou | Photo by Shanlian Mo

灰蜻属待定种2 *Orthetrum* sp. 2

【形态特征】雄性复眼蓝绿色,面部蓝白色;完全成熟后胸部和腹部覆盖蓝色粉霜,合胸侧面具1~2个小黄斑,翅透明。雌性主要黄褐色具黑色条纹;腹部第8节侧面具不发达的片状突起。本种与吕宋灰蜻相似,但翅痣色彩更深。【长度】体长 41~47 mm,腹长 28~32 mm,后翅 32~35 mm。【栖息环境】海拔 500~1500 m水草茂盛的湿地。【分布】贵州、浙江。【飞行期】5—9月。

[Identification] Male eyes bluish green, face bluish white. Fully mature male with thorax and abdomen covered by blue pruinosity, thorax laterally with 1-2 small yellow spots, wings hyaline. Female mainly yellowish brown with black markings. S8 with small lateral leaf-like flaps. The species is similar to *O. luzonicum* but the color of pterostigma darker. [Measurements] Total length 41-47 mm, abdomen 28-32 mm, hind wing 32-35 mm. [Habitat] Wetlands with plenty of emergent plants at 500-1500 m elevation. [Distribution] Guizhou, Zhejiang. [Flight Season] May to September.

灰蜻属待定种2 雌，贵州
Orthetrum sp. 2, female from Guizhou

灰蜻属待定种2 雄，贵州
Orthetrum sp. 2, male from Guizhou

灰蜻属待定种2 雄，贵州
Orthetrum sp. 2, male from Guizhou

灰蜻属待定种3 *Orthetrum* sp. 3

【形态特征】本种与褐肩灰蜻相似，但胸部褐色，雄性合胸背面无白色粉霜。【长度】体长 42～45 mm，腹长 27～29 mm，后翅 32～35 mm。【栖息环境】海拔 1500～3000 m的高山沼泽。【分布】云南（红河、大理、保山、德宏）。【飞行期】4—10月。

[Identification] The species is similar to *O. internum* but thoracic brown, male synthorax without white pruinosity on the dorsal part. [Measurements] Total length 42-45 mm, abdomen 27-29 mm, hind wing 32-35 mm. [Habitat] Marshes at 1500-3000 m elevation. [Distribution] Yunnan (Honghe, Dali, Baoshan, Dehong). [Flight Season] April to October.

灰蜻属待定种3 雌，云南（大理）
Orthetrum sp. 3, female from Yunnan (Dali)

灰蜻属待定种3 雄，云南（德宏）
Orthetrum sp. 3, male from Yunnan (Dehong)

灰蜻属待定种3 交尾，云南（德宏）
Orthetrum sp. 3, mating pair from Yunnan (Dehong)

灰蜻属待定种4 *Orthetrum* sp. 4

灰蜻属待定种4 雄，云南（大理）
Orthetrum sp. 4, male from Yunnan (Dali)

【形态特征】与鼎脉灰蜻非常相似，但是体色较淡，胸部褐色，腹部末端红色。是否为鼎脉灰蜻的变异型尚未明确。【长度】雄性体长 49～50 mm，腹长 31～33 mm，后翅 40～41 mm。【栖息环境】海拔 2000～3000 m的高山湿地。【分布】云南（大理）。【飞行期】不详。

[Identification] The species is similar to *O. triangulare* but body color paler, thorax brown and abdominal tip red. Not yet confirmed if this is just a variation of *O. triangulare*. [Measurements] Male total length 49-50 mm, abdomen 31-33 mm, hind wing 40-41 mm. [Habitat] Wetlands at 2000-3000 m elevation. [Distribution] Yunnan (Dali). [Flight Season] Unknown.

曲缘蜻属 Genus *Palpopleura* Rambur, 1842

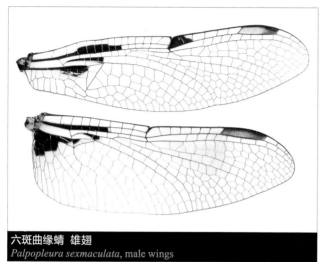

六斑曲缘蜻 雄翅
Palpopleura sexmaculata, male wings

本属全球已知7种，大多数分布在非洲。中国已知1种，南方广布。本属是一类体型短小但粗壮的蜻蜓；翅具黑色和琥珀色斑，翅痣较长，前翅的前缘脉显著弯曲，前翅最末端的结前横脉不完整，弓脉位于第1条与第2条结前横脉之间，盘区基方具2~3列翅室，基臀区具1条横脉，臀圈靴状。

本属蜻蜓栖息于浅水湿地。雄性会停落在水面附近的枝条顶端等待配偶，时而巡飞。它们飞行时腹部弯曲，形似蜜蜂。

The genus contains seven species and most of which are found in Africa. Only one species is recorded from China, widespread in the south. Species of the genus are short and robust. Wings tinted with black and amber spots, pterostigma long, costa in fore wings clearly curved, the distal antenodal in fore wings incomplete, arc between antenodals 1 and 2, discoidal field with 2-3 rows of cells basally, cubital space with one crossvein, anal loop boot-shaped.

Palpopleura species inhabit shallow wetlands. Males usually perch on top of branches above water and patrol sometimes. The abdomen is curved when flying thus looking like a bee.

六斑曲缘蜻 雄
Palpopleura sexmaculata, male

六斑曲缘蜻 *Palpopleura sexmaculata* (Fabricius, 1787)

六斑曲缘蜻 雄, 海南
Palpopleura sexmaculata, male from Hainan

六斑曲缘蜻 雄, 云南(德宏)
Palpopleura sexmaculata, male from Yunnan (Dehong)

六斑曲缘蜻 雌, 海南
Palpopleura sexmaculata, female from Hainan

六斑曲缘蜻 雌, 云南(西双版纳)
Palpopleura sexmaculata, female from Yunnan (Xishuangbanna)

【形态特征】雄性复眼蓝灰色, 面部黄色, 额具1个深蓝色金属斑; 胸部黄色具黑色斑纹, 前翅透明, 后翅大面积琥珀色, 具黑色斑纹; 腹部宽阔并覆盖浓密的蓝色粉霜。雌性大面积黄褐色, 腹部具黑色条纹。【长度】体长24~27 mm, 腹长14~16 mm, 后翅17~19 mm。【栖息环境】海拔 2500 m以下的浅水湿地、渗流地、沟渠和水稻田。【分布】中国南方广泛分布; 阿富汗、孟加拉国、不丹、印度、斯里兰卡、尼泊尔、缅甸、泰国、老挝、柬埔寨、越南。【飞行期】全年可见。

[Identification] Male eyes bluish grey, face yellow, frons with a metallic dark blue spot. Thorax yellow with black markings, fore wings hyaline, hind wings largely amber with black markings. Abdomen broad and covered by dense blue pruinosity. Female largely yellowish brown, abdomen with black markings. [Measurements] Total length 24-27 mm, abdomen 14-16 mm, hind wing 17-19 mm. [Habitat] Shallow wetlands, seepages, ditches and paddy fields below 2500 m elevation. [Distribution] Widespread in the south of China; Afghanistan, Bangladesh, Bhutan, India, Sri Lanka, Nepal, Myanmar, Thailand, Laos, Cambodia, Vietnam. [Flight Season] Throughout the year.

黄蜻属 Genus *Pantala* Hagen, 1861

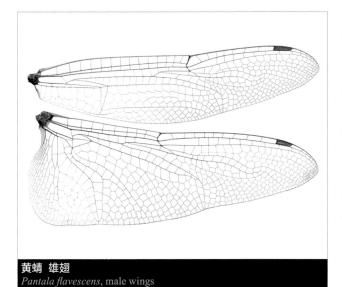

黄蜻 雄翅
Pantala flavescens, male wings

本属全球已知2种, 其中1种仅在美洲分布, 而另1种是著名的黄蜻, 除南极洲外全球广布。本属蜻蜓体中型, 复眼巨大; 翅透明, 后翅基方甚阔, 弓脉位于第1条与第2条结前横脉之间, 盘区具2~3列翅室, 基臀区具1~2条横脉, 臀圈靴状。

本属蜻蜓栖息于各种静水环境, 但偏爱浅水池塘和暂时性水潭。稚虫速生型。成虫经常集群捕食, 并可以飞行到城市的各个角落。中国的城市居民, 最容易遇见的蜻蜓即是黄蜻。雄性在水面上空来回飞行, 时而悬停。交尾在空中完成, 交尾后连结产卵。黄蜻具有迁飞的习性。

The genus contains two species, one of which is confined to Americas and the other, the famous *Pantala flavescens*, is found in all continents except Antarctica. Species of the genus are medium-sized, eyes large. Wings hyaline, hind wing bases broad, arc between antenodals 1 and 2, discoidal field with 2-3 rows of cells basally, cubital space with one to two crossveins, anal loop boot-shaped.

Pantala species inhabit standing water, preferring shallow ponds and temporary pools. Larval growth is very rapid. Individuals are usually seen on foraging in great numbers, seen from the every corner of urban area. *P. flavescens* is the most common species encountered by urbanites in China. Males patrol above water by flying constantly and short time hovering. Mating takes place in flight and they lay eggs in tandem. *P. flavescens* is a strong migrant.

黄蜻 交尾 | 金洪光 摄
Pantala flavescens, mating pair | Photo by Hongguang Jin

黄蜻群停落 | 金洪光 摄
Group of *Pantala flavescens* perching | Photo by Hongguang Jin

黄蜻 *Pantala flavescens* (Fabricius, 1798)

黄蜻 雄, 安徽 | 秦彧 摄
Pantala flavescens, male from Anhui | Photo by Yu Qin

黄蜻 雄, 吉林 | 金洪光 摄
Pantala flavescens, male from Jilin | Photo by Hongguang Jin

【形态特征】雄性复眼上方红褐色, 下方蓝灰色, 面部黄色; 胸部黄褐色, 翅透明, 后翅基方稍染黄褐色; 腹部背面红色, 具黑褐色斑, 其中第8~10节中央具较大的黑色斑。雌性身体主要黄褐色; 完全成熟后翅稍染褐色; 腹部土黄色, 腹面随年纪增长逐渐覆盖白色粉霜。【长度】体长 49~50 mm, 腹长 32~33 mm, 后翅 39~40 mm。【栖息环境】海拔 3500 m 以下的各类静水环境, 包括季节性水塘、渗流地、沟渠和水稻田。【分布】全国广布; 热带、亚热带和北美洲广布。分布于除南极洲以外的所有大陆地区。【飞行期】全年可见。

[Identification] Male eyes reddish brown above bluish grey below, face yellow. Thorax yellowish brown, wings hyaline, hind wing bases with yellowish brown tint. Abdomen red dorsally, with blackish brown markings, S8-S10 with

黄蜻 雄, 吉林 | 金洪光 摄
Pantala flavescens, male from Jilin | Photo by Hongguang Jin

黄蜻 雌，云南（德宏）
Pantala flavescens, female from Yunnan (Dehong)

large central black spots. Female mainly yellowish brown. Wings tinted with light brown when fully mature. Abdomen khaki yellow, and ventral part gradually covered by white pruinosity with age. **[Measurements]** Total length 49-50 mm, abdomen 32-33 mm, hind wing 39-40 mm. **[Habitat]** Standing water habitats below 3500 m elevation, including seasonal pools, seepages, ditches and paddy fields. **[Distribution]** Widespread in China; Widespread throughout the tropics and subtropics and in North America. Known from all continents except Antarctica. **[Flight Season]** Throughout the year.

长足蜻属 Genus *Phyllothemis* Fraser, 1935

沼长足蜻 雄
Phyllothemis eltoni, male

　　本属全球已知2种，分布于东南亚。中国已知1种，仅分布于云南西部。本属蜻蜓体型较小，体黑色具黄色斑纹，后足甚长；翅透明，结前横脉完整，数量较少，弓脉位于第2条和第3条结前横脉之间。

　　本属蜻蜓栖息于森林中的沼泽地。雄性停落在植物的叶片上占据领地。

The genus contains two species confined to Southeast Asia. One species is recorded from China, only found in the west of Yunnan. Species of the genus are small-sized, body fundamentally black with yellow markings, hind legs very long. Wings hyaline, antenodals complete but few in number, arc between antenodals 2 and 3.

Phyllothemis species inhabit marshes in forests. Territorial males perch on the plants near water.

沼长足蜻 *Phyllothemis eltoni* Fraser, 1935

【形态特征】雄性复眼蓝绿色，面部黄色，额具1个蓝黑色金属色斑；胸部和腹部黑色具丰富的黄色斑纹，翅透明，后足甚长。雌性与雄性相似。【长度】体长 30~31 mm，腹长 20~21 mm，后翅 23~24 mm。【栖息环境】海拔 500 m以下森林中的沼泽地。【分布】云南（德宏）；缅甸，泰国。【飞行期】9—12月。

[Identification] Male eyes bluish green, face yellow, frons with a metallic bluish black spot. Thorax and abdomen black with numerous yellow markings, wings hyaline, hind legs very long. Female similar to male. [Measurements] Total length 30-31 mm, abdomen 20-21 mm, hind wing 23-24 mm. [Habitat] Marshes in forest below 500 m elevation. [Distribution] Yunnan (Dehong); Myanmar, Thailand. [Flight Season] September to December.

沼长足蜻 雌，云南（德宏）
Phyllothemis eltoni, female from Yunnan (Dehong)

沼长足蜻 雄，云南（德宏）
Phyllothemis eltoni, male from Yunnan (Dehong)

狭翅蜻属 Genus *Potamarcha* Karsch, 1890

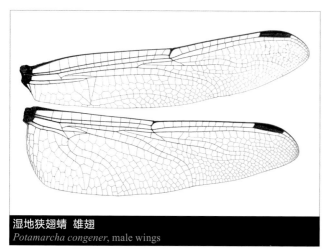

湿地狭翅蜻　雄翅
Potamarcha congener, male wings

本属全球仅2种，分布于亚洲和大洋洲。中国已知2种，主要分布于华南和西南地区。本属蜻蜓体中型；翅透明而狭长，翅痣很长，具有较多条结前横脉，弓脉位于第1条与第2条结前横脉之间，前翅三角室的上缘很短，IR 3呈明显的波浪状，基臀区具1条横脉，盘区基方具2～3列翅室，臀圈靴状。

本属蜻蜓栖息于浅水池塘和暂时性水潭。雄性在水面附近的枝头上停落，时而沿池塘边缘巡飞。交尾在空中进行，雄性护卫产卵。

The genus contains two species distributed in Asia and Oceania. Both are recorded from China, found in the South and Southwest. Species of the genus are medium-sized. Wings hyaline and narrow, pterostigma long, antenodals numerous, arc between antenodals 1 and 2, the anterior margin of triangle very short in fore wings, IR3 clearly waved, cubital space with one crossvein, discoidal field with 2-3 rows of cells, anal loop boot-shaped.

Potamarcha species inhabit in shallow ponds and temporary pools. Territorial males perch on tips of branches near water. Sometimes they patrol along the margin of ponds. Mating takes place in flight but is short in duration, male afterwards guards the female when laying eggs.

湿地狭翅蜻　雄
Potamarcha congener, male

湿地狭翅蜻 *Potamarcha congener* (Rambur, 1842)

湿地狭翅蜻 雄，云南（西双版纳）
Potamarcha congener, male from Yunnan (Xishuangbanna)

湿地狭翅蜻 雌，云南（西双版纳）
Potamarcha congener, female from Yunnan (Xishuangbanna)

湿地狭翅蜻 雄，云南（西双版纳）
Potamarcha congener, male from Yunnan (Xishuangbanna)

【形态特征】雄性复眼褐色，面部黄色，上额蓝黑色；胸部覆盖蓝色粉霜，翅透明；腹部第1~3节覆盖蓝色粉霜，第4~10节黑色具黄色条纹；雌性主要黄褐色具黑色条纹，腹部第8节侧缘具叶片状突起，年老后覆盖蓝灰色粉霜。【长度】体长43~45 mm，腹长28~30 mm，后翅32~33 mm。【栖息环境】海拔 1500 m以下的沼泽、水库和水稻田。【分布】云南、福建、广东、广西、海南、香港、台湾；南亚、东南亚、澳大利亚。【飞行期】全年可见。

[Identification] Male eyes brown, face yellow, antefrons bluish black. Thorax covered by blue pruinosity, wings hyaline. Abdomen covered by blue pruinosity on S1-S3, S4-S10 black with yellow stripes. Female mainly yellowish brown with black markings, S8 with small leaf-like flaps laterally, aged female body covered by bluish grey pruinosity. [Measurements] Total length 43-45 mm, abdomen 28-30 mm, hind wing 32-33 mm. [Habitat] Marshes, reservoirs and paddy fields below 1500 m elevation. [Distribution] Yunnan, Fujian, Guangdong, Guangxi, Hainan, Hong Kong, Taiwan; South and Southeast Asia, Australia. [Flight Season] Throughout the year.

玉带蜻属 Genus *Pseudothemis* Kirby, 1889

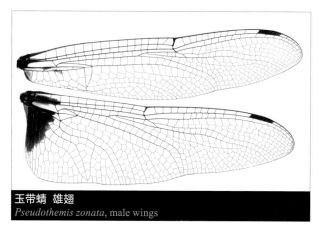

玉带蜻 雄翅
Pseudothemis zonata, male wings

本属全球已知2种，在亚洲东部和东南部广泛分布。中国已知1种，广泛分布。本属蜻蜓体中型，腹部基方白色或黄色；翅大面积透明，基方具色斑，翅痣较长，弓脉位于第1条与第2条结前横脉之间，最末端的结前横脉不完整，盘区基方具2~3列翅室，基臀区具1条横脉，臀圈靴状。

本属蜻蜓栖息于水面开阔的池塘和水库。雄性在边缘来回飞行，时而停落在水面的枝条上。交尾时间较短，在空中完成。雄性护卫产卵，雌性将卵产在水面漂浮的枝条上。

The genus contains two species widespread in the east and southeast of Asia. One species is recorded from China and widespread. Species of the genus are medium-sized, abdomen with basal half white or yellow. Wings largely hyaline with basal color spots, pterostigma long, arc between antenodals 1 and 2, distal antenodal incomplete, discoidal field with 2-3 rows of cells basally, cubital space with one crossvein, anal loop boot-shaped.

Pseudothemis species inhabit large ponds and reservoirs. Males usually patrol along the margin, and sometimes perch on the branches above water. Mating takes place in flight and is short in duration. Females lay eggs on floating branches on water's surface with hovering males guarding nearby.

玉带蜻 雄
Pseudothemis zonata, male

玉带蜻 *Pseudothemis zonata* (Burmeister, 1839)

【形态特征】雄性复眼褐色，面部黑色，额白色；胸部黑褐色，侧面具黄色细条纹，翅透明，后翅基方具甚大的黑褐色斑；腹部主要黑色，第2～4节白色。雌性与雄性相似，腹部第2～4节黄色，第5～7节侧面具黄斑。未熟的雄性与雌性色彩相似。【长度】体长 44～46 mm，腹长 29～31 mm，后翅 39～42 mm。【栖息环境】海拔 2000 m以下较大的池塘和水库。【分布】全国广布；朝鲜半岛、日本、老挝、越南。【飞行期】全年可见。

[Identification] Male eyes brown, face black, frons white. Thorax blackish brown, laterally with narrow yellow stripes, wings hyaline, hind wing bases with large blackish brown spots. Abdomen mainly black, S2-S4 white. Female similar to male, S2-S4 yellow, S5-S7 with lateral yellow spots. Immature male similar to female. [Measurements] Total length 44-46 mm, abdomen 29-31 mm, hind wing 39-42 mm. [Habitat] Large ponds and reservoirs below 2000 m elevation. [Distribution] Widespread in China; Korean peninsula, Japan, Laos, Vietnam. [Flight Season] Throughout the year.

玉带蜻 雌, 云南 (西双版纳)
Pseudothemis zonata, female from Yunnan (Xishuangbanna)

玉带蜻 雄, 云南 (西双版纳)
Pseudothemis zonata, male from Yunnan (Xishuangbanna)

玉带蜻 雌, 安徽 | 秦彧 摄
Pseudothemis zonata, female from Anhui | Photo by Yu Qin

红胭蜻属 Genus *Rhodothemis* Ris, 1909

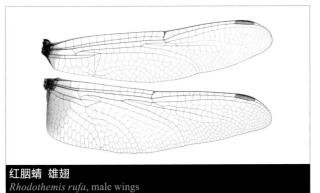

红胭蜻 雄翅
Rhodothemis rufa, male wings

本属全球已知5种，分布于亚洲的热带区域及大洋洲。中国已知1种，分布于华南和西南地区。本属蜻蜓体中型；翅狭长而透明，基方具色斑，翅痣较长，弓脉位于第1条与第2条结前横脉之间，前翅最末端的结前横脉不完整，盘区基方具1~3列翅室，基臀区具1条横脉，臀圈靴状。

本属蜻蜓栖息于水草茂盛的湿地，偏爱水葫芦滋生的池塘。雄性时常在水草上停落，时而飞行。交尾时间较短，雄性护卫产卵。红胭蜻与红蜻属种类相似，但其足上具长刺，且胸部背面具浅色条纹。

The genus contains five species distributed in tropical Asia and Oceania. One species is recorded from China, found in the South and Southwest. Species of the genus are medium-sized. Wings hyaline and narrow with basal color spots, pterostigma long, arc between antenodals 1 and 2, distal antenodal in fore wings incomplete, discoidal field with 1-3 rows of cells basally, cubital space with one crossvein, anal loop boot-shaped.

Species of the genus inhabit well vegetated wetlands, preferring ponds with plenty of water hyacinth. Males usually perch on plants, sometimes patrol. Mating duration is short and females lay eggs with male guarding. *Rhodothemis rufa* is similar to *Crocothemis* species but legs have longer spines, the dorsal part of thorax has pale stripes.

红胭蜻 雄
Rhodothemis rufa, male

红胭蜻 *Rhodothemis rufa* (Rambur, 1842)

红胭蜻 雄,云南(德宏)
Rhodothemis rufa, male from Yunnan (Dehong)

红胭蜻 雄,云南(德宏)
Rhodothemis rufa, male from Yunnan (Dehong)

红胭蜻 雌,广东 | 吴宏道 摄
Rhodothemis rufa, female from Guangdong | Photo by Hongdao Wu

　　【形态特征】雄性复眼红褐色,面部红色;胸部红褐色,合胸脊两侧具浅红色条纹,翅透明,后翅基方具红褐色斑;腹部鲜红色。雌性主要黄褐色,额具白色条纹;合胸脊两侧具黄白色条纹,翅透明,后翅基方的褐色斑较小;腹部黄褐色,第1~7节背面具黄白色斑。【长度】体长42~49 mm,腹长28~30 mm,后翅34~39 mm。【栖息环境】海拔1000 m以下水草茂盛的湿地。【分布】云南、广东、广西、海南、香港;南亚、东南亚广布。【飞行期】3—11月。

　　[Identification] Male eyes reddish brown, face red. Thorax reddish brown, dorsal carina with a pair of pale red stripes, wings hyaline, hind wing bases with reddish brown tints. Abdomen red. Female mainly yellowish brown, frons with white markings. Dorsal carina of thorax with a pair of yellowish white stripes, wings hyaline, hind wing bases with small brown tints. Abdomen yellowish brown, S1-S7 with yellowish white spots dorsally. [Measurements] Total length 42-49 mm, abdomen 28-30 mm, hind wing 34-39 mm. [Habitat] Wetlands with plenty of emergent plants below 1000 m elevation. [Distribution] Yunnan, Guangdong, Guangxi, Hainan, Hong Kong; Widespread in South and Southeast Asia. [Flight Season] March to November.

丽翅蜻属 Genus *Rhyothemis* Hagen, 1867

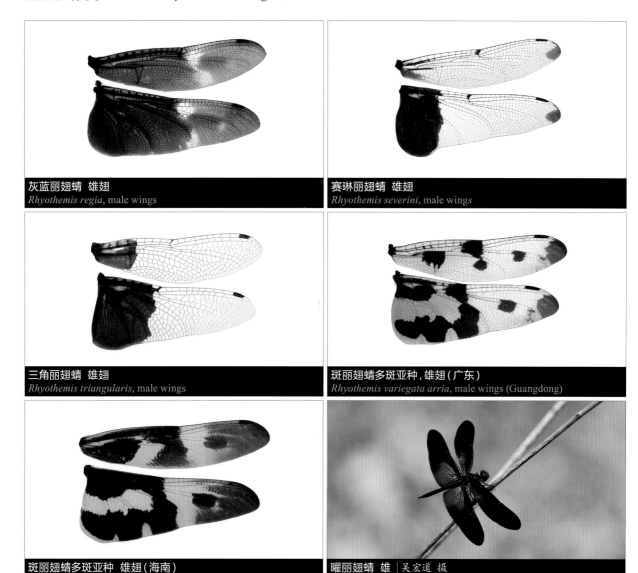

灰蓝丽翅蜻 雄翅
Rhyothemis regia, male wings

赛琳丽翅蜻 雄翅
Rhyothemis severini, male wings

三角丽翅蜻 雄翅
Rhyothemis triangularis, male wings

斑丽翅蜻多斑亚种,雄翅(广东)
Rhyothemis variegata arria, male wings (Guangdong)

斑丽翅蜻多斑亚种 雄翅(海南)
Rhyothemis variegata arria, male wings (Hainan)

曜丽翅蜻 雄 | 吴宏道 摄
Rhyothemis plutonia, male | Photo by Hongdao Wu

本属全球已知20余种, 分布于亚洲、非洲和大洋洲。本属蜻蜓体型小至中型, 身体暗色具金属光泽; 翅具艳丽的色彩和金属光泽, 后翅基方甚阔, 弓脉位于第1条与第2条结前横脉之间, 盘区具多列翅室, 臀圈靴状。

本属蜻蜓栖息于水草茂盛的湿地。雄性在水面附近的枝头和植物上停落, 护卫领地, 并以蝴蝶式的飞行姿态巡飞。雄性间经常展开争斗。交尾时间很短, 雄性护卫产卵。

The genus contains over 20 species distributed over Asia, Africa and Oceania. Species of the genus are small to medium sized, body dark with metallic reflections. Wings are colourful and with metallic reflections, hind wing base broad, arc between antenodals 1 and 2, discoidal field with many rows of cells basally, anal loop boot-shaped.

Rhyothemis species inhabit well vegetated wetlands. Territorial males usually perch on top of branches or plants and often patrol whose flight is suggestive of a butterfly. Males fight frequently. Mating is of short duration followed by the female laying eggs with the hovering male guarding afterward.

黑丽翅蜻 *Rhyothemis fuliginosa* Selys, 1883

【形态特征】通体蓝黑色具金属光泽；翅蓝黑色具蓝紫色或蓝绿色金属光色，前翅端方一半透明，后翅有时全黑色，有时端部有较小的透明区域。【长度】体长 31~36 mm，腹长 21~25 mm，后翅 31~36 mm。【栖息环境】海拔 500 m以下水草茂盛的湿地。【分布】北京、河北、河南、山东、江苏、湖北、安徽、浙江、福建、广东、香港、台湾；朝鲜半岛、日本。【飞行期】5—10月。

[Identification] Metallic bluish black throughout. Wings bluish black, shining metallic bluish purple or bluish green, fore wings with apical half hyaline, hind wings almost black or with small hyaline area apically. [Measurements] Total length 31-36 mm, abdomen 21-25 mm, hind wing 31-36 mm. [Habitat] Wetlands with plenty of emergent plants below 500 m elevation. [Distribution] Beijing, Hebei, Henan, Shandong, Jiangsu, Hubei, Anhui, Zhejiang, Fujian, Guangdong, Hong Kong, Taiwan; Korean peninsula, Japan. [Flight Season] May to October.

黑丽翅蜻 雄，湖北
Rhyothemis fuliginosa, male from Hubei

黑丽翅蜻 雄，湖北
Rhyothemis fuliginosa, male from Hubei

黑丽翅蜻 雌，北京 | 陈炜 摄
Rhyothemis fuliginosa, female from Beijing | Photo by Wei Chen

青铜丽翅蜻 *Rhyothemis obsolescens* Kirby, 1889

【形态特征】通体黑褐色；翅褐色具金属光泽，布满黄褐色的斑点。【长度】雄性体长 27 mm，腹长 18 mm，后翅 23 mm。【栖息环境】海拔 500 m 以下的湿地。【分布】云南（德宏）、海南；东南亚广布。【飞行期】4—8月。

青铜丽翅蜻 雄，云南（德宏）
Rhyothemis obsolescens, male from Yunnan (Dehong)

[Identification] Blackish brown throughout. Wings metallic brown with extensive yellowish brown markings. [Measurements] Male total length 27 mm, abdomen 18 mm, hind wing 23 mm. [Habitat] Wetlands below 500 m elevation. [Distribution] Yunnan (Dehong), Hainan; Widespread in Southeast Asia. [Flight Season] April to August.

青铜丽翅蜻 雄，云南（德宏）
Rhyothemis obsolescens, male from Yunnan (Dehong)

青铜丽翅蜻 雌，海南 ｜ Graham Reels 摄
Rhyothemis obsolescens, female from Hainan ｜ Photo by Graham Reels

臀斑丽翅蜻 *Rhyothemis phyllis* (Sulzer, 1776)

【形态特征】通体黑褐色；翅稍染金色，翅端具褐斑，后翅臀区具黄色和褐色斑。【长度】体长 30~41 mm，腹部 20~27 mm，后翅 33~37 mm。【栖息环境】海拔 500 m以下水草茂盛的湿地。【分布】云南（西双版纳）；广布于东南亚、澳新界和一些太平洋岛屿。【飞行期】10—11月。

[Identification] Blackish brown throughout. Wings slightly tinted with golden, tips with brown spots, hind wing anal field with yellow and brown marking. [Measurements] Total length 30-41 mm, abdomen 20-27 mm, hind wing 33-37 mm. [Habitat] Wetlands with plenty of emergent vegetation below 500 m elevation. [Distribution] Yunnan (Xishuangbanna); Widespread in Southeast Asia, Australasia and some Pacific Islands. [Flight Season] October to November.

臀斑丽翅蜻 雄，马来西亚
Rhyothemis phyllis, male from Malaysia

曜丽翅蜻 *Rhyothemis plutonia* Selys, 1883

【形态特征】通体蓝黑色；翅蓝黑色具蓝紫色或蓝绿色金属光色，雄性前翅端部有甚小的透明区域。雌性前后翅端部透明。本种与黑丽翅蜻相似，但前翅末端具较小的透明区域。【长度】体长 31~35 mm，腹长 19~23 mm，后翅 29~31 mm。【栖息环境】海拔 500 m以下水草茂盛的湿地。【分布】云南（西双版纳、德宏、红河）、广西、海南；南亚、东南亚广布。【飞行期】3—12月。

[Identification] Metallic bluish black throughout. Wings bluish black shining metallic bluish purple or bluish green, male fore wings with small area hyaline. Both wings of female with hyaline tips. The species is similar to *R.*

fuliginosa but fore wings with smaller hyaline tips. [Measurements] Total length 31-35 mm, abdomen 19-23 mm, hind wing 29-31 mm. [Habitat] Wetlands with plenty of emergent plants below 500 m elevation. [Distribution] Yunnan (Xishuangbanna, Dehong, Honghe), Guangxi, Hainan; Widespread in South and Southeast Asia. [Flight Season] March to December.

曝丽翅蜻 雄,云南(西双版纳)
Rhyothemis plutonia, male from Yunnan (Xishuangbanna)

曝丽翅蜻 雌,云南(西双版纳)
Rhyothemis plutonia, female from Yunnan (Xishuangbanna)

曝丽翅蜻 雄,云南(西双版纳)
Rhyothemis plutonia, male from Yunnan (Xishuangbanna)

灰黑丽翅蜻 *Rhyothemis regia* (Brauer, 1867)

【形态特征】通体黑褐色。雄性翅大面积蓝褐色具金属光色，亚翅端具透明区域。雌性翅大面积黑色具金属光泽，翅端透明，亚翅端具透明区域。【长度】体长 32~43 mm，腹长 22~28 mm，后翅 32~38 mm。【栖息环境】海拔 500 m以下的湿地。【分布】中国台湾；东南亚、新几内亚。【飞行期】4—10月。

[Identification] Blackish brown throughout. Male wings largely metallic bluish brown with transparent bands subapically. Female wings largely metallic black with hyaline tips and transparent bands subapically. [Measurements] Total length 32-43 mm, abdomen 22-28 mm, hind wing 32-38 mm. [Habitat] Wetlands below 500 m elevation. [Distribution] Taiwan of China; Southeast Asia, New Guinea. [Flight Season] April to October.

灰黑丽翅蜻 雌,台湾 | 嘎嘎 摄
Rhyothemis regia, female from Taiwan | Photo by Gaga

灰黑丽翅蜻 雄,台湾 | 嘎嘎 摄
Rhyothemis regia, male from Taiwan | Photo by Gaga

赛琳丽翅蜻 *Rhyothemis severini* Ris, 1913

赛琳丽翅蜻 雄,台湾 | 嘎嘎 摄
Rhyothemis severini, male from Taiwan | Photo by Gaga

【形态特征】通体蓝黑色具金属光泽。雄性翅基部和端部具金属蓝紫色斑。雌性翅具金属黑色斑。【长度】体长 41~44 mm,腹长 28~30 mm,后翅 40~42 mm。【栖息环境】海拔 500 m以下的湿地。【分布】中国台湾;日本、越南。【飞行期】4—9月。

[Identification] Metallic bluish black throughout. Male wing bases and tips with metallic bluish purple markings. Female wings with metallic black markings. [Measurements] Total length 41-44 mm, abdomen 28-30 mm, hind wing 40-42 mm. [Habitat] Wetlands below 500 m elevation. [Distribution] Taiwan of China; Japan, Vietnam. [Flight Season] April to September.

三角丽翅蜻 *Rhyothemis triangularis* Kirby, 1889

【形态特征】通体蓝黑色具金属光泽。雄性翅基部具金属蓝色斑。雌性翅基部具金属黑色斑。【长度】体长 25～28 mm，腹长 16～17 mm，后翅 23～25 mm。【栖息环境】海拔 1000 m 以下水草茂盛的湿地。【分布】云南（德宏）、广西、广东、海南、福建、香港、台湾；南亚、东南亚广布。【飞行期】4—11月。

[Identification] Metallic bluish black throughout. Male wing bases with metallic blue markings. Female wing bases with metallic black markings. [Measurements] Total length 25-28 mm, abdomen 16-17 mm, hind wing 23-25 mm. [Habitat] Wetlands with plenty of emergent vegatation below 1000 m elevation. [Distribution] Yunnan (Dehong), Guangxi, Guangdong, Hainan, Fujian, Hong Kong, Taiwan; Widespread in South and Southeast Asia. [Flight Season] April to November.

三角丽翅蜻 雄，广东 | 宋睿斌 摄
Rhyothemis triangularis, male from Guangdong | Photo by Ruibin Song

三角丽翅蜻 交尾，广东 | 吴宏道 摄
Rhyothemis triangularis, mating pair from Guangdong | Photo by Hongdao Wu

三角丽翅蜻 雄，广西
Rhyothemis triangularis, male from Guangxi

斑丽翅蜻多斑亚种 *Rhyothemis variegata arria* **Drury, 1773**

【形态特征】通体黑绿色具金属光泽。雄性翅琥珀色具丰富的黑色斑。雌性前翅端部透明。本种翅上的斑纹变异较大，海南个体翅大面积黑褐色。【长度】体长 37~42 mm，腹长 24~28 mm，后翅 35~39 mm。【栖息环境】海拔 1000 m 以下的湿地。【分布】云南（红河）、广西、广东、海南、福建、香港、台湾；日本、越南。【飞行期】4—11月。

[Identification] Metallic blackish green throughout. Male wings amber with extensive black markings. Female fore wings with apical area hyaline. A variable species, wings of individuals from Hainan largely blackish brown.

斑丽翅蜻多斑亚种 雄，云南（红河）
Rhyothemis variegata arria, male from Yunnan (Honghe)

斑丽翅蜻多斑亚种 雌，广东 | 莫善濂 摄
Rhyothemis variegata arria, female from Guangdong | Photo by Shanlian Mo

斑丽翅蜻多斑亚种 雄，广东 | 宋睿斌 摄
Rhyothemis variegata arria, male from Guangdong | Photo by Ruibin Song

斑丽翅蜻多斑亚种　雄,海南
Rhyothemis variegata arria, male from Hainan

斑丽翅蜻多斑亚种　雌,海南
Rhyothemis variegata arria, female from Hainan

[Measurements] Total length 37-42 mm, abdomen 24-28 mm, hind wing 35-39 mm. [Habitat] Wetlands below 1000 m elevation. [Distribution] Yunnan (Honghe), Guangxi, Guangdong, Hainan, Fujian, Hong Kong, Taiwan; Japan, Vietnam. [Flight Season] April to November.

斑丽翅蜻指名亚种 *Rhyothemis variegata variegata* (Linnaeus, 1763)

【形态特征】通体黑绿色具金属光泽。雄性翅琥珀色具稀疏的黑色斑。雌性前翅端部透明。本亚种和多斑亚种的区别在于翅上的黑色斑纹较少，体型更小。【长度】体长 31~37 mm，腹长 21~25 mm，后翅 32~35 mm。【栖息环境】海拔 500 m 以下的湿地。【分布】云南(临沧)；孟加拉国、印度、斯里兰卡、尼泊尔、缅甸、泰国、老挝、柬埔寨。【飞行期】不详。

斑丽翅蜻指名亚种 雄，云南(临沧)
Rhyothemis variegata variegata, male from Yunnan (Lincang)

斑丽翅蜻指名亚种 雌，云南(临沧)
Rhyothemis variegata variegata, female from Yunnan (Lincang)

[Identification] Metallic blackish green throughout. Male wings amber with sparse black markings. Female fore wings with apical area hyaline. Distinguished from *R. variegata arria* by fewer black markings of wings and smaller size. [Measurements] Total length 31-37 mm, abdomen 21-25 mm, hind wing 32-35 mm. [Habitat] Wetlands below 500 m elevation. [Distribution] Yunnan (Lincang); Bangladesh, India, Sri Lanka, Nepal, Myanmar, Thailand, Laos, Cambodia. [Flight Season] Unknown.

丽翅蜻属待定种 *Rhyothemis* sp.

丽翅蜻属待定种 雌，云南(西双版纳)
Rhyothemis sp., female from Yunnan (Xishuangbanna)

【形态特征】雌性翅琥珀色具稀疏的黑色斑。本种与臀斑丽翅蜻相似，但翅上的黑色斑点较多。【长度】雌性体长 32~33 mm，腹长 21~22 mm，后翅 33~35 mm。【栖息环境】海拔 500 m 以下的湿地。【分布】云南(西双版纳、德宏)。【飞行期】4—11月。

[Identification] Female wings amber with sparse black markings. The species is similar to *R. phyllis* but has more extensive black markings on wings. [Measurements] Female total length 32-33 mm, abdomen 21-22 mm, hind wing 33-35 mm. [Habitat] Wetlands below 500 m elevation. [Distribution] Yunnan (Xishuangbanna, Dehong). [Flight Season] April to November.

赛丽蜻属 Genus *Selysiothemis* Ris, 1897

本属全球仅1种，分布于亚洲、欧洲和撒哈拉的绿洲。赛丽蜻是体中型的蜻蜓。翅透明，翅脉稀疏，翅痣短而色彩较浅，具较少的结前横脉，弓脉位于第1条与第2条结前横脉之间，盘区基方具2列翅室，基臀区具1条横脉，臀圈靴状。

本属蜻蜓栖息于池塘和水库。雄性会停在水面上的树枝占据领地，并时而巡飞。

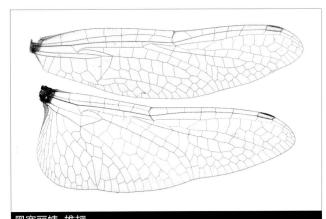

黑赛丽蜻 雄翅
Selysiothemis nigra, male wings

The genus contains a single species, distributed in Asia, Europe and oases of the Sahara. Species of the genus is medium-sized. Wings hyaline, venation sparse, pterostigma pale and short, antenodals sparse in number, arc between antenodals 1 and 2, discoidal field with 2 rows of cells basally, cubital space with one crossvein, anal loop boot-shaped.

Selysiothemis species inhabits in ponds and reserviors. Males perch on top of branches above water and patrol sometimes.

黑赛丽蜻 雄, 北京 | 陈炜 摄
Selysiothemis nigra, male from Beijing | Photo by Wei Chen

黑赛丽蜻 *Selysiothemis nigra* (Vander Linden, 1825)

黑赛丽蜻 雌,北京 | 陈炜 摄
Selysiothemis nigra, female from Beijing | Photo by Wei Chen

【形态特征】雄性复眼灰褐色,面部白色;胸部和腹部黑褐色,翅透明,翅脉灰白色。雌性黄褐色。【长度】体长 30~38 mm,腹长 21~26 mm,后翅 24~27 mm。【栖息环境】海拔 500 m以下的池塘和水库。【分布】北京、新疆、内蒙古、辽宁、河北;中亚、地中海和黑海地区、印度、撒哈拉。【飞行期】5—9月。

[Identification] Male eyes greyish brown, face white. Thorax and abdomen blackish brown, wings hyaline, venation greyish white. Female yellowish brown. [Measurements] Total length 30-38 mm, abdomen 21-26 mm, hind wing 24-27 mm. [Habitat] Ponds and reservoirs below 500 m elevation. [Distribution] Beijing, Xinjiang, Inner Mongolia, Liaoning, Hebei; Central Asia, the Mediterranean and the Black Sea region, India, Sahara. [Flight Season] May to September.

黑赛丽蜻 雄,北京 | 陈炜 摄
Selysiothemis nigra, male from Beijing | Photo by Wei Chen

赤蜻属 Genus *Sympetrum* Newman, 1833

旭光赤蜻 雄
Sympetrum hypomelas, male

本属已知超过60种，全球广布，但在较寒冷的区域繁盛。这是一类体小型至中型的蜻蜓；雄性通常红色；翅大面积透明，翅痣较长，弓脉位于第1条与第2条结前横脉之间，最末端的结前横脉不完整，盘区基方具1~3列翅室，基臀区具1条横脉，前翅的三角室具1条横脉，臀圈靴状。

本属蜻蜓栖息于水草茂盛的湿地和流速缓慢的溪流。雄性在水边停落占据领地，时而飞行。交尾时间较长，雌雄连结产卵。中国已知的赤蜻种类可以通过身体色彩、雄性钩片和肛附器的构造以及雌性下生殖板来区分。

The genus contains over 60 species distributed all over the world, but they are most speciose in colder regions. Species of the genus are small to medium sized. Males usually red. Wings larglely hyaline, pterostigma long, arc between antenodals 1 and 2, distal antenodal incomplete, discoidal field with 1-3 rows of cells basally, cubital space with one crossvein, triangle with one crossvein in forewings, anal loop boot-shaped.

Sympetrum species inhabit well vegetated wetlands and slow flowing streams. Males perch on plants, and sometimes fly around. Mating duration is long and they oviposit in tandem. The Chinese species can be distinguished by body color pattern, male hamulus and appendages, shape of female vulvar lamina.

大赤蜻指名亚种
Sympetrum baccha baccha

半黄赤蜻
Sympetrum croceolum

大理赤蜻
Sympetrum daliensis

夏赤蜻
Sympetrum darwinianum

扁腹赤蜻
Sympetrum depressiusculum

竖眉赤蜻指名亚种
Sympetrum eroticum eroticum

竖眉赤蜻多纹亚种
Sympetrum eroticum ardens

黄斑赤蜻
Sympetrum flaveolum

方氏赤蜻
Sympetrum fonscolombii

秋赤蜻
Sympetrum frequens

旭光赤蜻
Sympetrum hypomelas

褐顶赤蜻
Sympetrum infuscatum

赤蜻属 雄性肛附器
Genus *Sympetrum*, male anal appendages

南投赤蜻
Sympetrum nantouensis

姬赤蜻
Sympetrum parvulum

褐带赤蜻
Sympetrum pedemontanum

李氏赤蜻
Sympetrum risi risi

黄基赤蜻微斑亚种
Sympetrum speciosum haematoneura

黄基赤蜻指名亚种
Sympetrum speciosum speciosum

条斑赤蜻喜马亚种
Sympetrum striolatum commixtum

条斑赤蜻指名亚种
Sympetrum striolatum striolatum

大黄赤蜻
Sympetrum uniforme

赤蜻属 雄性肛附器
Genus *Sympetrum*, male anal appendages

肖氏赤蜻
Sympetrum xiaoi

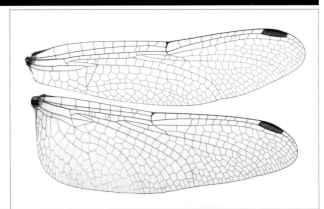

赤蜻属 雄性肛附器
Genus *Sympetrum*, male anal appendages

旭光赤蜻 雄翅
Sympetrum hypomelas, male wings

大赤蜻指名亚种 *Sympetrum baccha baccha* (Selys, 1884)

【形态特征】雄性面部红色；胸部黄褐色，侧面具黑色条纹，翅透明，末端具甚小的褐斑；腹部红色。雌性黄色具黑色条纹。【长度】体长 49～50 mm，腹长 33～34 mm，后翅 37～40 mm。【栖息环境】海拔 2000 m 以下的湿地。【分布】中国特有亚种，分布于河南、安徽、浙江、福建、贵州、四川、湖北、湖南、江西、台湾、广东。【飞行期】6—11月。

[Identification] Male face red. Thorax yellowish brown, laterally with black stripes, wings hyaline with small apical brown spots. Abdomen red. Female yellow with black markings. [Measurements] Total length 49-50 mm, abdomen 33-34 mm, hind wing 37-40 mm. [Habitat] Wetlands below 2000 m elevation. [Distribution] Endemic to China, recorded from Henan, Anhui, Zhejiang, Fujian, Guizhou, Sichuan, Hubei, Hunan, Jiangxi, Taiwan, Guangdong. [Flight Season] June to November.

大赤蜻指名亚种 雄, 湖北
Sympetrum baccha baccha, male from Hubei

大赤蜻指名亚种 雌, 安徽 | 秦彧 摄
Sympetrum baccha baccha, female from Anhui | Photo by Yu Qin

大赤蜻指名亚种 雄, 湖北
Sympetrum baccha baccha, male from Hubei

大赤蜻褐顶亚种 *Sympetrum baccha matutinum* Ris, 1911

大赤蜻褐顶亚种 雄, 吉林 │金洪光 摄
Sympetrum baccha matutinum, male from Jilin │ Photo by Hongguang Jin

【形态特征】雄性面部红色; 胸部红色, 侧面具黑色条纹, 翅透明, 端部具较大的黑褐色斑; 腹部红色。雌性黄色具黑色条纹。本亚种与指名亚种的明显区别在翅端部具较大的褐斑。【长度】体长 36~48 mm, 腹长 23~31 mm, 后翅 29~35 mm。【栖息环境】海拔 1000 m 以下的湿地。【分布】黑龙江、吉林; 朝鲜半岛、日本。【飞行期】6—9月。

[Identification] Male face red. Thorax red, laterally with black stripes, wings hyaline with large apical blackish brown spots. Abdomen red. Female yellow with black markings. The subspecies can be separated from nominate subspecies by the blackish brown wing tips. [Measurements] Total length 36-48 mm, abdomen 23-31 mm, hind wing 29-35 mm. [Habitat] Wetlands below 1000 m elevation. [Distribution] Heilongjiang, Jilin; Korean peninsula, Japan. [Flight Season] June to September.

大赤蜻褐顶亚种 雄, 吉林 │金洪光 摄
Sympetrum baccha matutinum, male from Jilin │ Photo by Hongguang Jin

大赤蜻褐顶亚种 雌, 吉林 │金洪光 摄
Sympetrum baccha matutinum, female from Jilin │ Photo by Hongguang Jin

长尾赤蜻 *Sympetrum cordulegaster* (Selys, 1883)

【形态特征】雄性面部黄白色；胸部黄褐色，翅透明；腹部红色，侧缘具较小的黑斑，第7～9节向腹面隆起。雌性土黄色，下生殖板呈锥状，伸达腹部末端。【长度】体长 29～42 mm，腹长 17～28 mm，后翅 23～31 mm。【栖息环境】海拔 1000 m以下的湿地。【分布】黑龙江、吉林、辽宁、台湾；朝鲜半岛、日本、俄罗斯远东。【飞行期】6—10月。

长尾赤蜻 雌,吉林 | 金洪光 摄
Sympetrum cordulegaster, female from Jilin | Photo by Hongguang Jin

长尾赤蜻 交尾,吉林 | 金洪光 摄
Sympetrum cordulegaster, mating pair from Jilin | Photo by Hongguang Jin

[Identification] Male face yellowish white. Thorax yellowish brown, wings hyaline. Abdomen red with lateral small black markings, S7-S9 expanded ventrally. Female mainly khaki, ovipositor elongated and extending beyond tip of abdomen. [Measurements] Total length 29-42 mm, abdomen 17-28 mm, hind wing 23-31 mm. [Habitat] Wetlands below 1000 m elevation. [Distribution] Heilongjiang, Jilin, Liaoning, Taiwan; Korean peninsula, Japan, Russian Far East. [Flight Season] June to October.

长尾赤蜻 雄,吉林 | 金洪光 摄
Sympetrum cordulegaster, male from Jilin | Photo by Hongguang Jin

半黄赤蜻 *Sympetrum croceolum* Selys, 1883

半黄赤蜻 雄，黑龙江
Sympetrum croceolum, male from Heilongjiang

半黄赤蜻 雌，黑龙江
Sympetrum croceolum, female from Heilongjiang

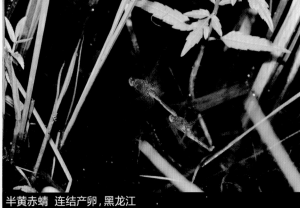

半黄赤蜻 连结产卵，黑龙江
Sympetrum croceolum, laying eggs in tandem from Heilongjiang

　　【形态特征】雄性头部、胸部和翅金褐色；腹部红色。雌性腹部黄褐色，下生殖板较突出。【长度】体长 37～48 mm，腹长 24～32 mm，后翅 28～35 mm。【栖息环境】海拔 1500 m 以下的湿地。【分布】除西北地区、云南、广西、海南和台湾外全国广布；朝鲜半岛、日本。【飞行期】6—11月。

[Identification] Male head, thorax and wings golden brown. Abdomen red. Female abdomen yellowish brown, vulvar lamina projecting. [Measurements] Total length 37-48 mm, abdomen 24-32 mm, hind wing 28-35 mm. [Habitat] Wetlands below 1500 m elevation. [Distribution] Widespread except Northwest China, Yunnan, Guangxi, Hainan and Taiwan; Korean peninsula, Japan. [Flight Season] June to November.

半黄赤蜻 雄，黑龙江
Sympetrum croceolum, male from Heilongjiang

大理赤蜻 *Sympetrum daliensis* Zhu, 1999

　　【形态特征】雄性面部黄色；胸部黄褐色，侧面具黑色条纹，翅透明；腹部红色，侧缘具黑色斑纹。雌性多型，腹部红色或黄色。【长度】体长 40~44 mm，腹长 27~29 mm，后翅 33~34 mm。【栖息环境】海拔 1500~2500 m 的湿地。【分布】中国云南（大理）特有。【飞行期】6—11月。

大理赤蜻 雌，黄色型，云南（大理）
Sympetrum daliensis, female, the yellow morph from Yunnan (Dali)

大理赤蜻 雌，红色型，云南（大理）
Sympetrum daliensis, female, the red morph from Yunnan (Dali)

大理赤蜻 雄，云南（大理）
Sympetrum daliensis, male from Yunnan (Dali)

[Identification] Male face yellow. Thorax yellowish brown, laterally with black stripes, wings hyaline. Abdomen red with black markings laterally. Female polymorphic, abdomen red or yellow. [Measurements] Total length 40-44 mm, abdomen 27-29 mm, hind wing 33-34 mm. [Habitat] Wetlands at 1500-2500 m elevation. [Distribution] Endemic to Yunnan (Dali) of China. [Flight Season] June to November.

黑赤蜻 *Sympetrum danae* (Sulzer, 1776)

【形态特征】雄性完全成熟后全身黑褐色,翅透明。雌性黄色,胸部侧面具黑色条纹,腹部侧缘具黑色条纹。【长度】体长 29～34 mm, 腹长 18～26 mm, 后翅 20～30 mm。【栖息环境】海拔 2500 m以下的湿地。【分布】内蒙古、黑龙江、吉林、四川; 欧亚大陆北部和北美洲广布。【飞行期】6—9月。

[Identification] Male blackish brown throughout when fully mature, wings hyaline. Female yellow, thorax with black stripes laterally, abdomen with black stripes laterally. [Measurements] Total length 29-34 mm, abdomen 18-26 mm, hind wing 20-30 mm. [Habitat] Wetlands below 2500 m elevation. [Distribution] Inner Mongolia, Heilongjiang, Jilin, Sichuan; Widespread across the north of Eurasia and North America. [Flight Season] June to September.

黑赤蜻 雄,内蒙古 | 吕非 摄
Sympetrum danae, male from Inner Mongolia | Photo by Fei Lv

黑赤蜻 雄,芬兰 | Matti Hämäläinen 摄
Sympetrum danae, male from Finland | Photo by Matti Hämäläinen

黑赤蜻 雌,芬兰 | Matti Hämäläinen 摄
Sympetrum danae, female from Finland | Photo by Matti Hämäläinen

夏赤蜻 *Sympetrum darwinianum* Selys, 1883

夏赤蜻 雄, 重庆
Sympetrum darwinianum, male from Chongqing

夏赤蜻 雄, 重庆
Sympetrum darwinianum, male from Chongqing

夏赤蜻 雌, 湖北 | 宋睿斌 摄
Sympetrum darwinianum, female from Hubei | Photo by Ruibin Song

【形态特征】雄性面部红色；胸部背面红色，侧面黄褐色具黑色条纹，翅透明；腹部红色。雌性腹部橙红色，侧缘具黑斑。北方的雄性胸部完全红色。【长度】体长 37～42 mm，腹长 25～28 mm，后翅 29～32 mm。【栖息环境】海拔 2000 m 以下的湿地。【分布】除西北地区外全国广布；朝鲜半岛、日本。【飞行期】6—11月。

[Identification] Male face red. Thorax red dorsally, yellowish brown laterally with black stripes, wings hyaline. Abdomen red. Female abdomen orange red with black markings laterally. Individuals from the north with entirely red thorax. [Measurements] Total length 37-42 mm, abdomen 25-28 mm, hind wing 29-32 mm. [Habitat] Wetlands below 2000 m elevation. [Distribution] Widespread in China except Northwest China; Korean peninsula, Japan. [Flight Season] June to November.

扁腹赤蜻 *Sympetrum depressiusculum* (Selys, 1841)

扁腹赤蜻 雄,黑龙江
Sympetrum depressiusculum, male from Heilongjiang

扁腹赤蜻 雌,黑龙江
Sympetrum depressiusculum, female from Heilongjiang

扁腹赤蜻 雌,黑龙江
Sympetrum depressiusculum, female from Heilongjiang

　　【形态特征】雄性面部黄色;胸部黄褐色,侧面具黑色条纹,翅透明;腹部红色。雌性多型,腹部橙红色或土黄色,侧缘具较小的褐色斑。【长度】体长 27～40 mm,腹长 17～27 mm,后翅 22～30 mm。【栖息环境】海拔 1000 m以下的湿地。【分布】黑龙江、吉林、辽宁、内蒙古、北京、台湾;朝鲜半岛、日本、欧洲、西伯利亚。【飞行期】6—10月。

　　[Identification] Male face yellow. Thorax yellowish brown with black stripes laterally, wings hyaline. Abdomen red. Female polymorphic, abdomen orange red or khaki yellow with small black markings laterally. [Measurements] Total length 27-40 mm, abdomen 17-27 mm, hind wing 22-30 mm. [Habitat] Wetlands below 1000 m elevation. [Distribution] Heilongjiang, Jilin, Liaoning, Inner Mongolia, Beijing, Taiwan; Korean peninsula, Japan, Europe, Siberia. [Flight Season] June to October.

竖眉赤蜻指名亚种 *Sympetrum eroticum eroticum* (Selys, 1883)

【形态特征】雄性面部褐色，前额具1对黑色斑点；胸部初熟时黄色，老熟以后红褐色，具宽阔的肩条纹，翅透明；腹部红色。雌性多型，腹部有红色和黄褐色两种；翅透明，有时端部具褐斑；下生殖板较长。【长度】体长 33~40 mm，腹长 23~28 mm，后翅 25~30 mm。【栖息环境】海拔 1000 m以下的湿地。【分布】黑龙江、吉林、辽宁、内蒙古、北京、河北、山西、山东、河南；俄罗斯远东、朝鲜半岛、日本。【飞行期】6—10月。

[Identification] Male face brown, antefrons with a pair of black spots. Thorax yellow when young and reddish brown when aged, with broad black humeral stripes, wings hyaline. Abdomen red. Female polymorphic, abdomen red or yellowish brown. Wings hyaline or with apical brown spots. Vulvar lamina long. [Measurements] Total length 33-40 mm, abdomen 23-28 mm, hind wing 25-30 mm. [Habitat] Wetlands below 1000 m elevation. [Distribution] Heilongjiang, Jilin, Liaoning, Inner Mongolia, Beijing, Hebei, Shanxi, Shandong, Henan; Russian Far East, Korean peninsula, Japan. [Flight Season] June to October.

竖眉赤蜻指名亚种 雄，黑龙江
Sympetrum eroticum eroticum, male from Heilongjiang

竖眉赤蜻指名亚种 雄，北京
Sympetrum eroticum eroticum, male from Beijing

竖眉赤蜻指名亚种 雌，山东
Sympetrum eroticum eroticum, female from Shandong

竖眉赤蜻指名亚种 交尾，黑龙江
Sympetrum eroticum eroticum, mating pair from Heilongjiang

竖眉赤蜻多纹亚种 *Sympetrum eroticum ardens* (McLachlan, 1894)

【形态特征】与指名亚种非常近似，但合胸侧面具较多的黑色条纹。【长度】体长 40~44 mm，腹长 27~31 mm，后翅 31~32 mm。【栖息环境】海拔 2000 m 以下的湿地。【分布】安徽、湖北、湖南、四川、重庆、云南、贵州、浙江、福建、广东、台湾；越南。【飞行期】6—12月。

[Identification] Similar to nominate subspecies, but thorax with black stripes laterally. [Measurements] Total length 40-44 mm, abdomen 27-31 mm, hind wing 31-32 mm. [Habitat] Wetlands below 2000 m elevation. [Distribution] Anhui, Hubei, Hunan, Sichuan, Chongqing, Yunnan, Guizhou, Zhejiang, Fujian, Guangdong, Taiwan; Vietnam. [Flight Season] June to December.

竖眉赤蜻多纹亚种 雄，贵州
Sympetrum eroticum ardens, male from Guizhou

竖眉赤蜻多纹亚种 雄，贵州
Sympetrum eroticum ardens, male from Guizhou

竖眉赤蜻多纹亚种 雌，贵州
Sympetrum eroticum ardens, female from Guizhou

黄斑赤蜻 *Sympetrum flaveolum* (Linnaeus, 1758)

黄斑赤蜻 雄, 内蒙古 | 吕非 摄
Sympetrum flaveolum, male from Inner Mongolia | Photo by Fei Lv

【形态特征】雄性面部红色；胸部红褐色，前翅前缘从基方至翅结处具琥珀色斑，后翅基方具1个甚大琥珀色斑；腹部红色具黑色条纹。雌性土黄色，腹部具黑色条纹。【长度】体长 32~37 mm，腹长 19~27 mm，后翅 23~32 mm。【栖息环境】海拔 500 m 以下的湿地。【分布】黑龙江、吉林、内蒙古；欧洲西部至日本广布。【飞行期】7—9月。

[Identification] Male face red. Thorax reddish brown, fore wings with amber markings from base to nodus, hind wing bases with a large amber markings. Abdomen red with black markings. Female khaki yellow, abdomen with black markings. [Measurements] Total length 32-37 mm, abdomen 19-27 mm, hind wing 23-32 mm. [Habitat] Wetlands below 500 m elevation. [Distribution] Heilongjiang, Jilin, Inner Mongolia; Widespread from western Europe to Japan. [Flight Season] July to September.

黄斑赤蜻 雌, 吉林 | 金洪光 摄
Sympetrum flaveolum, female from Jilin | Photo by Hongguang Jin

黄斑赤蜻 雌, 内蒙古 | 吕非 摄
Sympetrum flaveolum, female from Inner Mongolia | Photo by Fei Lv

方氏赤蜻 *Sympetrum fonscolombii* (Selys, 1840)

【形态特征】雄性面部红色；胸部红褐色，侧面具2条黄色条纹，翅透明，后翅基方具橙黄色斑；腹部红色，末端具黑色斑点。雌性黄色具黑色条纹。【长度】体长 35～41 mm，腹长 24～39 mm，后翅 26～32 mm。【栖息环境】海拔 2500 m以下的湿地。【分布】除西北地区外全国广布；亚洲、欧洲、非洲广布。【飞行期】5—12月。

方氏赤蜻 雄，云南（红河）
Sympetrum fonscolombii, male from Yunnan (Honghe)

方氏赤蜻 雌，辽宁
Sympetrum fonscolombii, female from Liaoning

[Identification] Male face red. Thorax reddish brown with two yellow stripes laterally, wings hyaline, hind wing bases with orange markings. Abdomen red, the tip with black markings. Female yellow with black markings. [Measurements] Total length 35-41 mm, abdomen 24-39 mm, hind wing 26-32 mm. [Habitat] Wetlands below 2500 m elevation. [Distribution] Widespread except Northwest China; Widespread in Asia, Europe, Africa. [Flight Season] May to December.

方氏赤蜻 雄，云南（红河）
Sympetrum fonscolombii, male from Yunnan (Honghe)

秋赤蜻 *Sympetrum frequens* (Selys, 1883)

【形态特征】雄性面部黄色；胸部黄褐色，侧面具黑色细条纹，翅透明；腹部红色。雌性多型，腹部橙红色或土黄色，侧缘具较小的褐色斑。本种与扁腹赤蜻相似，但身体的黑色条纹更显著，雄性的肛附器和雌性的下生殖板构造都与扁腹赤蜻不同。【长度】体长 34~42 mm，腹长 23~28 mm，后翅 27~31 mm。【栖息环境】海拔 1000 m以下的湿地和水稻田。【分布】黑龙江、吉林、辽宁；俄罗斯远东、朝鲜半岛、日本。【飞行期】6—10月。

[Identification] Male face yellow. Thorax yellowish brown with black stripes laterally, wings hyaline. Abdomen red. Female polymorphic, abdomen orange red or khaki yellow with small brown markings laterally. The species is similar to *S. depressiusculum*, but the black stripes are broader, male anal appendages and female vulvar lamina are different. [Measurements] Total length 34-42 mm, abdomen 23-28 mm, hind wing 27-31 mm. [Habitat] Wetlands and paddy fields below 1000 m elevation. [Distribution] Heilongjiang, Jilin, Liaoning; Russian Far East, Korean peninsula, Japan. [Flight Season] June to October.

秋赤蜻 雄，黑龙江
Sympetrum frequens, male from Heilongjiang

秋赤蜻 雌，吉林 | 金洪光 摄
Sympetrum frequens, female from Jilin | Photo by Hongguang Jin

旭光赤蜻 *Sympetrum hypomelas* (Selys, 1884)

旭光赤蜻 雄,云南(普洱)
Sympetrum hypomelas, male from Yunnan (Pu'er)

旭光赤蜻 雌,黄色型,云南(普洱)
Sympetrum hypomelas, female, the yellow morph from Yunnan (Pu'er)

旭光赤蜻 雌,红色型,云南(普洱)
Sympetrum hypomelas, female, the red morph from Yunnan (Pu'er)

【形态特征】雄性面部红色;胸部背面红色,侧面具2条宽阔的黄色条纹,翅透明;腹部红色,侧缘具黑色斑纹。雌性多型,腹部橙红色或土黄色,侧面具黑色条纹。【长度】体长 35～41 mm,腹长 23～27 mm,后翅 29～33 mm。【栖息环境】海拔 1500～2500 m的湿地。【分布】云南;孟加拉国、印度、尼泊尔、缅甸。【飞行期】7—12月。

[Identification] Male face red. Thorax red dorsally, sides with two yellow stripes, wings hyaline. Abdomen red with black markings laterally. Female polymorphic, abdomen orange red or khaki yellow with black markings laterally. [Measurements] Total length 35-41 mm, abdomen 23-27 mm, hind wing 29-33 mm. [Habitat] Wetlands at 1500-2500 m elevation. [Distribution] Yunnan; Bangladesh, India, Nepal, Myanmar. [Flight Season] July to December.

褐顶赤蜻 *Sympetrum infuscatum* (Selys, 1883)

【形态特征】雄性身体深褐色，合胸侧面和腹部具黑色条纹；翅透明，翅端具褐斑；年长的个体腹部红褐色。雌性体色较浅；胸部大面积黄色；腹部黄褐色。【长度】体长 42~47 mm，腹长 29~32 mm，后翅 32~37 mm。【栖息环境】海拔 1500 m以下的湿地。【分布】除西北地区、海南和台湾外全国广布；俄罗斯远东、朝鲜半岛、日本。【飞行期】6—10月。

褐顶赤蜻 雄，黑龙江
Sympetrum infuscatum, male from Heilongjiang

褐顶赤蜻 雌，吉林 ｜金洪光 摄
Sympetrum infuscatum, female from Jilin ｜ Photo by Hongguang Jin

[Identification] Male dark brown throughout, sides of thorax and abdomen with black stripes. Wings hyaline with brown apical spots. Abdomen reddish brown when aged. Female paler. Thorax largely yellow. Abdomen yellowish brown. [Measurements] Total length 42-47 mm, abdomen 29-32 mm, hind wing 32-37 mm. [Habitat] Wetlands below 1500 m elevation. [Distribution] Widespread in China except Northwest China, Hainan and Taiwan; Russian Far East, Korean peninsula, Japan. [Flight Season] June to October.

褐顶赤蜻 雄，重庆
Sympetrum infuscatum, male from Chongqing

小黄赤蜻 *Sympetrum kunckeli* (Selys, 1884)

【形态特征】雄性面部白色，稍染绿色；胸部黄褐色，合胸脊黑色，肩条纹黑色，甚阔，侧面具小黑斑，翅透明；腹部红色。雌性黄褐色具黑色斑纹。【长度】体长 29~40 mm，腹长 17~25 mm，后翅 20~28 mm。【栖息环境】海拔 1000 m 以下的湿地。【分布】吉林、辽宁、北京、河北、山西、山东、河南、安徽、陕西、湖北、台湾；俄罗斯远东、朝鲜半岛、日本。【飞行期】6—9月。

[Identification] Male face white, slightly tinted with green. Thorax yellowish brown, dorsal carina black, humeral stripes black and broad, sides with small black spots, wings hyaline. Abdomen red. Female yellowish brown with black markings. [Measurements] Total length 29-40 mm, abdomen 17-25 mm, hind wing 20-28 mm. [Habitat] Wetlands below 1000 m elevation. [Distribution] Jilin, Liaoning, Beijing, Hebei, Shanxi, Shandong, Henan, Anhui, Shaanxi, Hubei, Taiwan; Russian Far East, Korean peninsula, Japan. [Flight Season] June to September.

小黄赤蜻 雄, 湖北
Sympetrum kunckeli, male from Hubei

小黄赤蜻 雄, 湖北
Sympetrum kunckeli, male from Hubei

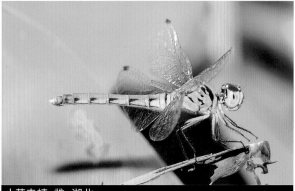

小黄赤蜻 雌, 湖北
Sympetrum kunckeli, female from Hubei

南投赤蜻 *Sympetrum nantouensis* Tang, Yeh & Chen, 2013

南投赤蜻 雄，台湾 | 嘎嘎 摄
Sympetrum nantouensis, male from Taiwan | Photo by Gaga

【形态特征】雄性面部白色；胸部淡褐色，侧面具黑色条纹，翅透明；腹部红色，侧缘具黑色斑纹。【长度】雄性体长 35 mm，腹长 24 mm，后翅 26 mm。【栖息环境】海拔 300~800 m 的杂草沼泽。【分布】中国台湾特有。【飞行期】9—11月。

[Identification] Male face white. Thorax pale brown, sides with black stripes, wings hyaline. Abdomen red with black markings laterally. [Measurements] Male total length 35 mm, abdomen 24 mm, hind wing 26 mm. [Habitat] Grassy marshes at 300-800 m elevation. [Distribution] Endemic to Taiwan of China. [Flight Season] September to November.

姬赤蜻 *Sympetrum parvulum* (Bartenev, 1913)

【形态特征】雄性面部白色；胸部黄褐色，合胸脊黑色，肩条纹黑色，甚阔，翅透明；腹部红色。雌性腹部橙红色。本种与竖眉赤蜻相似，但雄性肛附器和雌性下生殖板的构造不同。【长度】体长 32~34 mm，腹长 22~23 mm，后翅 25~26 mm。【栖息环境】海拔 500~1500 m 水草茂盛的湿地。【分布】河南、湖北、湖南、贵州、重庆、福建、广东；俄罗斯远东、朝鲜半岛、日本。【飞行期】6—10月。

姬赤蜻 雄，重庆
Sympetrum parvulum, male from Chongqing

姬赤蜻 雌，广东｜吴宏道 摄
Sympetrum parvulum, female from Guangdong ｜ Photo by Hongdao Wu

[Identification] Male face white. Thorax yellowish brown, dorsal carina black, humeral stripes black and broad, wings hyaline. Abdomen red. Female abdomen orange red. The species is similar to *Sympetrum eroticum* but can be distinguished from male anal appendages and female vulvar lamina. [Measurements] Total length 32-34 mm, abdomen 22-23 mm, hind wing 25-26 mm. [Habitat] Wetlands with plenty of emergent vegetation at 500-1500 m elevation. [Distribution] Henan, Hubei, Hunan, Guizhou, Chongqing, Fujian, Guangdong; Russian Far East, Korean peninsula, Japan. [Flight Season] June to October.

姬赤蜻 雄，广东｜宋黎明 摄
Sympetrum parvulum, male from Guangdong ｜ Photo by Liming Song

褐带赤蜻 *Sympetrum pedemontanum* (Müller, 1766)

【形态特征】雄性面部红色；胸部红褐色，翅在近翅端处具较宽的褐带；腹部红色，末端具黑斑。雌性黄褐色。【长度】体长 28~35 mm，腹长 18~24 mm，后翅 21~28 mm。【栖息环境】海拔 500~1500 m水草茂盛的湿地和流速缓慢的开阔小溪。【分布】新疆、内蒙古、黑龙江、吉林、辽宁、北京；广布于从欧洲至日本的欧亚大陆温带区域。【飞行期】6—9月。

[Identification] Male face red. Thorax reddish brown, wings with broad brown bands subapically. Abdomen red, tip with black spots. Female largely yellowish brown. [Measurements] Total length 28-35 mm, abdomen 18-24 mm, hind wing 21-28 mm. [Habitat] Wetlands with plenty of emergent vegetation and slow flowing streams at 500-1500 m elevation. [Distribution] Xinjiang, Inner Mongolia, Heilongjiang, Jilin, Liaoning, Beijing; Widespread in the temperate area of Eurasia from Europe to Japan. [Flight Season] June to September.

褐带赤蜻 雄，黑龙江
Sympetrum pedemontanum, male from Heilongjiang

褐带赤蜻 雄，黑龙江
Sympetrum pedemontanum, male from Heilongjiang

褐带赤蜻 雌，黑龙江
Sympetrum pedemontanum, female from Heilongjiang

李氏赤蜻 *Sympetrum risi risi* Bartenev, 1914

【形态特征】雄性面部黄色；胸部黄褐色，侧面具黑色条纹，翅透明，端部具褐斑；腹部红色，侧缘具黑色斑纹。雌性多型，腹部红色或黄色。本种与大理赤蜻相似，但翅端具褐斑。【长度】体长 37~45 mm，腹长 25~31 mm，后翅 29~35 mm。【栖息环境】海拔 2000 m 以下的湿地。【分布】黑龙江、吉林、辽宁、四川、贵州、湖北、湖南、浙江、福建、广东；俄罗斯远东、朝鲜半岛、日本。【飞行期】6—12月。

[Identification] Male face yellow. Thorax yellowish brown, sides with black stripes, wings hyaline with brown tips. Abdomen red with black markings laterally. Female polymorphic, abdomen red or yellow. Similar to *S. daliensis* but wings with brown tips. [Measurements] Total length 37-45 mm, abdomen 25-31 mm, hind wing 29-35 mm. [Habitat] Wetlands below 2000 m elevation. [Distribution] Heilongjiang, Jilin, Liaoning, Sichuan, Guizhou, Hubei, Hunan, Zhejiang, Fujian, Guangdong; Russian Far East, Korean peninsula, Japan. [Flight Season] June to December.

李氏赤蜻 雄，贵州
Sympetrum risi risi, male from Guizhou

李氏赤蜻 雄，黑龙江
Sympetrum risi risi, male from Heilongjiang

李氏赤蜻 雌，黄色型，贵州
Sympetrum risi risi, female, the yellow morph from Guizhou

李氏赤蜻 雌,红色型,黑龙江
Sympetrum risi risi, female, the red morph from Heilongjiang

李氏赤蜻 连结产卵,贵州
Sympetrum risi risi, laying eggs in tandem from Guizhou

血红赤蜻 *Sympetrum sanguineum* (Müller, 1764)

血红赤蜻 雄,芬兰 | Sami Karjalainen 摄
Sympetrum sanguineum, male from Finland | Photo by Sami Karjalainen

血红赤蜻 交尾,芬兰 | Sami Karjalainen 摄
Sympetrum sanguineum, mating pair from Finland | Photo by Sami Karjalainen

血红赤蜻 雌,芬兰 | Sami Karjalainen 摄
Sympetrum sanguineum, female from Finland | Photo by Sami Karjalainen

【形态特征】雄性面部红色;胸部红褐色,翅透明,基方稍染黄褐色;腹部红色,末端具黑斑。雌性多型,腹部红色或黄色。【长度】体长 34~39 mm,腹部 20~26 mm,后翅 23~31 mm。【栖息环境】水草茂盛的湿地。【分布】新疆;欧洲至中国新疆地区广布。【飞行期】7—9月。

[Identification] Male face red. Thorax reddish brown, wings hyaline, bases slightly tinted with yellowish brown. Abdomen red, tip with black markings. Female polymorphic, abdomen red or yellow. [Measurements] Total length 34-39 mm, abdomen 20-26 mm, hind wing 23-31 mm. [Habitat] Wetlands with plenty of emergent plants. [Distribution] Xinjiang; Widespread ocurring from Europe to Xinjiang of China. [Flight Season] July to September.

黄基赤蜻微斑亚种 *Sympetrum speciosum haematoneura* **Fraser, 1924**

【形态特征】本亚种与指名亚种近似但后翅基部具甚小的琥珀色斑或缺如。【长度】体长 42~44 mm，腹长27~28 mm，后翅 35~36 mm。【栖息环境】海拔 1500~3000 m挺水植物匮乏的池塘和水库。【分布】西藏、云南（大理、保山、德宏）；印度、尼泊尔。【飞行期】5—11月。

[Identification] The subspecies is similar to nominate subspecies, but the amber spots on hind wing bases are very small or absent. [Measurements] Total length 42-44 mm, abdomen 27-28 mm, hind wing 35-36 mm. [Habitat] Ponds without emergent plants and reservoirs at 1500-3000 m elevation. [Distribution] Tibet, Yunnan (Dali, Baoshan, Dehong); India, Nepal. [Flight Season] May to November.

黄基赤蜻微斑亚种 雄，云南（德宏）
Sympetrum speciosum haematoneura, male from Yunnan (Dehong)

黄基赤蜻指名亚种 *Sympetrum speciosum speciosum* Oguma, 1915

【形态特征】雄性面部红色；胸部大面积红色具黑色条纹，翅大面积透明，后翅基方具甚大的琥珀色斑；腹部红色。雌性胸部黄褐色；腹部橙红色具黑色斑点。【长度】体长 42~44 mm，腹长 26~28 mm，后翅 34~35 mm。【栖息环境】海拔 2000 m 以下挺水植物匮乏的池塘和水库。【分布】除西北地区外全国广布；朝鲜半岛、日本、越南。【飞行期】6—10月。

黄基赤蜻指名亚种 雄，贵州
Sympetrum speciosum speciosum, male from Guizhou

[Identification] Male face red. Thorax largely red with black stripes, wings largely hyaline, hind wing bases with large amber spots. Abdomen red. Female thorax yellowish brown. Abdomen orange red with black spots. [Measurements] Total length 42-44 mm, abdomen 26-28 mm, hind wing 34-35 mm. [Habitat] Ponds without emergent plants and reservoirs below 2000 m elevation. [Distribution] Widespread in China except for in the Northwest; Korean peninsula, Japan, Vietnam. [Flight Season] June to October.

黄基赤蜻指名亚种 雄，贵州
Sympetrum speciosum speciosum, male from Guizhou

黄基赤蜻指名亚种 雌，湖北｜宋睿斌 摄
Sympetrum speciosum speciosum, female from Hubei ｜ Photo by Ruibin Song

黄基赤蜻台湾亚种 *Sympetrum speciosum taiwanum* Asahina, 1951

【形态特征】本亚种与指名亚种近似，但后翅基部具甚小的琥珀色斑或缺如。本亚种与微斑亚种的差异不显著，但两者的分布区域不重叠。【长度】体长40~45 mm，腹长 25~28 mm，后翅 33~35 mm。【栖息环境】海拔 500~3500 m 挺水植物匮乏的池塘和水库。【分布】中国台湾特有。【飞行期】4—12月。

黄基赤蜻台湾亚种 雄，台湾│嘎嘎 摄
Sympetrum speciosum taiwanum, male from Taiwan | Photo by Gaga

[Identification] This subspecies is similar to nominate subspecies, but the amber spots on hind wing bases very small or absent. The subspecies is not clearly distinct from *S. speciosum haematoneura*, but their distributions do not overlap. [Measurements] Total length 40-45 mm, abdomen 25-28 mm, hind wing 33-35 mm. [Habitat] Ponds without emergent plants and reservoirs at 500-3500 m elevation. [Distribution] Endemic to Taiwan of China. [Flight Season] April to December.

条斑赤蜻喜马亚种 *Sympetrum striolatum commixtum* (Selys, 1884)

条斑赤蜻喜马亚种 雄，云南（大理）
Sympetrum striolatum commixtum, male from Yunnan (Dali)

【形态特征】本亚种与指名亚种相似,但翅脉深褐色。【长度】体长 40~43 mm,腹长 27~29 mm,后翅 31~33 mm。【栖息环境】海拔 1500~3000 m水草茂盛的湿地和山区的开阔小溪。【分布】云南(大理、保山); 不丹、印度、尼泊尔。【飞行期】6—12月。

[Identification] This subspecies is similar to nominate subspecies but venation dark brown. [Measurements] Total length 40-43 mm, abdomen 27-29 mm, hind wing 31-33 mm. [Habitat] Wetlands with plenty of emergent vegetation and exposed montane streams at 1500-3000 m elevation. [Distribution] Yunnan (Dali, Baoshan); Bhutan, India, Nepal. [Flight Season] June to December.

条斑赤蜻喜马亚种 雌,云南(大理)
Sympetrum striolatum commixtum, female from Yunnan (Dali)

条斑赤蜻喜马亚种 连结产卵,云南(大理)
Sympetrum striolatum commixtum, laying eggs in tandem from Yunnan (Dali)

条斑赤蜻喜马亚种 雄,云南(大理)
Sympetrum striolatum commixtum, male from Yunnan (Dali)

条斑赤蜻指名亚种 *Sympetrum striolatum striolatum* (Charpentier, 1840)

条斑赤蜻指名亚种 雌,黑龙江
Sympetrum striolatum striolatum, female from Heilongjiang

条斑赤蜻指名亚种 交尾,吉林｜金洪光 摄
Sympetrum striolatum striolatum, mating pair from Jilin ｜ Photo by Hongguang Jin

条斑赤蜻指名亚种 雄,黑龙江
Sympetrum striolatum striolatum, male from Heilongjiang

【形态特征】雄性面部红色；胸部红褐色，翅透明，翅脉金黄色；腹部红色，末端具黑斑。雌性多型，腹部红色或黄色。【长度】体长 36～45 mm，腹部 22～31 mm，后翅 25～32 mm。【栖息环境】海拔 2000 m以下水草茂盛的湿地和山区的开阔小溪。【分布】黑龙江、吉林、辽宁、内蒙古、新疆、北京、山东、山西、河南、陕西、四川；广布于从欧洲至日本的欧亚大陆温带区域。【飞行期】6—10月。

[Identification] Male face red. Thorax reddish brown, wings hyaline, venation golden. Abdomen red, tip with black markings. Female polymorphic, abdomen red or yellow. [Measurements] Total length 36-45 mm, abdomen 22-31 mm, hind wing 25-32 mm. [Habitat] Wetlands with plenty of emergent vegetation and exposed montane streams below 2000 m elevation. [Distribution] Heilongjiang, Jilin, Liaoning, Inner Mongolia, Xinjiang, Beijing, Shandong, Shanxi, Henan, Shaanxi, Sichuan; Widespread in the temperate regions of Eurasia from Europe to Japan. [Flight Season] June to October.

大黄赤蜻 *Sympetrum uniforme* (Selys, 1883)

【形态特征】整个身体金黄色,仅翅痣红色,但色彩随年纪增长而逐渐变暗。本种与半黄赤蜻相似,但翅的色彩均匀,半黄赤蜻翅基方色彩加深,中央透明。【长度】体长 42~47 mm,腹长 29~31 mm,后翅 30~34 mm。【栖息环境】海拔 1000 m 以下的湿地。【分布】黑龙江、吉林、辽宁、内蒙古、北京、河北、河南、山东、山西、陕西;俄罗斯远东、朝鲜半岛、日本。【飞行期】6—10月。

大黄赤蜻 雄,黑龙江
Sympetrum uniforme, male from Heilongjiang

[Identification] Golden throughout, only the pterostigma red, but its color darkened with age. The species is similar to *S. croceolum*, but the color of wings uniform, wings of *S. croceolum* darkened at bases and hyaline centrally. [Measurements] Total length 42-47 mm, abdomen 29-31 mm, hind wing 30-34 mm. [Habitat] Wetlands below 1000 m elevation. [Distribution] Heilongjiang, Jilin, Liaoning, Inner Mongolia, Beijing, Hebei, Henan, Shandong, Shanxi, Shaanxi; Russian Far East, Korean peninsula, Japan. [Flight Season] June to October.

大黄赤蜻 雄,黑龙江
Sympetrum uniforme, male from Heilongjiang

大黄赤蜻 雌,吉林 | 金洪光 摄
Sympetrum uniforme, female from Jilin | Photo by Hongguang Jin

普赤蜻 *Sympetrum vulgatum* (Linnaeus, 1758)

【形态特征】雄性面部黄褐色；胸部褐色，具白色的肩前条纹，翅透明，翅脉褐色；腹部红色，末端具黑斑。雌性黄褐色具黑色条纹；前翅具淡琥珀色斑；下生殖板向下伸出。本种与条斑赤蜻相似，但雄性钩片较短，雌性的下生殖板更突出。【长度】体长 35~40 mm，腹部 23~28 mm，后翅 24~29 mm。【栖息环境】海拔 1000 m以下水草茂盛的湿地。【分布】黑龙江、吉林、内蒙古；广布于从欧洲至日本的欧亚大陆温带区域。【飞行期】6—10月。

[Identification] Male face yellowish brown. Thorax brown with white antehumeral stripes, wings hyaline, venation brown. Abdomen red, tip with black markings. Female mainly yellowish brown with black markings. Fore wings with light amber markings. Vulvar lamina projected ventrally. The species is similar to *S. striolatum* but hook of hamule shorter, female vulvar lamina more protruding. [Measurements] Total length 35-40 mm, abdomen 23-28 mm, hind wing 24-29 mm. [Habitat] Wetlands with plenty of emergent plants below 1000 m elevation. [Distribution] Heilongjiang, Jilin, Inner Mongolia; Widespread in the temperate area of Eurasia from Europe to Japan. [Flight Season] June to October.

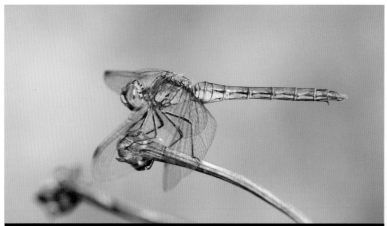

普赤蜻 雌,吉林 | 金洪光 摄
Sympetrum vulgatum, female from Jilin | Photo by Hongguang Jin

普赤蜻 交尾,吉林 | 金洪光 摄
Sympetrum vulgatum, mating pair from Jilin | Photo by Hongguang Jin

普赤蜻 雄,吉林 | 金洪光 摄
Sympetrum vulgatum, male from Jilin | Photo by Hongguang Jin

肖氏赤蜻 *Sympetrum xiaoi* Han & Zhu, 1997

【形态特征】雄性面部黄色；胸部黄色具黑色条纹，翅透明，端部具褐斑；腹部红色，侧缘具黑色条纹。雌性胸部浅黄色；腹部土黄色。【长度】体长 37~40 mm，腹长 25~27 mm，后翅 32~34 mm。【栖息环境】海拔 500~2000 m的湿地。【分布】中国特有，分布于山西、贵州、湖南。【飞行期】7—10月。

肖氏赤蜻 雄，湖南 | 吴宏道 摄
Sympetrum xiaoi, male from Hunan | Photo by Hongdao Wu

[Identification] Male face yellow. Thorax yellow with black stripes, wings hyaline with brown apical spots. Abdomen red with lateral black stripes. Female thorax light yellow. Abdomen khaki yellow. [Measurements] Total length 37-40 mm, abdomen 25-27 mm, hind wing 32-34 mm. [Habitat] Wetlands at 500-2000 m elevation. [Distribution] Endemic to China, recorded from Shanxi, Guizhou, Hunan. [Flight Season] July to October.

肖氏赤蜻 交尾，湖南 | 吴宏道 摄
Sympetrum xiaoi, mating pair from Hunan | Photo by Hongdao Wu

赤蜻属待定种1 *Sympetrum* sp. 1

【形态特征】雄性面部黄色；胸部黄褐色具黑色条纹，翅透明；腹部红色，侧缘具黑斑。雌性与雄性相似。【长度】体长 38 mm，腹长 26 mm，后翅 30 mm。【栖息环境】海拔 1500～2000 m的高山湿地。【分布】云南（大理）。【飞行期】7—9月。

[Identification] Male face yellow. Thorax yellowish brown with black stripes, wings hyaline. Abdomen red with black stripes laterally. Female similar to male. [Measurements] Total length 38 mm, abdomen 26 mm, hind wing 30 mm. [Habitat] Wetlands at 1500-2000 m elevation. [Distribution] Yunnan (Dali). [Flight Season] July to September.

赤蜻属待定种1 雄，云南（大理）
Sympetrum sp. 1, male from Yunnan (Dali)

赤蜻属待定种2 *Sympetrum* sp. 2

【形态特征】雄性面部黄色；胸部黄色具黑色条纹，翅透明，基方具琥珀色斑；腹部黑色具黄色斑点。本种与日本分布的蓝灰赤蜻相似。【长度】未测量。【栖息环境】海拔 500～1500 m的湿地和水稻田。【分布】湖北、浙江、贵州。【飞行期】6—10月。

[Identification] Male face yellow. Thorax yellow with black stripes, wings hyaline, bases with amber markings.

Abdomen black with yellow spots. The species is similar to *S. gracile* from Japan. [Measurements] Not measured. [Habitat] Wetlands and paddy fields at 500-1500 m elevation. [Distribution] Hubei, Zhejiang, Guizhou. [Flight Season] June to October.

赤蜻属待定种2 雌，湖北
Sympetrum sp. 2, female from Hubei

赤蜻属待定种3 *Sympetrum* sp. 3

【形态特征】雄性完全成熟后身体蓝黑色；翅透明，端部具黑褐色斑。雌性身体黄色具黑色条纹。【长度】雄性体长 36 mm，腹长 23 mm，后翅 29 mm。【栖息环境】海拔 500～1000 m 的湿地和水稻田。【分布】浙江、安徽。【飞行期】6—9月。

[Identification] Male bluish black when fully mature, wings hyaline with blackish brown spots apically. Female body yellow with black stripes. [Measurements] Male total length 36 mm, abdomen 23 mm, hind wing 29 mm. [Habitat] Wetlands and paddy fields at 500-1000 m elevation. [Distribution] Zhejiang, Anhui. [Flight Season] June to September.

赤蜻属待定种3 雌，浙江 | 莫善濂 摄
Sympetrum sp. 3, female from Zhejiang | Photo by Shanlian Mo

赤蜻属待定种3 雄，浙江
Sympetrum sp. 3, male from Zhejiang

赤蜻属待定种4 *Sympetrum* sp. 4

【形态特征】与竖眉赤蜻相似，但本种面部白色，身体条纹、雄性肛附器和次生殖器也略有差异。【长度】体长 39~42 mm，腹长 26~30 mm，后翅 30~32 mm。【栖息环境】海拔 1000 m 以下的湿地。【分布】广东。【飞行期】5—11月。

[Identification] Similar to *S. eroticum*, but face white, body maculation, male anal appendages and secondary genitalia slightly different. [Measurements] Total length 39-42 mm, abdomen 26-30 mm, hind wing 30-32 mm. [Habitat] Wetlands below 1000 m elevation. [Distribution] Guangdong. [Flight Season] May to November.

赤蜻属待定种4 雄，广东 | 宋睿斌 摄
Sympetrum sp. 4, male from Guangdong | Photo by Ruibin Song

赤蜻属待定种4 雌，广东 | 宋睿斌 摄
Sympetrum sp. 4, female from Guangdong | Photo by Ruibin Song

方蜻属 Genus *Tetrathemis* Brauer, 1868

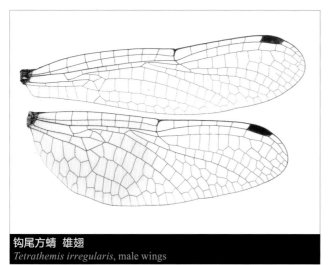

钩尾方蜻 雄翅
Tetrathemis irregularis, male wings

本属全球已知10余种，分布于亚洲和非洲。中国已知2种，分布于华南和西南地区。本属蜻蜓体型较小；体黑色具黄色斑纹，腹部较短；翅透明，后翅基方具琥珀色斑，结前横脉完整，通常6~9条，弓脉位于第1条和第2条结前横脉之间，盘区基方仅有1列翅室，无臀圈。本属中国已知的2种雄性可以通过肛附器的构造区分。

本属蜻蜓栖息于森林中的林荫池塘。雄性停落在水边的植物上占据领地。雌性产卵时腹部弯曲，将卵粘附在水面上方悬挂的树枝上。

The genus contains over ten species, distributed in Asia and Africa. Two species are recorded from China, found in the South and Southwest. Species of the genus are small-sized. Body fundamentally black with yellow markings, abdomen short. Wings hyaline, hind wings with amber brown at bases, antenodals complete, usually 6-9 in number, arc between antenodals 1 and 2, discoidal field with only one row of cells, anal loop absent. Males of Chinese species can be distinguished by anal appendages.

Tetrathemis species inhabit shady ponds in forests. Territorial males perch on plants near water. Females lay the eggs by curving the abdomen and attaching the eggs to the branches hanging above water.

钩尾方蜻 雄
Tetrathemis irregularis, male

宽翅方蜻
Tetrathemis platyptera

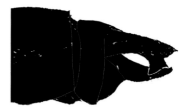

钩尾方蜻
Tetrathemis irregularis

方蜻属 雄性肛附器
Genus *Tetrathemis*, male anal appendages

钩尾方蜻 *Tetrathemis irregularis* **Brauer, 1868**

钩尾方蜻 雄,云南(德宏)
Tetrathemis irregularis, male from Yunnan (Dehong)

【形态特征】雄性复眼蓝绿色,面部黄色,额具1个金属蓝黑色斑;胸部黑色具肩前条纹,侧面具2条宽阔的黄条纹,翅透明,后翅基方具甚大的琥珀色斑;腹部黑色具黄色斑点。雌性与雄性相似,但后翅的琥珀色斑面积更大;腹部更粗壮。【长度】体长 25~27 mm,腹长 16~17 mm,后翅 19~21 mm。【栖息环境】海拔 1000 m以下的水潭和池塘。【分布】云南(德宏、临沧、西双版纳);东南亚、澳大利亚。【飞行期】3—12月。

[Identification] Male eyes bluish green, face yellow, frons with a metallic bluish black spot. Thorax black with antehumeral stripe, sides with two broad yellow stripes, wings hyaline, hind wing bases with large amber spots. Abdomen black with yellow markings. Female similar to male, hind wing bases with larger amber spots. Abdomen stouter. [Measurements] Total length 25-27 mm, abdomen 16-17 mm, hind wing 19-21 mm. [Habitat] Pools and ponds below 1000 m elevation. [Distribution] Yunnan (Dehong, Lincang, Xishuangbanna); Southeast Asia, Australia. [Flight Season] March to December.

钩尾方蜻 雌,云南(德宏)
Tetrathemis irregularis, female from Yunnan (Dehong)

钩尾方蜻 雌,产卵,云南(德宏)
Tetrathemis irregularis, female laying eggs from Yunnan (Dehong)

宽翅方蜻 *Tetrathemis platyptera* Selys, 1878

【形态特征】本种与钩尾方蜻相似,但雄性上肛附器的构造不同。本种雌性的肩前条纹稍短,腹部第7节侧面无黄色斑点,可与钩尾方蜻区分。【长度】体长 27~30 mm,腹长 17~19 mm,后翅 21~24 mm。【栖息环境】海拔1500 m以下的水潭和池塘。【分布】云南(红河)、安徽、江苏、浙江、福建、广东、广西、海南;印度、缅甸、泰国、老挝、柬埔寨、越南、马来半岛、印度尼西亚。【飞行期】5—10月。

宽翅方蜻 雌,广东
Tetrathemis platyptera, female from Guangdong

[Identification] The species is similar to *T. irregularis*, but male superior appendages different. Female can be distinguished from *T. irregularis* by shorter antehumeral stripe and the absence of yellow spots on S7. [Measurements] Total length 27-30 mm, abdomen 17-19 mm, hind wing 21-24 mm. [Habitat] Pools and ponds below 1500 m elevation. [Distribution] Yunnan (Honghe), Anhui, Jiangsu, Zhejiang, Fujian, Guangdong, Guangxi, Hainan; India, Myanmar, Thailand, Laos, Cambodia, Vietnam, Peninsular Malaysia, Indonesia. [Flight Season] May to October.

宽翅方蜻 雄,云南(红河)
Tetrathemis platyptera, male from Yunnan (Honghe)

宽翅方蜻 雌,广东
Tetrathemis platyptera, female from Guangdong

云斑蜻属 Genus *Tholymis* Hagen, 1867

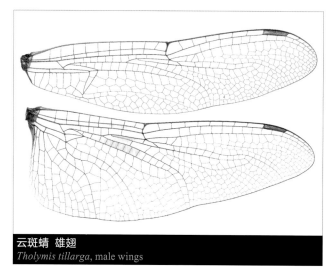

云斑蜻 雄翅
Tholymis tillarga, male wings

本属全球已知2种，分布于南美洲、亚洲、非洲和大洋洲。中国已知1种，分布于华南和西南地区。本属蜻蜓体中型；复眼十分发达，在头顶交会呈一条直线；翅大面积透明，后翅具色斑，弓脉位于第1条与第2条结前横脉之间，R3末端较曲折，盘区基方具2~3列翅室，基臀区具1条横脉，臀圈开放。

本属蜻蜓栖息于水草茂盛的湿地。它们仅在黄昏时活跃，并飞行到光线很暗的时间。雄性在水面上来回飞行，时而悬停。交尾时间很短，空中进行。

The genus contains two species distributed in South America, Asia, Africa and Oceania. One species is recorded from China, found in the South and Southwest. Species of the genus are medium-sized. Eyes large and broadly confluent above head. Wings largely hyaline, hind wings with spots, arc between antenodals 1 and 2, R3 waved distally, discoidal field with 2-3 rows of cells basally, cubital space with one crossvein, anal loop open.

Tholymis species inhabit well vegetated wetlands. They are active only at twilight, flying up to just before night time. Males fly rapidly above water and sometimes hover. Mating duration is short and takes place in flight.

云斑蜻 雄
Tholymis tillarga, male

云斑蜻 *Tholymis tillarga* (Fabricius, 1798)

云斑蜻 雄,海南｜莫善濂 摄
Tholymis tillarga, male from Hainan | Photo by Shanlian Mo

【形态特征】雄性身体红色；翅大面积透明，后翅具1个乳白色斑和1个褐色斑。雌性黄褐色，后翅具1个褐色斑。【长度】体长 42~47 mm，腹长 29~32 mm，后翅 33~35 mm。【栖息环境】海拔 1500 m以下的湿地。【分布】云南、福建、广东、广西、海南、香港、台湾；亚洲、非洲、大洋洲广布。【飞行期】全年可见。

云斑蜻 雄,海南｜莫善濂 摄
Tholymis tillarga, male from Hainan | Photo by Shanlian Mo

云斑蜻 雌,云南(西双版纳)
Tholymis tillarga, female from Yunnan (Xishuangbanna)

[Identification] Male body red. Wings largely hyaline, hing wings with a white spot and a brown spot. Female yellowish brown, hind wings with a brown spot. [Measurements] Total length 42-47 mm, abdomen 29-32 mm, hind wing 33-35 mm. [Habitat] Wetlands below 1500 m elevation. [Distribution] Yunnan, Fujian, Guangdong, Guangxi, Hainan, Hong Kong, Taiwan; Widespread in Asia, Africa, Oceania. [Flight Season] Throughout the year.

斜痣蜻属 Genus *Tramea* Hagen, 1861

本属全球已知20余种，分布于美洲、亚洲、非洲和大洋洲。中国已知5种及亚种，主要分布在南方。本属是一类较大型的蜻科物种。翅大面积透明但后翅基方具甚阔深色斑，翅痣较短，弓脉位于第1条与第2条结前横脉之间，盘区基具3～4列翅室，基臀区具1～2条横脉，臀圈靴状。本属中国的已知种可以通过后翅基方色斑的形状区分。

本属蜻蜓栖息于水草茂盛的湿地。雄性在水面上来回飞行，巡视领地。交尾时停落，交尾结束后雌雄连结产卵，雌性仅在点水瞬间被放开，然后立刻再连结。有时雄性将雌性放置于安全地点产卵，在旁边悬停护卫。本属的许多种类有迁飞的习性。

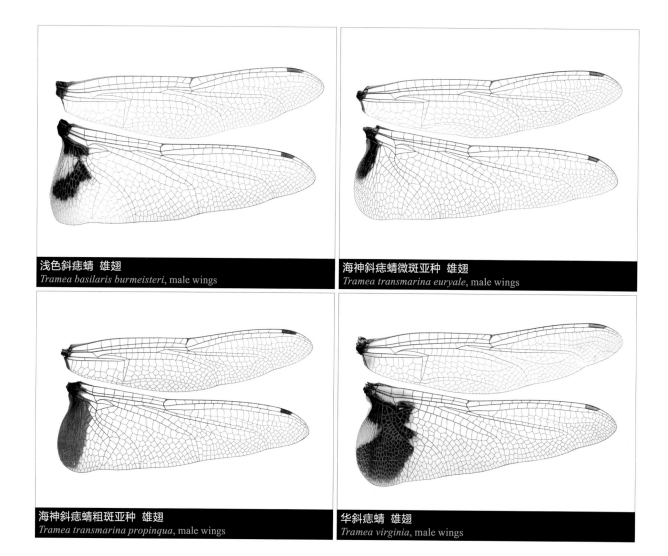

浅色斜痣蜻 雄翅
Tramea basilaris burmeisteri, male wings

海神斜痣蜻微斑亚种 雄翅
Tramea transmarina euryale, male wings

海神斜痣蜻粗斑亚种 雄翅
Tramea transmarina propinqua, male wings

华斜痣蜻 雄翅
Tramea virginia, male wings

The genus contains over 20 species, widespread in the Americas, Asia, Africa and Oceania. Five species and subspecies are recorded from China, mainly found in the south. Species of the genus are moderately large-sized libellulids. Wings largely hyaline, hind wing bases broad with dark spots, pterostigma relatively short, arc between antenodals 1 and 2, discoidal field with 3-4 rows of cells basally, cubital space with 1-2 crossvein, anal loop boot-shaped. The Chinese species can be distinguished by the shape of dark spots of hind wing bases.

Tramea species inhabit well vegetated wetlands. Males usually constantly patrol above water. Mating pair perch on

海神斜痣蜻微斑亚种　连结产卵
Tramea transmarina euryale, laying eggs in tandem

trees. They lay eggs in tandem, but the pair disconnect just before the female descends to deposit her eggs into water, after which they connect again. Sometimes the female lays eggs with a hovering male guarding. Many species of the genus are strong migrants.

浅色斜痣蜻 *Tramea basilaris burmeisteri* **Kirby,1889**

【形态特征】雄性面部红色；胸部浅褐色，后翅臀区具1个甚大的红褐色斑；腹部红色，各节末端具黑色环纹。雌性黄褐色，后翅基方具1个甚大的褐色斑。【长度】体长 49~50 mm，腹长 34~35 mm，后翅 41~43 mm。【栖息环境】海拔 1000 m以下的杂草池塘。【分布】云南（德宏、西双版纳、红河）、广西；印度、东南亚、日本。【飞行期】全年可见。

[Identification] Male face red. Thorax light brown, hind wing anal field with a large reddish brown spot. Abdomen red with black apical rings in all segments. Female yellowish brown, hind wing anal field with a large brown spot. [Measurements] Total length 49-50 mm, abdomen 34-35 mm, hind wing 41-43 mm. [Habitat]

浅色斜痣蜻　雄，云南（西双版纳）
Tramea basilaris burmeisteri, male from Yunnan (Xishuangbanna)

浅色斜痣蜻 雌，云南（德宏）
Tramea basilaris burmeisteri, female from Yunnan (Dehong)

浅色斜痣蜻 雄，云南（德宏）
Tramea basilaris burmeisteri, male from Yunnan (Dehong)

Grassy ponds below 1000 m elevation. [Distribution] Yunnan (Dehong, Xishuangbanna, Honghe), Guangxi; India, Southeast Asia, Japan. [Flight Season] Throughout the year.

缘环斜痣蜻 *Tramea limbata similata* (Rambur, 1842)

【形态特征】雌性身体黄褐色；后翅基方几乎透明；腹部各节末端具黑色环纹。【长度】雌性体长 53 mm，腹长 36 mm，后翅 45 mm。【栖息环境】海拔 1000 m 以下的杂草池塘。【分布】广西；印度、尼泊尔、斯里兰卡、泰国。【飞行期】5月。

[Identification] Female body yellowish brown. Hind wing bases almost hyaline. Abdomen with black apical rings. [Measurements] Female total length 53 mm, abdomen 36 mm, hind wing 45 mm. [Habitat] Grassy ponds below 1000 m elevation. [Distribution] Guangxi; India, Nepal, Sri Lanka, Thailand. [Flight Season] May.

缘环斜痣蜻 雌，广西
Tramea limbata similata, female from Guangxi

海神斜痣蜻微斑亚种 *Tramea transmarina euryale* (Selys, 1878)

【形态特征】雄性面部红褐色，额具较小的黑色斑；胸部深褐色，后翅基方具1个较小的黑褐色斑；腹部红色，第8~10节具黑斑。雌性与雄性相似，但面部黄褐色；后翅基方色斑甚小；腹部橙红色。【长度】体长 50~52 mm，腹长 33~34 mm，后翅 41~43 mm。【栖息环境】海拔 1000 m 以下的湿地。【分布】云南、广东、海南、香港、台湾；东南亚、日本。【飞行期】全年可见。

海神斜痣蜻微斑亚种 交尾，云南（德宏）
Tramea transmarina euryale, mating pair from Yunnan (Dehong)

[Identification] Male face reddish brown, frons with small black spot. Thorax dark brown, hind wing anal field with a small blackish brown spot. Abdomen red, S8-S10 with black spots. Female similar to male but face yellowish brown. Hind wing anal field with small spots. Abdomen orange red. [Measurements] Total length 50-52 mm, abdomen 33-34 mm, hind wing 41-43 mm. [Habitat] Wetlands below 1000 m elevation. [Distribution] Yunnan, Guangdong, Hainan, Hong Kong, Taiwan; Southeast Asia, Japan. [Flight Season] Throughout the year.

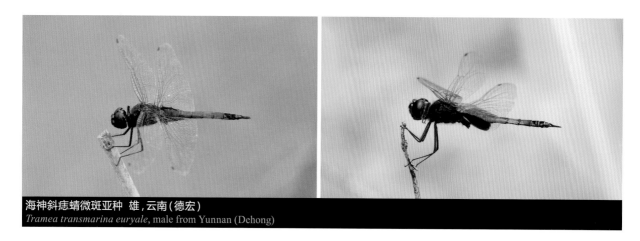

海神斜痣蜻微斑亚种 雄，云南（德宏）
Tramea transmarina euryale, male from Yunnan (Dehong)

海神斜痣蜻粗斑亚种 *Tramea transmarina propinqua* Lieftinck, 1942

【形态特征】本亚种与微斑亚种相似，但后翅臀区的黑褐色斑面积更大。【长度】雄性体长 50 mm，腹长 33 mm，后翅 40 mm。【栖息环境】海拔 500 m以下的湿地。【分布】中国台湾；澳大利亚、密克罗尼西亚、印度尼西亚、斐济、日本、马绍尔群岛、马里亚纳群岛北部、皮特凯恩、帕劳、巴布亚新几内亚、泰国。【飞行期】4—12月。

海神斜痣蜻粗斑亚种 雄，台湾
Tramea transmarina propinqua, male from Taiwan

[Identification] The subspecies is similar to *Tramea transmarina euryale* but hind wing anal field with larger blackish brown spot. [Measurements] Male total length 50 mm, abdomen 33 mm, hind wing 40 mm. [Habitat] Wetlands below 500 m elevation. [Distribution] Taiwan of China; Australia, Micronesia, Indonesia, Fiji, Japan, Marshall Islands, Northern Mariana Islands, Pitcairn, Palau, Papua New Guinea, Thailand. [Flight Season] April to December.

华斜痣蜻 *Tramea virginia* (Rambur, 1842)

华斜痣蜻 雄，贵州
Tramea virginia, male from Guizhou

华斜痣蜻 雄, 贵州
Tramea virginia, male from Guizhou

【形态特征】雄性面部红褐色；胸部褐色，翅稍染琥珀色，后翅臀区具较大的红褐色斑；腹部红色，第8～10节具黑斑。雌性身体黄褐色，后翅臀区具较大的褐色斑。【长度】体长 53～56 mm，腹长 36～38 mm，后翅43～48 mm。【栖息环境】海拔 1500 m 以下的湿地。【分布】除西北地区外全国广布；朝鲜半岛、日本、东南亚。【飞行期】全年可见。

华斜痣蜻 雌, 广东 | 吴宏道 摄
Tramea virginia, female from Guangdong | Photo by Hongdao Wu

[Identification] Male face reddish brown. Thorax brown, wings slightly tinted with amber, hind wing anal field with large reddish brown spots. Abdomen red, S8-S10 with black spots. Female mainly yellowish brown, hind wing anal field with large brown spots. [Measurements] Total length 53-56 mm, abdomen 36-38 mm, hind wing 43-48 mm. [Habitat] Wetlands below 1500 m elevation. [Distribution] Widespread in China except Northwest China; Korean peninsula, Japan, Southeast Asia. [Flight Season] Throughout the year.

褐蜻属 Genus *Trithemis* Brauer, 1868

在本属全球已知的40余种中，大多数在非洲分布，少数种类分布于亚洲、欧洲和大洋洲。中国已知3种，南方较常见。本属蜻蜓体色艳丽，体中型；翅大面积透明，基方具色斑，翅痣较短，弓脉位于第1条与第2条结前横脉之间，前翅最末端的结前横脉不完整，盘区基方具2~3列翅室，基臀区具1条横脉，臀圈靴状。

本属蜻蜓栖息于各类静水环境和流速缓慢的溪流。雄性通常停落在水边植物的枝头上并经常巡飞。交尾在空中完成，时间较短，雄性护卫产卵。

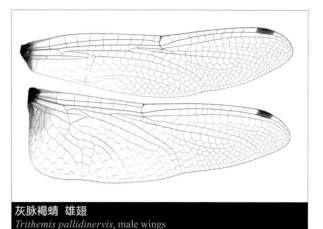

灰脉褐蜻 雄翅
Trithemis pallidinervis, male wings

The genus contains over 40 species with most of them distributed in Africa, a few species are found in Asia, Europe and Oceania. Three species are recorded from China, common in the south. Species of the genus are colorful, medium-sized dragonflies. Wings largely hyaline with basal spots, pterostigma short, arc between antenodals 1 and 2, distal antenodal in fore wings incomplete, fore wings discoidal field with 2-3 rows of cells basally, cubital space with one crossvein, anal loop boot-shaped.

Trithemis species inhabit both standing water and slow flowing streams. Males usually perch on branches above water and fly frequently. Mating duration is short and takes place in flight. Females lay eggs with males guarding around.

晓褐蜻 雄
Trithemis aurora, male

晓褐蜻 *Trithemis aurora* (Burmeister, 1839)

【形态特征】雄性身体紫红色；翅脉紫红色，后翅基方具甚大的褐色斑。雌性黄色具黑色条纹；后翅基方具黄褐色斑。【长度】体长 33~35 mm，腹长 22~24 mm，后翅 27~29 mm。【栖息环境】海拔 2000 m 以下的湿地和流速缓慢的河流。【分布】除东北地区外全国广布；日本、南亚、东南亚。【飞行期】全年可见。

[Identification] Male body purple red. Venation purple red, hind wing bases with large brown spots. Female yellow with black markings. Hind wing bases with yellowish brown spots. [Measurements] Total length 33-35 mm, abdomen 22-24 mm, hind wing 27-29 mm. [Habitat] Wetlands and slow flowing rivers below 2000 m elevation. [Distribution] Widespread in China except Northeast China; Japan, South and Southeast Asia. [Flight Season] Throughout the year.

晓褐蜻 雄, 云南 (西双版纳)
Trithemis aurora, male from Yunnan (Xishuangbanna)

晓褐蜻 雌, 云南 (德宏)
Trithemis aurora, female from Yunnan (Dehong)

晓褐蜻 雄, 云南 (德宏)
Trithemis aurora, male from Yunnan (Dehong)

庆褐蜻 *Trithemis festiva* (Rambur, 1842)

【形态特征】雄性面部褐色，额蓝黑色具金属光泽；胸部覆盖蓝色粉霜，翅透明，后翅基方具褐斑；腹部大面积黑色，第1～3节背面覆盖蓝色粉霜，第4～6节背面有时具黄色斑点。雌性黄褐色具黑色条纹；后翅基方稍染琥珀色。【长度】体长 36～38 mm，腹长 24～26 mm，后翅 30～32 mm。【栖息环境】海拔 2500 m以下的湿地和流速缓慢的河流。【分布】华南和西南地区广布；从欧洲东南部经南亚和东南亚至新几内亚广布。【飞行期】全年可见。

庆褐蜻 雄，云南（德宏）
Trithemis festiva, male from Yunnan (Dehong)

[Identification] Male face brown, frons metallic bluish black. Thorax covered by blue pruinosity, wings hyaline, hind wing bases with brown spots. Abdomen largely black, S1-S3 covered by blue pruinosity, S4-S6 sometimes with yellow spots. Female mainly yellowish brown with black markings, wing bases slightly tinted with amber. [Measurements] Total length 36-38 mm, abdomen 24-26 mm, hind wing 30-32 mm. [Habitat] Wetlands and slow flowing streams below 2500 m elevation. [Distribution] Widespread in the south and southwest of China; Widespread from Southeast Europe throughout South and Southeast Asia to New Guinea. [Flight Season] Throughout the year.

庆褐蜻 雄，云南（德宏）
Trithemis festiva, male from Yunnan (Dehong)

庆褐蜻 雌，广东 | 吴宏道 摄
Trithemis festiva, female from Guangdong | Photo by Hongdao Wu

灰脉褐蜻 *Trithemis pallidinervis* (Kirby, 1889)

灰脉褐蜻 雄, 云南 (西双版纳)
Trithemis pallidinervis, male from Yunnan (Xishuangbanna)

灰脉褐蜻 雄, 云南 (西双版纳)
Trithemis pallidinervis, male from Yunnan (Xishuangbanna)

灰脉褐蜻 雌, 云南 (西双版纳)
Trithemis pallidinervis, female from Yunnan (Xishuangbanna)

【形态特征】雄性面部红褐色, 额蓝黑色具金属光泽; 胸部褐色, 侧面具黑色条纹, 翅透明, 后翅基方具较小的褐斑; 腹部黑褐色具黄褐色斑。雌性黄色具黑色条纹; 翅稍染琥珀色, 基方具较小的黄褐色斑; 腹部腹面具白色粉霜。【长度】体长 38~42 mm, 腹长 26~30 mm, 后翅 31~32 mm。【栖息环境】海拔 1000 m 以下的湿地、水库和流速缓慢的河流。【分布】云南、广东、广西、海南、香港、台湾; 南亚、东南亚。【飞行期】4—12月。

[Identification] Male face reddish brown, frons metallic bluish black. Thorax brown, sides with black stripes, wings hyaline, hind wing bases with small brown spots. Abdomen blackish brown with yellowish brown markings. Female yellow with black markings. Wings slightly tinted with amber, bases with small yellowish brown spots. Abdomen with ventral part covered by white pruinosity. [Measurements] Total length 38-42 mm, abdomen 26-30 mm, hind wing 31-32 mm. [Habitat] Wetlands, reservoirs and slow flowing streams below 1000 m elevation. [Distribution] Yunnan, Guangdong, Guangxi, Hainan, Hong Kong, Taiwan; South and Southeast Asia. [Flight Season] April to December.

曲钩脉蜻属 Genus *Urothemis* Brauer, 1868

本属全球已知10种,分布于亚洲、非洲和大洋洲,在热带区域较常见。中国已知1种包括3个亚种,分布于华南和西南地区。本属蜻蜓体中型,腹部较宽阔。翅大面积透明但基方具深色斑,具较少的结前横脉,翅痣较长,弓脉位于第1条与第2条结前横脉之间,盘区基方具2列翅室,基臀区具1条横脉,臀圈靴状。

本属蜻蜓栖息于水草茂盛的湿地,在水葫芦滋生的池塘较常见。雄性通常停在叶片顶端,经常巡飞。交尾在空中进行,时间较短,雄性护卫产卵。

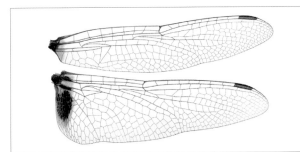

赤斑曲钩脉蜻微斑亚种 雄翅
Urothemis signata insignata, male wings

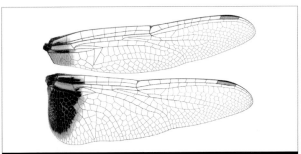

赤斑曲钩脉蜻指名亚种 雄翅
Urothemis signata signata, male wings

The genus contains ten species, distributed in Asia, Africa and Oceania, they are common in tropical regions. Three subspecies of one species are recorded from China, found in the South and Southwest. Species of the genus are medium-sized, abdomen flattened and broad. Wings largely hyaline, hind wing bases have dark spots, antenodals few in number, pterostigma long, arc between antenodals 1 and 2, discoidal field with 2 rows of cells basally, cubital space with one crossvein, anal loop boot-shaped.

Urothemis species inhabit well vegetated wetlands, they are commonly observed in ponds with dense coverage of water hyacinth. Males usually perch on top of branches, and patrol frequently. Mating takes place in flight, but lasts for only a few seconds. Females lay eggs with males guarding.

赤斑曲钩脉蜻微斑亚种 雄
Urothemis signata insignata, male

赤斑曲钩脉蜻微斑亚种 *Urothemis signata insignata* (Selys, 1872)

【形态特征】本亚种与指名亚种相似，但后翅基方的红褐色斑较小。【长度】体长 41～43 mm，腹长 26～28 mm，后翅 33～34 mm。【栖息环境】海拔 1000 m以下水草茂盛的湿地和流速缓慢、河岸水葫芦滋生的河流。【分布】云南（德宏）；马来群岛。【飞行期】4—11月。

赤斑曲钩脉蜻微斑亚种 雄，云南（德宏）
Urothemis signata insignata, male from Yunnan (Dehong)

赤斑曲钩脉蜻微斑亚种 雌，云南（德宏）
Urothemis signata insignata, female from Yunnan (Dehong)

[Identification] The subspecies is similar to nominate subspecies but the reddish brown spots on hind wing bases much smaller. [Measurements] Total length 41-43 mm, abdomen 26-28 mm, hind wing 33-34 mm. [Habitat] Wetlands with plenty of emergent vegetation and rivers with marginal water hyacinth below 1000 m elevation. [Distribution] Yunnan (Dehong); Malay Archipelago. [Flight Season] April to November.

赤斑曲钩脉蜻微斑亚种 雄，云南（德宏）
Urothemis signata insignata, male from Yunnan (Dehong)

赤斑曲钩脉蜻指名亚种 *Urothemis signata signata* (Rambur, 1842)

赤斑曲钩脉蜻指名亚种　雄, 海南
Urothemis signata signata, male from Hainan

【形态特征】雄性面部红褐色；胸部红褐色，翅大面积透明，翅基方具红褐色斑；腹部红色，第8~9节背面具黑斑。雌性多型；面部黄色；胸部黄色或黄褐色，翅基方具褐色或黄色斑；腹部黄色或橙红色，背面具黑色斑。【长度】体长 47~48 mm，腹长 31~32 mm，后翅 40~41 mm。【栖息环境】海拔 500 m 以下水草茂盛的湿地和流速缓慢、河岸水葫芦滋生的河流。【分布】广东、广西、海南、香港；南亚、东南亚。【飞行期】3—11月。

赤斑曲钩脉蜻指名亚种　雄，海南
Urothemis signata signata, male from Hainan

[Identification] Male face reddish brown. Thorax reddish brown, wings largely hyaline, bases with reddish brown spots. Abdomen red, S8-S9 with black spots. Female polymorphic. Face yellow. Thorax yellow or yellowish brown, wing bases with brown or yellow markings. Abdomen yellow or orange red with dorsal black spots. [Measurements] Total

赤斑曲钩脉蜻指名亚种　雌，黄色型，广东
Urothemis signata signata, female, the yellow morph from Guangdong

赤斑曲钩脉蜻指名亚种　雌，橙色型，广东｜吴宏道 摄
Urothemis signata signata, female, the orange morph from Guangdong｜Photo by Hongdao Wu

length 47-48 mm, abdomen 31-32 mm, hind wing 40-41 mm. **[Habitat]** Wetlands with plenty of emergent vegetation and rivers with marginal water hyacinth below 500 m elevation. **[Distribution]** Guangdong, Guangxi, Hainan, Hong Kong; South and Southeast Asia. **[Flight Season]** March to November.

赤斑曲钩脉蜻台湾亚种 *Urothemis signata yiei* **Asahina, 1972**

【形态特征】本亚种被认为是台湾特有，但与指名亚种未见显著差异。【长度】雄性体长 48 mm，腹长 31 mm，后翅 40 mm。【栖息环境】海拔 500 m以下水草茂盛的湿地和流速缓慢、河岸水葫芦滋生的河流。【分布】中国台湾特有。【飞行期】4—11月。

[Identification] The subspecies is considered to be endemic to Taiwan, but shows no clear difference from the nominate subspecies. **[Measurements]** Male total length 48 mm, abdomen 31 mm, hind wing 40 mm. **[Habitat]** Wetlands with plenty of emergent plants and rivers with marginal water hyacinth below 500 m elevation. **[Distribution]** Endemic to Taiwan of China. **[Flight Season]** April to November.

赤斑曲钩脉蜻台湾亚种 雄，台湾 | 嘎嘎 摄
Urothemis signata yiei, male from Taiwan | Photo by Gaga

虹蜻属 Genus *Zygonyx* Hagen, 1867

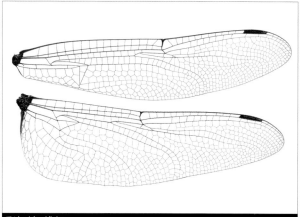

彩虹蜻 雄翅
Zygonyx iris insignis, male wings

本属全球已知20余种，主要分布于亚洲、欧洲南端、非洲和大洋洲。中国已知3种，分布于华南和西南地区。本属蜻蜓体大型，体色较暗；翅大面积透明而狭长，具较多的结前横脉，弓脉位于第1条与第2条结前横脉之间，盘区基方具2~3列翅室，基臀区具2条横脉，臀圈靴状。

本属蜻蜓栖息于开阔溪流和大型河流，喜欢水流速较快的河段。雄性在水面上方长时间定点悬停等待雌性。交尾时停落，交尾时间较长，部分种类连结产卵。有些种类将卵产在具渗流的陡峭石壁上。

The genus contains over 20 species, distributed in Aisa, southernmost Europe, Africa and Oceania. Three species are recorded from China, found in the South and Southwest. Species of the genus are large-sized libellulids, body dark. Wings largely hyaline and long, antenodals abundant, arc between antenodals 1 and 2, discoidal field with 2-3 rows of cells basally, cubital space with two crossveins, anal loop boot-haped.

Zygonyx species inhabit exposed streams and rivers, preferring rapid section of streams. Males patrol above water by continually hovering. They perch on trees when mating, mating duration is long, some species oviposit in tandem. Some females lay eggs in trickles within the steep precipiles.

彩虹蜻 连结产卵
Zygonyx iris insignis, laying eggs in tandem

朝比奈虹蜻 *Zygonyx asahinai* Matsuki & Saito, 1995

朝比奈虹蜻 雌，广东
Zygonyx asahinai, female from Guangdong

【形态特征】雄性面部蓝黑色具金属光泽；胸部蓝黑色具黄色的肩条纹，侧面具2条黄色条纹，翅透明；腹部黑色具黄色细纹。雌性翅基部和端部具琥珀色斑；腹部黑色具甚细的黄色条纹。【长度】体长 56～65 mm，腹长

朝比奈虹蜻 雄，云南（红河）
Zygonyx asahinai, male from Yunnan (Honghe)

朝比奈虹蜻 雄，云南（红河）
Zygonyx asahinai, male from Yunnan (Honghe)

40～47 mm，后翅 45～52 mm。【栖息环境】海拔 1000 m以下森林中的瀑布和狭窄小溪。【分布】湖北、云南、贵州、福建、广东、广西、香港；越南。【飞行期】4—10月。

[Identification] Male face metallic bluish black. Thorax bluish black with yellow antehumeral stripes, sides with two yellow stripes, wings hyaline. Abdomen black with narrow yellow stripes. Female wing bases and tips with amber spots. Abdomen black with narrow yellow markings. [Measurements] Total length 56-65 mm, abdomen 40-47 mm, hind wing 45-52 mm. [Habitat] Waterfalls and narrow streams in forest below 1000 m elevation. [Distribution] Hubei, Yunnan, Guizhou, Fujian, Guangdong, Guangxi, Hong Kong; Vietnam. [Flight Season] April to October.

彩虹蜻 *Zygonyx iris insignis* **Kirby, 1900**

【形态特征】面部具蓝紫色金属光泽；胸部绿黑色具金属光泽，合胸具黄色的肩条纹，侧面具2条黄色条纹，随年纪增长，黄条纹逐渐加深变成褐色，翅稍染褐色；腹部黑色，第1～3节侧面具黄斑，第3～7节背面中央具黄条纹。本种胸部色彩随年纪增长而逐渐变暗，与其他2种不同。【长度】体长 57～61 mm，腹部 38～43 mm，后翅 48～52 mm。【栖息环境】海拔 1500 m 以下的开阔溪流。【分布】云南、贵州、福建、广东、广西、海南、香港；南亚、东南亚。【飞行期】全年可见。

[Identification] Face metallic bluish purple. Thorax metallic greenish black with yellow antehumeral stripes, sides with two yellow stripes, gradually darken to brown with age, wings tinted with light brown. Abdomen black, S1-S3 with lateral yellow spots, S3-S7 with yellow stripes mid-dorsally. The thoracic yellow stripes darken with age, differs from other two species. [Measurements] Total length 57-61 mm, abdomen 38-43 mm, hind wing 48-52 mm. [Habitat] Exposed streams below 1500 m elevation. [Distribution] Yunnan, Guizhou, Fujian, Guangdong, Guangxi, Hainan, Hong Kong; South and Southeast Asia. [Flight Season] Throughout the year.

彩虹蜻 交尾，广西
Zygonyx iris insignis, mating pair from Guangxi

彩虹蜻 雄，广东｜宋睿斌 摄
Zygonyx iris insignis, male from Guangdong | Photo by Ruibin Song

彩虹蜻 雌，广西
Zygonyx iris insignis, female from Guangxi

高砂虹蜻 *Zygonyx takasago* Asahina, 1966

【形态特征】面部具蓝紫色金属光泽；胸部蓝黑色，合胸具黄色的肩条纹，有时甚短，侧面具2条黄色条纹，翅透明，有时翅端具褐斑。腹部黑色，基方具黄斑。本种与彩虹蜻相似但腹部更粗壮，雄性腹部背面通常仅在基方具黄条纹。【长度】体长 55～62 mm，腹部 38～43 mm，后翅 45～52 mm。【栖息环境】海拔 1500 m以下的开阔溪流和河流。【分布】云南、贵州、浙江、广东、广西、海南、台湾；老挝、越南。【飞行期】4—9月。

[Identification] Face metallic bluish purple. Thorax bluish black with yellow antehumeral stripes, sometimes very short, sides with two yellow stripes, wings hyaline, sometimes with apical brown spots. Abdomen black with basal yellow spots. The species is similar to *Z. iris* but the abdomen stouter, dorsum with yellow mid-dorsal stripes in male confined to basal segments. [Measurements] Total length 55-62 mm, abdomen 38-43 mm, hind wing 45-52 mm. [Habitat] Exposed streams and rivers below 1500 m elevation. [Distribution] Yunnan, Guizhou, Zhejiang, Guangdong, Guangxi, Hainan, Taiwan; Laos, Vietnam. [Flight Season] April to September.

高砂虹蜻 雄，广东｜宋睿斌 摄
Zygonyx takasago, male from Guangdong | Photo by Ruibin Song

高砂虹蜻 雌，云南（德宏）
Zygonyx takasago, female from Yunnan (Dehong)

高砂虹蜻 连结产卵，广西
Zygonyx takasago, laying eggs in tandem from Guangxi

开臀蜻属 Genus *Zyxomma* Rambur, 1842

本属全球已知6种,分布于亚洲和非洲,在热带区域较常见。本属蜻蜓体中型,复眼非常发达,在头顶交汇呈一条直线,面部较窄;翅大面积透明,弓脉位于第1条与第2条结前横脉之间,盘区基方具2~3列翅室,基臀区具1条横脉,臀圈开放。

本属蜻蜓栖息于有树荫遮蔽的池塘。雄性在黄昏时非常活跃,飞行快速并时而悬停。交尾时间较短,在空中完成。

The genus contains six species distributed in Asia and Africa, common in tropical area. Species of the genus are medium-sized, eyes large and broadly confluent above head, face narrow. Wings largely hyaline, arc between antenodals 1 and 2, discoidal field with 2-3 rows of cells basally, cubital space with one crossvein, anal loop open.

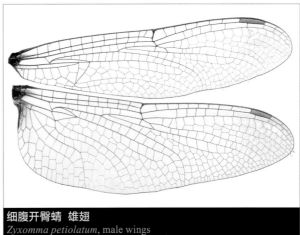

细腹开臀蜻 雄翅
Zyxomma petiolatum, male wings

Zyxomma species inhabit shady ponds. Males are active only at twilight, flying rapidly above water and sometimes hovering. Mating duration is short and takes place in flight.

细腹开臀蜻 雄
Zyxomma petiolatum, male

霜白开臀蜻 *Zyxomma obtusum* Albarda, 1881

【形态特征】雄性通体覆盖白色粉霜，翅端具褐斑。雌性身体褐色，翅端具褐斑。【长度】体长 48~49 mm，腹长 35~36 mm，后翅 36~38 mm。【栖息环境】海拔 500 m 以下的池塘。【分布】中国台湾；南亚、东南亚、日本。【飞行期】全年可见。

[Identification] Male covered by white pruinosity throughout, wing tips with brown spots. Female body brown, wing tips with brown spots. [Measurements] Total length 48-49 mm, abdomen 35-36 mm, hind wing 36-38 mm. [Habitat] Ponds below 500 m elevation. [Distribution] Taiwan of China; South and Southeast Asia, Japan. [Flight Season] Throughout the year.

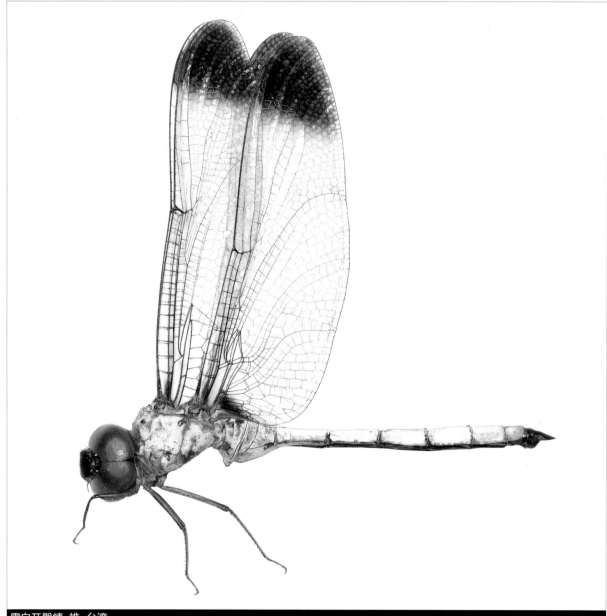

霜白开臀蜻 雄，台湾
Zyxomma obtusum, male from Taiwan

细腹开臀蜻 *Zyxomma petiolatum* **Rambur, 1842**

细腹开臀蜻 雄，云南（西双版纳）
Zyxomma petiolatum, male from Yunnan (Xishuangbanna)

细腹开臀蜻 雄，云南（西双版纳）
Zyxomma petiolatum, male from Yunnan (Xishuangbanna)

【形态特征】复眼黄绿色；胸部深褐色，翅褐色，有时翅端具深褐斑；腹部第1~3节较宽阔，第4~10节甚细。【长度】体长 50~52 mm，腹长 38~39 mm，后翅 32~33 mm。【栖息环境】海拔 500 m 以下的池塘。【分布】云南、福建、广东、广西、海南、香港、台湾；南亚、东南亚、大洋洲。【飞行期】3—12月。

[Identification] Eyes yellowish green. Thorax dark brown, wings brown, sometimes with dark brown apical spots. S1-S3 broad, S4-S10 slim. [Measurements] Total length 50-52 mm, abdomen 38-39 mm, hind wing 32-33 mm. [Habitat] Ponds below 500 m elevation. [Distribution] Yunnan, Fujian, Guangdong, Guangxi, Hainan, Hong Kong, Taiwan; South and Southeast Asia, Oceania. [Flight Season] March to December.

细腹开臀蜻 雌，云南（西双版纳）
Zyxomma petiolatum, female from Yunnan (Xishuangbanna)

双纹缅春蜓 雄
Burmagomphus arvalis, male

云南棘尾春蜓 雄
Trigomphus yunnanensis, male

脊纹环尾春蜓 雄
Lamelligomphus teinus, male

短角亚春蜓 雄
Asiagomphus giza, male

间翅亚目

SUBORDER ANISOZYGOPTERA

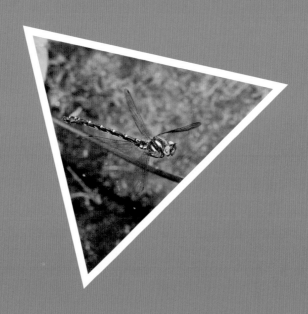

1 昔蜓科 Family Epiophlebiidae

昔蜓属 Genus *Epiophlebia* Calvert, 1903

我们可以把蜻蜓划分为四大类：第一类是蜻蜓，即属于差翅亚目体型粗壮的种类；第二类是豆娘，即属于束翅亚目具有纤细体态的种类；第三类是昔蜓，被认为是古代蜻蜓在现代唯一的后裔；第四类是化石蜻蜓。某些化石蜻蜓是迄今为止最古老体型最大的昆虫。尽管人类拥有大量的化石蜻蜓，昔蜓的地位在上个世纪一直存在争议。1906年Handlirsch为昔蜓建立了新的亚目，即间翅亚目，在一段时期内被广泛接受。然而也有很多学者质疑，并把昔蜓放在差翅亚目或束翅亚目中，认为昔蜓仅是这两类蜻蜓中的一个特例。在2013年Dijkstra等人建立的最新分类系统中，昔蜓仍然被认为是独立的亚目。

昔蜓的显著形态特征包括：成虫后翅基方具有发达的翅柄，臀脉和后翅边缘之间只有1列或2列翅室，上唇向下方膨大加阔，触角5节，梗节长而阔，额顶具一个屋檐形凸起，雌性第8节腹面末端中央具刺状突起等。稚虫的构造也非常特殊，触角5节，并可以通过足和腹部之间的摩擦发出声音。

目前昔蜓属包括4个已知种：喜马拉雅昔蜓，分布于印度、尼泊尔和不丹；日本昔蜓，分布于日本；川昔蜓，分布于中国四川；中华昔蜓，分布于中国东北和朝鲜。除了日本昔蜓，其他种的成虫非常稀有，这是由于它们神秘的行为、特殊的栖息地和短暂的飞行期等方面的限制。在中国已知的2种昔蜓中，川昔蜓至今仅知稚虫的模式标本，成虫未知；中华昔蜓也仅有模式标本记录，而且仅有原始文献的第一作者见过模式标本，合作者和其他研究人员未曾见过和检查过模式标本，故使这种昔蜓的身份存疑。

作者于2012年在云南地区考察期间，曾获得一种昔蜓稚虫。目前保存有4头标本，包括1头末龄稚虫。经过比对，本种和已知种稚虫在形态上有差异，然而稚虫的分类体系尚不完善，很难完全确定其身份，这种昔蜓也有可能是未被描述的新种。我们正在努力地寻找成虫，希望不久可以确定其真实身份。

日本昔蜓 雌雄连结，日本 | 酒井正次 摄
Epiophlebia superstes, pair in tandem from Japan | Photo by Shoji Sakai

日本昔蜓 雌，日本 | 酒井正次 摄
Epiophlebia superstes, female from Japan | Photo by Shoji Sakai

Traditionally Odonata includes Anisoptera (dragonflies), Zygoptera (damselflies), the genus *Epiophlebia* Calvert, 1903 and a diverse array of fossil forms. Some related pre-odonate fossil species, are both the largest known insects and among the oldest known insects. However, despite its extensive fossil record, the systematic position of *Epiophlebia* has changed dramatically over the last century. In 1906, Handlirsch proposed a new suborder, the Anisozygoptera, for this enigmatic odonate, which for a long time was widely accepted. Nevertheless *Epiophlebia* continued to be considered either a zygopteran or an anisopteran by various authors. Dijkstra et al. (2013) reviewed the classification of Odonata, with *Epiophlebia* being retained in Anisozygoptera.

The autapomorphies of *Epiophlebia* include: petiole of hind wing well developed, hind wing with one or two cell rows between A and wing margin, labrum widened distally, antennae five segmented with pedicel elongate and flattened, post frons with transverse shield-like intraocellar ridge, female abdominal segment 8 with midventral apical spur. *Epiophlebia* larvae are also unique in that the antennae are five segmented, and they can produce sound by rubbing the inner apex of the femora against lateral abdominal files.

The genus *Epiophlebia* currently includes four described species: *E. laidlawi* Tillyard, 1921 from India, Nepal and Bhutan; *E. superstes* (Selys, 1889) from Japan; *E. diana* Carle, 2012 from Sichuan, China and *E. sinensis* Li & Nel, 2012 described from northeastern China and also reported from North Korea. With the exception of *E. superstes*, adult *Epiophlebia* specimens are very rare in collections due to their short flight seasons, isolated habitats, and unusual lifestyles. The two species from China are very poorly known and not taxonomically well supported. *E. diana* is known only from the larval stage. *E. sinensis* is known only by the first author of the species and unavailable for examination by other workers, a circumstance which makes the status of this claimed species very much open to question.

In 2012, the present author conducted fieldwork to high mountains in Yunnan province, and discovered the larvae of an *Epiophlebia* species. Four larvae were collected including a final stage one. It is very difficult to identify them from larval morphology although some difference have been found from the described larvae. Our dragonfly team is currently searching for the adult so that its identity can be established.

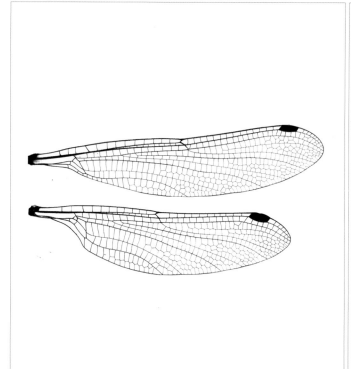

日本昔蜓 雄翅
Epiophlebia superstes, male wings

一种采自中国云南的昔蜓稚虫
Larva of *Epiophlebia* sp. from Yunnan, China

束翅亚目

SUBORDER ZYGOPTERA

1 德丽螅科 Family Devadattidae

　　本科全世界仅1属，主要分布于中南半岛和婆罗洲。本科豆娘体中型，色彩灰暗，翅的显著特征是具较长的翅柄，翅痣基边倾斜。

　　This oriental family contains only one genus with most of the species occurring in Indochina and Borneo. Species of the family are medium-sized, body dark colored, wings are characterized by having a long stalk and peculiar skewed pterostigma.

褐顶德丽螈 雄
Devadatta ducatrix, male

褐顶德丽螈 雌
Devadatta ducatrix, female

德丽螅属 Genus *Devadatta* Kirby, 1890

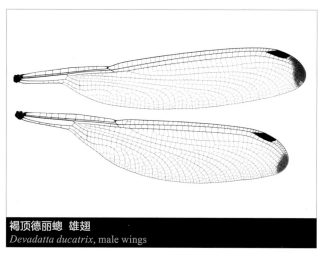

褐顶德丽螅 雄翅
Devadatta ducatrix, male wings

本属全世界已知13种，中国已知1种，分布于广西和云南的中越边境地区。

本属豆娘主要栖息于森林中的林荫小溪和渗流地，雌雄都可在小溪附近的阴暗环境中找到，通常停落在地面低处的枝条上，偶尔飞行。

The genus contains 13 species but only one species is recorded from China, found near the Vietnamese border in Guangxi and Yunnan.

Devadatta species inhabit shady streamlets and seepages in forest, both sexes can be found perching inconspicuously in low vegetation close to water, occasionally making short flights.

褐顶德丽螅 *Devadatta ducatrix* Lieftinck, 1969

【形态特征】雄性头部黑褐色；胸部蓝黑色具金属光泽，侧面轻微覆盖粉霜，翅透明，闪烁蓝绿色或蓝紫色光泽，翅端褐色；腹部蓝黑色。雌性与雄性色彩相似，胸部和腹部具白斑。【长度】体长 42～50 mm，腹长 33～40 mm，后翅 33～35 mm。【栖息环境】海拔 1000 m 以下森林中的林荫小溪和渗流地。【分布】云南（红河）、广西；老挝、越南。【飞行期】4—6月。

褐顶德丽螅 雌，云南（红河）
Devadatta ducatrix, female from Yunnan (Honghe)

[Identification] Male head blackish brown. Thorax metallic bluish black, slightly pruinosed laterally, wings hyaline, shining bluish green or bluish purple, wing tips brown. Abdomen bluish black. Female similarly colored to male, thorax and abdomen with white spots. **[Measurements]** Total length 42-50 mm, abdomen 33-40 mm, hind wing 33-35 mm. **[Habitat]** Shady streams and seepages in forest below 1000 m elevation. **[Distribution]** Yunnan (Honghe), Guangxi; Laos, Vietnam. **[Flight Season]** April to June.

褐顶德丽螅 雄，云南（红河）
Devadatta ducatrix, male from Yunnan (Honghe)

2 色蟌科 Family Calopterygidae

　　本科全球已知21属180余种，除大洋洲和太平洋群岛广泛分布，中国已知12属40余种，广布全国。本科是中至大型豆娘，许多种类身体具金属光泽，腹部细长，翅较宽阔，翅脉密集。如科名含义所示，本科豆娘具"艳丽的翅"且体态优美，十分引人注目。

　　本科豆娘主要栖息于山区溪流和低海拔河流。雄性通常停落在水边的植物或水面的岩石上占据领地。很多种类的雌性具有潜水产卵的能力。

The family contains 21 genera with over 180 species worldwide, widely distributed except Oceania and the Pacific Islands. Over 40 species in 12 genera are recorded from China. They are medium to large damselflies. Many species possess metallic color on almost the whole body, abdomen slim, wings broad with dense venation. They are among the most spectacular odonates and possess colored wings (the family name means beautiful wing).

Species of the family inhabit montane streams and lowland rivers. Males usually perch on marginal plants or rocks above water. Many females lay eggs underwater.

赤基色蟌 雄 | 宋睿斌 摄
Archineura incarnata,
male | Photo by Ruibin Song

褐带暗色蟌 雄
Atrocalopteryx fasciata, male

基色蟌属 Genus *Archineura* Kirby, 1894

赤基色蟌 雄翅
Archineura incarnata, male wings

本属豆娘，是全世界体型最大的色蟌，全球已知3种，分布于中国、老挝和越南。中国已知2种，广泛分布于中国南方。雄性翅基方具有红色或霜白色斑，雌性翅基部无斑但整个翅染有淡褐色。两性均具翅痣，基室具横脉。

本属豆娘栖息于茂盛森林中的开阔溪流。雄性具领域行为，通常停立在溪流中的岩石上。雄性会为争夺领地而争斗，有时会追逐飞行至高空。雌性栖息于河岸带的岩石上。交尾较难遇见，交尾结束后雄性护卫产卵。雌性可以半身潜水将卵产于溪流中的朽木或水草的茎干中。

Species of the genus are magnificent damselflies, the largest in their family, three species are presently recognized, distributed in China, Laos and Vietnam. Two species are recorded from China, widely distributed in the southern part. The wing bases in males have red or creamy white color, absent in females. The wings of females are tinted light brown. Both sexes have a pterostigma on the wings and median space with crossveins.

Archineura species inhabit exposed streams in forest. Males usually perch on rocks in streams defending territories. Two males will sometimes engage in distinctive high-flying territorial contest. Females perch on nearby rocks. Mating behavior has seldom been seen, but after mating the male often guards the ovipositing female. Female inserts her eggs into the deadwood or aquatic plants while partly submerging her body under the surface.

霜基色蟌 雄
Archineura hetaerinoides, male

霜基色螅 *Archineura hetaerinoides* (Fraser, 1933)

霜基色螅 雄，广西
Archineura hetaerinoides, male from Guangxi

【形态特征】雄性面部墨绿色，上唇具黄色斑点；胸部和腹部墨绿色具金属光泽，翅透明，基方具霜白色斑，老熟以后胸部侧面覆盖粉霜。雌性翅稍染琥珀色。本种模式产地老挝的雄性仅后翅基方具乳白色斑，与在云南西部的个体吻合，广西、云南东部和越南的个体前后翅均具乳白色斑，体型稍大。【长度】体长 80~87 mm，腹长 64~69 mm，后翅 49~54 mm。【栖息环境】海拔 1200 m 以下森林中布满岩石的开阔溪流。【分布】广西、云南（红河、普洱）；老挝、越南。【飞行期】4—7月。

[Identification] Male face dark green, labrum with yellow spots. Thorax and abdomen metallic dark green, wings hyaline with cream white patches at base, sides of synthorax with slight pruinescence in old males. Female wings slightly tinted with amber. Males from the type locality in Laos have cream white patches only on hind wings, identical to the population in the west of Yunnan, males from Guangxi, the east of Yunnan and Vietnam have cream white patches on both wings, size slightly larger. [Measurements] Total length 80-87 mm, abdomen 64-69 mm, hind wing 49-54 mm. [Habitat] Exposed and rocky streams in forested area below 1200 m elevation. [Distribution] Guangxi, Yunnan (Honghe, Pu'er); Laos, Vietnam. [Flight Season] April to July.

霜基色蟌 交尾,广西
Archineura hetaerinoides, mating pair from Guangxi

霜基色蟌 雄,云南(普洱)
Archineura hetaerinoides, male from Yunnan (Pu'er)

霜基色螅 雌，云南（普洱）
Archineura hetaerinoides, female from Yunnan (Pu'er)

霜基色螅 雄，云南（普洱）
Archineura hetaerinoides, male from Yunnan (Pu'er)

赤基色蟌 *Archineura incarnata* (Karsch, 1892)

【形态特征】雄性面部墨绿色，上唇具黄色斑点；胸部和腹部墨绿色具金属光泽，翅透明，基方具深红色斑，老熟以后身体色彩略带红色，侧面稍微覆盖粉霜。雌性比雄性略大，翅稍染琥珀色。【长度】体长 75～85 mm，腹长 61～67 mm，后翅 45～52 mm。【栖息环境】海拔 300～1500 m森林中布满岩石的开阔溪流。【分布】中国特有，分布于四川、重庆、贵州、安徽、湖北、湖南、江西、浙江、福建、广西、广东。【飞行期】4—10月。

赤基色蟌 雄，贵州
Archineura incarnata, male from Guizhou

赤基色蟌 雄，广东 | 蒋先兰 摄
Archineura incarnata, male from Guangdong | Photo by Xianlan Jiang

赤基色螅 雌,贵州
Archineura incarnata, female from Guizhou

赤基色螅 交尾,广东 | 莫善濂 摄
Archineura incarnata, mating pair from Guangdong | Photo by Shanlian Mo

[Identification] Male face dark green, labrum with yellow spots. Thorax and abdomen metallic dark green, wings hyaline with bases broadly carmine red, body color darker with slight reddish tint in old males. Female a little larger, wings slightly tinted amber. [Measurements] Total length 75-85 mm, abdomen 61-67 mm, hind wing 45-52 mm. [Habitat] Exposed and rocky streams in forested area at 300-1500 m elevation. [Distribution] Endemic to China, recorded from Sichuan, Chongqing, Guizhou, Anhui, Hubei, Hunan, Jiangxi, Zhejiang, Fujian, Guangxi, Guangdong. [Flight Season] April to October.

暗色蟌属 Genus *Atrocalopteryx* Dumont, Vanfleteren, De Jonckheere & Weekers, 2005

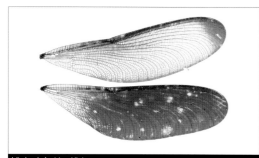

越南暗色蟌 雄翅
Atrocalopteryx coomani, male wings

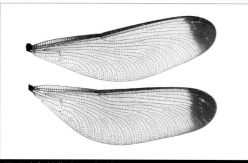

黑顶暗色蟌指名亚种 雄翅
Atrocalopteryx melli melli, male wings

本属仅在亚洲分布，全球已知8种，其中6种分布于中国。本属豆娘体型较大，通体墨绿色具金属光泽，不同种类的翅染有不同的色彩，有些完全黑色，有些具宽阔的深色斑纹。翅脉十分密集，基室无横脉。有些种类的雌性具白色的伪翅痣。

本属豆娘以溪流生境为主，有些可以生活在较宽阔的河流中，如黑暗色蟌，但具有树荫遮蔽的山区溪流是它们最理想的生境。本属雄性具领域行为，通常是停立在溪流中的岩石或者挺水植物上。雌性栖息于溪流附近的森林中，交尾和产卵时才会接近水面。它们可以半身潜水产卵。卵产在溪流中的朽木和水生植物的茎干中。

Presently eight species are placed in this genus, which is confined to Asia, six species are recorded from China. Species of the genus are moderately large damselflies, body metallic dark green, wings tinted with different colors according to species, some with entirely black wings, and others with broad dark bands centrally or apically. The venation is dense, basal space without crossveins. Females with or without white pseudopterostigma depending on species.

Atrocalopteryx species mainly inhabit streams, some species, such as *A. atrata* occur also in large lowland rivers, but in general, montane streams with shady forest are their preferred habitats. Males exhibit territorial behavior, and usually perch on the rocks or emergent plants. Females usually remain in nearby forest and only approach water for copulation and oviposition. They often oviposit while partly submerged. Eggs are inserted into deadwood or the stalks of aquatic plants.

黑顶暗色蟌指名亚种 雄
Atrocalopteryx melli melli, male

黑暗色蟌 *Atrocalopteryx atrata* (Selys, 1853)

　　【形态特征】雄性头部黑褐色；胸部和腹部深绿色具金属光泽，翅黑色。雌性通体黑褐色。【长度】体长 47~58 mm，腹长 38~48 mm，后翅 31~38 mm。【栖息环境】海拔 1500 m以下的溪流和河流。【分布】除西北地 区外全国广布；朝鲜半岛、日本、俄罗斯远东。【飞行期】4—10月。

[Identification] Male head blackish brown. Thorax and abdomen metallic dark green, wings black. Female overall blackish brown. [Measurements] Total length 47-58 mm, abdomen 38-48 mm, hind wing 31-38 mm. [Habitat] Streams and rivers below 1500 m elevation. [Distribution] Widespread throughout China except the Northwest region; Korean peninsula, Japan, Russian Far East. [Flight Season] April to October.

黑暗色蟌 交尾，北京
Atrocalopteryx atrata, mating pair from Beijing

黑暗色蟌 雌，北京
Atrocalopteryx atrata, female from Beijing

黑暗色蟌 雄，北京
Atrocalopteryx atrata, male from Beijing

黑蓝暗色鏓 *Atrocalopteryx atrocyana* (Fraser, 1935)

【形态特征】雄性头部黑褐色；胸部和腹部深绿色具金属光泽，翅黑色具蓝紫色金属光泽，甚阔。雌性体色稍暗；腹部黑褐色。【长度】体长 51～61 mm，腹长 42～50 mm，后翅 40～41 mm。【栖息环境】海拔 1000 m以下的林荫小溪。【分布】贵州、广东；越南。【飞行期】5—11月。

[Identification] Male head blackish brown. Thorax and abdomen metallic dark green, wings broad, black and shining bluish purple. Female body darker. Abdomen blackish brown. [Measurements] Total length 51-61 mm, abdomen 42-50 mm, hind wing 40-41 mm. [Habitat] Shady forest streams below 1000 m elevation. [Habitat] Guizhou, Guangdong; Vietnam. [Flight Season] May to November.

黑蓝暗色鏓 雄，广东 | 宋黎明 摄
Atrocalopteryx atrocyana, male from Guangdong | Photo by Liming Song

黑蓝暗色鏓 雌，越南 | Matti Hämäläinen 摄
Atrocalopteryx atrocyana, female from Vietnam | Photo by Matti Hämäläinen

黑蓝暗色鏓 雄，广东 | 宋黎明 摄
Atrocalopteryx atrocyana, male from Guangdong | Photo by Liming Song

越南暗色蟌 *Atrocalopteryx coomani* (Fraser, 1935)

越南暗色蟌 雄，云南（红河）
Atrocalopteryx coomani, male from Yunnan (Honghe)

【形态特征】雄性复眼上方黑褐色，下方绿色；胸部和腹部深绿色具金属光泽，后胸具黄色条纹，后翅黑褐色，仅在基方半透明，前翅半透明，前缘和端部褐色。雌性体色稍暗；翅具白色的伪翅痣；腹部黑褐色。【长度】体长 72～80 mm，腹长 60～65 mm，后翅 46～47 mm。【栖息环境】海拔 1000 m以下的林荫小溪。【分布】云南（红河）；老挝、越南。【飞行期】5—12月。

[Identification] Male eyes blackish brown above, green below. Thorax and abdomen metallic dark green, metathorax with yellow stripes, hind wings blackish brown, except base sub-hyaline, fore wings sub-hyaline, with costal margin and tips brown. Female with darker body color. Wings with white pseudopterostigma. Abdomen blackish brown. [Measurements] Total length 72-80 mm, abdomen 60-65mm, hind wing 46-47 mm. [Habitat] Shady forest streams below 1000 m elevation. [Distribution] Yunnan (Honghe); Laos, Vietnam. [Flight Season] May to December.

越南暗色鲽 雌，越南 | Matti Hämäläinen 摄
Atrocalopteryx coomani, female from Vietnam | Photo by Matti Hämäläinen

褐带暗色蟌 *Atrocalopteryx fasciata* Yang, Hämäläinen & Zhang, 2014

褐带暗色蟌 雄，云南（德宏）
Atrocalopteryx fasciata, male from Yunnan (Dehong)

【形态特征】雄性复眼黑褐色；胸部和腹部深绿色具金属光泽，后胸侧面具黄色条纹，翅烟色，半透明，中央具显著的深褐色带。雌性体色稍淡；翅染有琥珀色，无伪翅痣。【长度】体长 58~68 mm，腹长 45~55 mm，后翅 38~41 mm。【栖息环境】海拔 1000~1500 m 的开阔溪流和河流。【分布】云南（德宏、保山、普洱）；老挝。【飞行期】5—11月。

[Identification] Male eyes blackish brown. Thorax and abdomen metallic dark green, metathorax with yellow stripes, wings lightly tinted with amber brown, with distinct dark brown bands centrally. Female with slightly paler body color. Wings tinted with amber, pseudopterostigma absent. [Measurements] Total length 58-68 mm, abdomen 45-55 mm, hind wing 38-41mm. [Habitat] Exposed streams and rivers at 1000-1500 m elevation. [Distribution] Yunnan (Dehong, Baoshan, Pu'er); Laos. [Flight Season] May to November.

褐带暗色蟌 雄，云南（德宏）
Atrocalopteryx fasciata, male from Yunnan (Dehong)

褐带暗色蟌 雌，云南（普洱）
Atrocalopteryx fasciata, female from Yunnan (Pu'er)

黑顶暗色鳋指名亚种 *Atrocalopteryx melli melli* Ris, 1912

【形态特征】雄性复眼上方黑褐色，下方绿色；胸部和腹部深绿色具金属光泽，后胸侧面具黄色条纹，翅染黑褐色，半透明，翅端黑色。雌性体色稍暗；翅具白色的伪翅痣；腹部褐色。本种与越南暗色鳋相似，但本种后翅大面积半透明，而后者后翅是大面积黑褐色。【长度】体长 64~80 mm，腹长 53~66 mm，后翅 45~48 mm。【栖息环境】海拔 1000 m 以下的林荫小溪。【分布】中国特有，分布于浙江、福建、广东、广西。【飞行期】4—11月。

黑顶暗色鳋指名亚种　雄，广东
Atrocalopteryx melli melli, male from Guangdong

黑顶暗色鳋指名亚种　雌，广东｜宋睿斌 摄
Atrocalopteryx melli melli, female from Guangdong | Photo by Ruibin Song

[Identification] Male eyes blackish brown above and green below. Thorax and abdomen metallic dark green, metathorax with yellow stripes, wings sub-hyaline, brownish tinted, wing tips opaque black. Female with darker body. Wings with white pseudopterostigma present. Abdomen brown. Similar to *A. coomani* but hind wings largely sub-hyaline, hind wings of *A. coomani* largely blackish brown. [Measurements] Total length 64-80 mm, abdomen 53-66 mm, hind wing 45-48 mm. [Habitat] Shady forest streams below 1000 m elevation. [Distribution] Endemic to China, recorded from Zhejiang, Fujian, Guangdong, Guangxi. [Flight Season] April to November.

黑顶暗色鳋指名亚种　雄，广西
Atrocalopteryx melli melli, male from Guangxi

黑顶暗色螅海南亚种 *Atrocalopteryx melli orohainani* Guan, Han & Dumont, 2012

【形态特征】本亚种依靠分子鉴定的差异而建立。本亚种比指名亚种体型稍大，无显著的形态差异。【长度】雄性正模体长 79 mm，腹长 66 mm，后翅 51 mm。【栖息环境】海拔 1000 m 以下的林荫小溪。【分布】中国海南特有。【飞行期】4—11月。

[Identification] This subspecies was erected largely based on molecular data. A little larger in size than the nominate subspecies, but without clear morphological differences. [Measurements] Holotype male total length 79 mm, abdomen 66 mm, hind wing 51 mm. [Habitat] Shady forest streams below 1000 m elevation. [Distribution] Endemic to Hainan of China. [Flight Season] April to November.

透顶暗色螅 *Atrocalopteryx oberthueri* (McLachlan, 1894)

【形态特征】雄性复眼黑褐色；胸部和腹部深绿色具金属光泽，后胸侧面具黄色条纹，翅大面积黑色，仅翅端透明。雌性体色稍淡；翅半透明，具长而窄的白色伪翅痣；腹部褐色。【长度】体长 57~70 mm，腹长 46~51 mm，后翅 40~41 mm。【栖息环境】海拔 1000~2000 m 的山区溪流。【分布】四川、云南；老挝的分布记录存疑。【飞行期】6—9月。

[Identification] Male eyes blackish brown. Thorax and abdomen metallic dark green, metathorax with yellow stripe, wings largely opaque black with the tips hyaline. Female with paler color of body. Wings sub-hyaline with long, white pseudopterostigma. Abdomen brown. [Measurements] Total length 57-70 mm, abdomen 46-51 mm, hind wing 40-41mm. [Habitat] Montane streams at 1000-2000 m elevation. [Distribution] Sichuan, Yunnan; doubtable record from Laos. [Flight Season] June to September.

透顶暗色螅 雄，云南（丽江）
Atrocalopteryx oberthueri, male from Yunnan (Lijiang)

透顶暗色蟌 雌，云南（丽江）
Atrocalopteryx oberthueri, female from Yunnan (Lijiang)

透顶暗色蟌 雄，云南（丽江）
Atrocalopteryx oberthueri, male from Yunnan (Lijiang)

闪色蟌属 Genus *Caliphaea* Hagen, 1859

本属仅在亚洲分布，全球已知的5种都在中国分布。本属豆娘在本科中体型相对较小；身体铜色，后胸具有黄色条纹，翅透明，基室无横脉，具翅痣，翅基方具翅柄；腹部末端具粉霜。本属雄性可以通过下肛附器的形状区分。

本属豆娘主要栖息于具有一定海拔高度的山区，有些可以生活在海拔超过 2500 m 的高山环境。通常山区的狭窄小溪和沟渠是本属青睐的一类生境，有些种类也会生活在小型瀑布附近的细小水流环境。雄性具领域行为，停立在小型岩石或者水草上。雌性将卵产在溪流中的朽木或水草上。

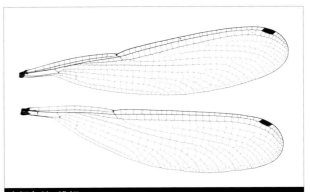

亮闪色蟌 雄翅
Caliphaea nitens, male wings

The genus is confined to Asia, all of the five described species occur in China. Species of the genus are relatively small compared with other members of the family. The body is usually coppery, metathorax with yellow markings. Wings hyaline, median space without crossveins, pterostigma present in both sexes, and wing bases stalked. Tip of abdomen pruinosed. Males of different *Caliphaea* species can be distinguished from each other by the shape of the inferior appendages.

紫闪色蟌 雄
Caliphaea consimilis, male

Caliphaea species mostly occur at moderate altitudes, but some species live in high mountains above 2500 m elevation. Narrow streams and ditches are ideal habitats for them and sometimes they frequent trickles from small waterfalls. Males exhibit territorial behavior and perch on small rocks or emergent plants. Females lay their eggs into deadwood or aquatic plants.

昂卡闪色螅
Caliphaea angka

绿闪色螅
Caliphaea confusa

紫闪色螅
Caliphaea consimilis

亮闪色螅
Caliphaea nitens

泰国闪色螅
Caliphaea thailandica

闪色螅属待定种
Caliphaea sp.

闪色螅属 雄性肛附器
Genus *Caliphaea*, male anal appendages

昂卡闪色螅 *Caliphaea angka* Hämäläinen, 2003

【形态特征】雄性面部、胸部和腹部主要铜色具金属光泽；雌性与雄性相似，但腹部较短。【长度】体长 44～50 mm，腹长 35～40 mm，后翅 30～32 mm。【栖息环境】海拔 1500～2500 m的沟渠和狭窄小溪。【分布】云南（大理）；泰国。【飞行期】6—9月。

[Identification] Male face, thorax and abdomen mainly metallic coppery. Female similar to male, abdomen shorter. [Measurements] Total length 44-50 mm, abdomen 35-40 mm, hind wing 30-32 mm. [Habitat] Ditches and narrow streams at 1500-2500 m elevation. [Distribution] Yunnan (Dali); Thailand. [Flight Season] June to September.

昂卡闪色螅 雄, 云南（大理）
Caliphaea angka, male from Yunnan (Dali)

昂卡闪色螅 雌, 云南（大理）
Caliphaea angka, female from Yunnan (Dali)

绿闪色蟌 *Caliphaea confusa* Hagen, 1859

【形态特征】雄性面部、胸部和腹部青铜色具金属光泽。【长度】雄性体长 47～48 mm，腹长 38～39 mm，后翅 29～30 mm。【栖息环境】海拔 1500～2500 m的沟渠、狭窄小溪和小型瀑布。【分布】云南（红河）；不丹、印度、尼泊尔、缅甸、老挝、越南。【飞行期】4—8月。

[Identification] Male face, thorax and abdomen metallic coppery green. [Measurements] Male total length 47-48 mm, abdomen 38-39 mm, hind wing 29-30 mm. [Habitat] Ditches, narrow streams and small waterfalls at 1500-2500 m elevation. [Distribution] Yunnan (Honghe); Bhutan, India, Nepal, Myanmar, Laos, Vietnam. [Flight Season] April to August.

绿闪色蟌 雄，云南（红河）
Caliphaea confusa, male from Yunnan (Honghe)

紫闪色蟌 *Caliphaea consimilis* McLachlan, 1894

【形态特征】雄性面部、胸部和腹部铜褐色具紫红色光泽，后胸随年纪增长逐渐覆盖粉霜。雌性与雄性相似。【长度】体长 44～48 mm，腹长 35～39 mm，后翅 28～31 mm。【栖息环境】海拔 500～2500 m的沟渠、狭窄小溪和小型瀑布。【分布】中国特有，分布于甘肃、四川、重庆、广西。【飞行期】6—9月。

[Identification] Male face, thorax and abdomen coppery brown and shining purple red, metathorax becomes gradually pruinosed with age. Female similar to male. [Measurements] Total length 44-48 mm, abdomen 35-39 mm, hind wing 28-31 mm. [Habitat] Ditches, narrow streams and small waterfalls at 500-2500 m elevation. [Distribution] Endemic to China, recorded from Gansu, Sichuan, Chongqing, Guangxi. [Flight Season] June to September.

紫闪色螅 雄, 重庆
Caliphaea consimilis, male from Chongqing

紫闪色螅 雌, 重庆
Caliphaea consimilis, female from Chongqing

亮闪色蟌 *Caliphaea nitens* Navás, 1934

【形态特征】雄性面部、胸部和腹部青铜色具金属光泽，后胸随年纪增长逐渐覆盖粉霜。雌性与雄性相似。
【长度】体长 43～47 mm，腹长 34～38 mm，后翅 29～32 mm。【栖息环境】海拔 500～2000 m的沟渠和狭窄小溪。
【分布】中国特有，分布于甘肃、四川、重庆、湖北、湖南、贵州、江西、浙江、福建、广西、广东。【飞行期】6—9月。

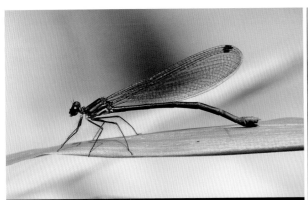

亮闪色蟌 雌，重庆
Caliphaea nitens, female from Chongqing

亮闪色蟌 雄，重庆
Caliphaea nitens, male from Chongqing

[Identification] Male face, thorax and abdomen metallic coppery green, metathorax gradually pruinosed with age. Female similar to male. [Measurements] Total length 43-47 mm, abdomen 34-38 mm, hind wing 29-32 mm. [Habitat] Ditches and narrow streams at 500-2000 m elevation. [Distribution] Endemic to China, recorded from Gansu, Sichuan, Chongqing, Hubei, Hunan, Guizhou, Jiangxi, Zhejiang, Fujian, Guangxi, Guangdong. [Flight Season] June to September.

亮闪色蟌 交尾，重庆
Caliphaea nitens, mating pair from Chongqing

泰国闪色螅 *Caliphaea thailandica* Asahina, 1976

【形态特征】雄性面部、胸部和腹部青铜色具金属光泽。【长度】雄性体长 47 mm，腹长 39 mm，后翅 30 mm。【栖息环境】海拔 1000~1500 m的狭窄小溪。【分布】云南（普洱）；老挝、泰国、越南。【飞行期】5—7月。

[Identification] Male face, thorax and abdomen metallic coppery green. [Measurements] Male total length 47 mm, abdomen 39 mm, hind wing 30 mm. [Habitat] Narrow streams at 1000-1500 m elevation. [Distribution] Yunnan (Pu'er); Laos, Thailand, Vietnam. [Flight Season] May to July.

泰国闪色螅 雄，云南（普洱）
Caliphaea thailandica, male from Yunnan (Pu'er)

闪色螅属待定种 *Caliphaea* sp.

【形态特征】雄性面部、胸部和腹部铜色具金属光泽。雌性与雄性相似。【长度】体长 40~43 mm，腹长 32~36 mm，后翅 27~30 mm。【栖息环境】海拔 1000~2500 m的狭窄小溪。【分布】云南（德宏、保山）。【飞行期】5—7月。

[Identification] Male face, thorax and abdomen metallic coppery. Female similar to male. [Measurements] Total length 40-43 mm, abdomen 32-36 mm, hind wing 27-30 mm. [Habitat] Narrow streams at 1000-2500 m elevation. [Distribution] Yunnan (Dehong, Baoshan). [Flight Season] May to July.

闪色螅属待定种　雄，云南（保山）
Caliphaea sp, male from Yunnan (Baoshan)

闪色螅属待定种　雌，云南（保山）
Caliphaea sp, female from Yunnan (Baoshan)

色蟌属 Genus *Calopteryx* Leach, 1815

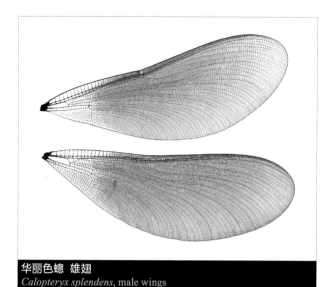

华丽色蟌 雄翅
Calopteryx splendens, male wings

本属全球已知16种，主要分布于北美洲、欧洲和西亚地区。中国已知2种，其中一个西部种分布于新疆，另一个分布于东北地区。本属豆娘和暗色蟌属近似，但体型略小。它们通体墨绿色具金属光泽，不同种和亚种的翅上染有不同色彩，有些完全黑色，有些具宽阔的深色斑纹；翅脉十分密集，基室无横脉；雌性具白色的伪翅痣。

本属豆娘主要栖息于海拔较低的溪流和河流，行为与暗色蟌属相似。

The genus contains 16 species and most of them occur in North America, Europe and the western part of Asia. Two species are recorded from China, the range of one western species reaches Xinjiang and that of the other species reaches Northeast China. Species of the genus are similar to *Atrocalopteryx* but smaller. The body is metallic dark green, wing color pattern varies in different species and subspecies ranging from totally black wings, wings with opaque bands centrally to entirely opaque wings. Venation dense, basal space without crossveins. Female with white pseudopterostigma.

Calopteryx species live in lowland streams and rivers. Behavior similar to *Atrocalopteryx* species.

日本色蟌 交尾 | 金洪光 摄
Calopteryx japonica, mating pair | Photo by Hongguang Jin

日本色蟌 *Calopteryx japonica* Selys, 1869

　　【形态特征】雄性面部、胸部和腹部主要墨绿色具金属光泽；翅黑色具深蓝色光泽；腹部末端腹面具白色斑。雌性体色灰暗；翅棕色，半透明，具白色的伪翅痣。【长度】体长 55～63 mm，腹长 40～48 mm，后翅 31～40 mm。【栖息环境】海拔 500 m 以下的河流和开阔溪流。【分布】黑龙江、吉林、辽宁；朝鲜半岛、日本、俄罗斯远东。【飞行期】6—8月。

日本色蟌 雄，黑龙江｜莫善濂 摄
Calopteryx japonica, male from Heilongjiang | Photo by Shanlian Mo

[Identification] Male face, thorax and abdomen metallic dark green. Wings black opaque with dark bluish lustre. Underside of distal abdominal segments white. Female with darker body color. Wings brown and translucent, with white pseudopterostigma. [Measurements] Total length 55-63 mm, abdomen 40-48 mm, hind wing 31-40 mm. [Habitat] Rivers and exposed streams below 500 m elevation. [Distribution] Heilongjiang, Jilin, Liaoning; Korean peninsula, Japan, Russian Far East. [Flight Season] June to August.

日本色蟌 雄，吉林｜金洪光 摄
Calopteryx japonica, male from Jilin | Photo by Hongguang Jin

日本色蟌 雌，吉林｜金洪光 摄
Calopteryx japonica, female from Jilin | Photo by Hongguang Jin

华丽色蟌 *Calopteryx splendens* (Harris, 1780)

华丽色蟌 护卫产卵,芬兰 | Sami Karjalainen 摄
Calopteryx splendens, female laying eggs with male guarding from Finland | Photo by Sami Karjalainen

【形态特征】雄性面部、胸部和腹部主要墨绿色具金属光泽;翅具1个甚大的不透明蓝色斑,大概为翅长的2/3。雌性体色较淡;翅透明具绿色光泽,或者与雄性翅色彩相似,具白色的伪翅痣。【长度】体长 45~48 mm,腹长 33~41 mm,后翅 27~36 mm。【栖息环境】河流和开阔溪流。【分布】欧洲和亚洲西部地区广布,在亚洲东部地区分布于中国新疆、蒙古和雅库特西南地区。【飞行期】5—7月。

华丽色蟌 雄,芬兰 | Sami Karjalainen 摄
Calopteryx splendens, male from Finland | Photo by Sami Karjalainen

[Identification] Male face, thorax and abdomen metallic dark green. Wings with a very large opaque bluish patch covering nearly two thirds of the wing length. Female paler. Wings either hyaline with greenish sheen, or with similar patches to male, with white pseudopterostigma. [Measurements] Total length 45-48 mm, abdomen 33-41 mm, hind wing 27-36 mm. [Habitat] Rivers and exposed streams. [Distribution] Widely distributed in Europe and western Asia, in the east its range extends to Xinjiang, Mongolia and southwestern Yakutia. [Flight Season] May to July.

亮翅色螅属 Genus *Echo* Selys, 1853

本属全世界已知仅5种，分布于东洋界。中国已知的3个珍稀种仅分布于云南和西藏。本属豆娘身体色彩较暗并具金属光泽，翅透明并闪烁绿紫色霓虹光泽或具深色不透明斑纹。翅脉较密集，基室具横脉，两性具翅痣。

本属豆娘主要栖息于植被茂盛的森林中的狭窄小溪、渗流地等细流生境。雄性停落在距水面较低处的植物上，雌性产卵于水草的茎干内。

The genus contains only five known Oriental species. Three of them are recorded from China, all of them are rare species confined to Yunnan or Tibet. Species of the genus have a metallic dark green body, wings either hyaline with greenish violet iridescence, or with dark or opaque markings, venation dense, median space with crossveins, pterostigma present in both sexes.

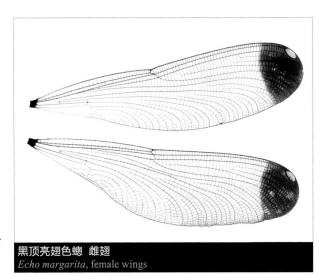

黑顶亮翅色螅 雌翅
Echo margarita, female wings

Echo species live in narrow streams and seepages in dense forests. Males perch on the leaves very low above the water, females lay eggs in the stalks of aquatic plants.

白背亮翅色螅 雄
Echo candens, male

白背亮翅色螅 *Echo candens* Zhang, Hämäläinen & Cai, 2015

白背亮翅色螅 雌，云南（德宏）
Echo candens, female from Yunnan (Dehong)

【形态特征】雄性面部具白色粉霜；胸部深绿色具金属光泽，合胸背面覆盖浓密的白色粉霜，侧面微染粉霜，翅稍染褐色，闪烁蓝绿色光泽，翅痣黑褐色；腹部深绿色。雌性身体深绿色具金属光泽；翅褐色。【长度】体长 52~57 mm，腹长 41~45 mm，后翅 35~38 mm。【栖息环境】海拔 800~1500 m 雨林中的狭窄小溪、沟渠和渗流地。【分布】云南（德宏）；缅甸。【飞行期】5—7月。

[Identification] Male face with whitish pruinosed patch. Thorax metallic dark green, dorsal part of synthorax heavily whitish pruinosed, wings slightly tinted with brown, and shining bluish green, pterostigma blackish brown. Abdomen dark green. Female body metallic dark green. Wings tinted with brown. [Measurements] Total length 52-57 mm, abdomen 41-45 mm, hind wing 35-38 mm. [Habitat] Seepages, ditches and narrow streams in rain forest from 800-1500 m elevation. [Distribution] Yunnan (Dehong); Myanmar. [Flight Season] May to July.

白背亮翅色螅 雄，云南（德宏）
Echo candens, male from Yunnan (Dehong)

黑顶亮翅色蟌 *Echo margarita* Selys, 1853

黑顶亮翅色蟌 雄，云南（德宏）
Echo margarita, male from Yunnan (Dehong)

【形态特征】雄性头部黑褐色；胸部金属黑绿色，翅大面积透明闪烁蓝紫色，端部具黑褐色斑，翅痣白色；腹部黑褐色。雌性与雄性相似，但腹部较短较粗壮。【长度】体长 49～55 mm，腹长 39～44 mm，后翅 34～39 mm。【栖息环境】海拔 500～1500 m雨林中的狭窄小溪、沟渠和渗流地。【分布】云南（德宏）；印度、缅甸。【飞行期】6—11月。

[Identification] Male head blackish brown. Thorax metallic blackish green, wings largely hyaline and shining bluish purple, wing tips with blackish brown spots, pterostigma white. Abdomen blackish brown. Female similar to male, but abdomen shorter and stouter. [Measurements] Total length 49-55 mm, abdomen 39-44 mm, hind wing 34-39 mm. [Habitat] Seepages, ditches and narrow streams in rain forest at 500-1500 m elevation. [Distribution] Yunnan (Dehong); India, Myanmar. [Flight Season] June to November.

黑顶亮翅色蟌 雌，云南（德宏）
Echo margarita, female from Yunnan (Dehong)

华丽亮翅色蟌 *Echo perornata* Yu & Hämäläinen, 2012

华丽亮翅色蟌 雄,西藏｜吴超 摄
Echo perornata, male from Tibet｜Photo by Chao Wu

华丽亮翅色蟌 雌,西藏｜吴超 摄
Echo perornata, female from Tibet｜Photo by Chao Wu

【形态特征】雄性头部黑褐色；胸部金属墨绿色,翅稍染烟色,翅结以上具1条显著的褐带,端部具同样的深色斑,翅痣稍覆盖白色粉霜；腹部黑褐色。雌性与雄性相似,翅结以上的褐带较阔,翅端的褐斑宽阔,翅痣白色。【长度】体长 51～53 mm,腹长 40～43 mm,后翅 36～37 mm。【栖息环境】海拔 800～1200 m的山区溪流。【分布】中国西藏特有。【飞行期】6—9月。

[Identification] Male head blackish brown. Thorax metallic dark green, slightly brownish tinted wings have a distinct dark brown transverse band beyond nodus and wing apex with the same dark color, pterostigma slightly tinted with white pruinescence. Abdomen blackish brown. Female similar to male, the brown band beyond nodus is wider, wing apex more broadly darkened, pterostigma cream white. [Measurements] Total length 51-53 mm, abdomen 40-43 mm, hind wing 36-37 mm. [Habitat] Forest streams at 800-1200 m elevation. [Distribution] Endemic to Tibet of China. [Flight Season] June to September.

单脉色蟌属 Genus *Matrona* Selys, 1853

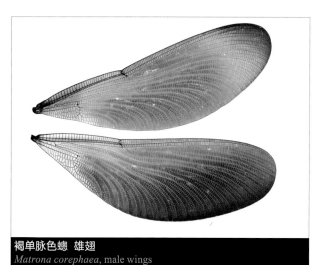

褐单脉色蟌 雄翅
Matrona corephaea, male wings

本属是仅分布于亚洲东部的大型豆娘,全世界已知的9种中有7种在中国分布。它们的身体通体是具金属光泽的墨绿色；翅完全黑色或棕色,翅脉十分密集,基室具横脉,雌性具白色的伪翅痣。

本属以溪流生境为主,有些可以生活在较宽阔的河流中,如透顶单脉色蟌。本属的雄性具领域行为,通常停立在溪流中的岩石或者挺水植物上。雌性栖息于溪流附近的森林中,它们可以潜水产卵。卵产在溪流中的朽木和水生植物的茎干中。

Species of the genus are rather large damselflies confined to eastern Asia, seven of the nine presently recognized species occur in China. Their body is metallic dark green. Wings entirely black or brown, venation dense, basal space with crossveins, female has white pseudopterostigma.

Most species of the genus inhabit streams in forested areas, and some species, such as *Matrona basilaris*, can even live in large rivers. Males exhibit territorial behavior, typically perching on rocks or emergent plants. Females usually rest in nearby forest, they can lay eggs underwater. Eggs are inserted into deadwood or the stalks of aquatic plants.

黑单脉色螅 雄
Matrona nigripectus, male

安妮单脉色螅 *Matrona annina* Zhang & Hämäläinen, 2012

【形态特征】雄性上唇具1对黄色斑点；胸部和腹部深绿色具金属光泽，后胸具黄色条纹，翅红褐色，翅结以下翅脉蓝白色；腹部第8~10节腹面黄褐色。雌性胸部金属铜褐色，翅浅褐色，具白色的伪翅痣；腹部深褐色。【长度】体长 63~70 mm，腹长 51~59 mm，后翅 40~44 mm。【栖息环境】海拔 500 m以下的半阴小溪。【分布】中国特有，分布于广西、广东。【飞行期】8—11月。

[Identification] Male labrum with a pair of yellow spots. Thorax and abdomen metallic dark green, metathorax with yellow stripes, wings reddish brown with bluish white reticulationat the base. Underside of S8-S10 yellowish brown. Female with metallic coppery brown thorax, wings light brown, with white pseudopterostigma. Abdomen dark brown. **[Measurements]** Total length 63-70 mm, abdomen 51-59 mm, hind wing 40-44 mm. **[Habitat]** Semi-shady streams below 500 m elevation. **[Distribution]** Endemic to China, recorded from Guangxi, Guangdong. **[Flight Season]** August to November.

安妮单脉色蟌 雄, 广东 | 吴宏道 摄
Matrona annina, male from Guangdong | Photo by Hongdao Wu

安妮单脉色蟌 雌, 广东 | 吴宏道 摄
Matrona annina, female from Guangdong | Photo by Hongdao Wu

透顶单脉色蟌 *Matrona basilaris* Selys, 1853

【形态特征】雄性面部金属绿色；胸部深绿色具金属光泽，后胸具黄色条纹，翅黑色，翅脉在基方1/2处蓝色；腹部第8～10节腹面黄褐色。雌性胸部青铜色，翅深褐色，具白色的伪翅痣；腹部褐色。北方雄性翅正面几乎完全深蓝色，南方雄性仅在基方不足1/2处蓝色。【长度】中国南方：体长 63～70 mm，腹长 51～57 mm，后翅 38～48 mm。中国北方：体长 56～62 mm，腹长 46～51 mm，后翅 34～43 mm。【栖息环境】海拔 1500 m以下的开阔溪流和河流。【分布】除西北地区外全国广布；老挝、越南。【飞行期】5—11月。

透顶单脉色蟌 雄，北京
Matrona basilaris, male from Beijing

透顶单脉色蟌 雌，产卵，北京
Matrona basilaris, female laying eggs from Beijing

[Identification] Male face metallic green. Thorax metallic dark green, metathorax with yellow stripes, wings black with bluish white reticulation at basal half. Underside of S8-S10 distinctly yellowish brown. Female with coppery green thorax, wings dark brown, with white pseudopterostigma. Abdomen brown. Males from the north have most of wings

透顶单脉色蟌 雄，北京
Matrona basilaris, male from Beijing

透顶单脉色蟌 交尾，贵州
Matrona basilaris, mating pair from Guizhou

透顶单脉色蟌 雌，贵州
Matrona basilaris, female from Guizhou

dark bluish, males from the south have only blue reticulation at basal half. **[Measurements]** Individuals from the south of China: Total length 63-70 mm, abdomen 51-57 mm, hind wing 38-48 mm. Individuals from the north of China: Total length 56-62 mm, abdomen 46-51 mm, hind wing 34-43 mm. **[Habitat]** Exposed streams and rivers below 1500 m elevation. **[Distribution]** Widespread throughout China except the Northwest region; Laos, Vietnam. **[Flight Season]** May to November.

透顶单脉色蟌 雄
Matrona basilaris, male

褐单脉色蟌 *Matrona corephaea* Hämäläinen, Yu & Zhang, 2011

【形态特征】雄性面部金属绿色，上唇淡绿色；胸部和腹部深绿色具金属光泽，后胸具黄色条纹，翅红棕色，后翅色彩较深。雌性胸部青铜色，翅红棕色，后翅色彩较深，具白色的伪翅痣；腹部褐色。【长度】体长 62~70 mm，腹长 50~55 mm，后翅 38~44 mm。【栖息环境】海拔 1500 m以下的开阔溪流和林荫小溪。【分布】中国特有，分布于贵州、重庆、湖北、湖南、浙江。【飞行期】6—9月。

[Identification] Male face metallic green, labrum pale green, thorax and abdomen metallic dark green, metathorax with yellow stripes, wings reddish umber, hind wings clearly darker colored. Female with coppery green thorax, wings reddish umber, with white pseudopterostigma, hind wings darker in color. Abdomen brown. [Measurements] Total length 62-70 mm, abdomen 50-55 mm, hind wing 38-44 mm. [Habitat] Exposed and shady streams below 1500 m

褐单脉色蟌 雄，重庆
Matrona corephaea, male from Chongqing

elevation. [Distribution] Endemic to China, recorded from Guizhou, Chongqing, Hubei, Hunan, Zhejiang. [Flight Season] June to September.

褐单脉色蟌 雄，重庆
Matrona corephaea, male from Chongqing

褐单脉色蟌 雌，产卵，重庆
Matrona corephaea, female laying eggs from Chongqing

台湾单脉色蟌 *Matrona cyanoptera* Hämäläinen & Yeh, 2000

【形态特征】雄性面部金属绿色；胸部和腹部深绿色具金属光泽，翅黑色，翅脉蓝白色。雌性胸部青铜色，翅深褐色，具白色的伪翅痣；腹部褐色。【长度】体长 62～67 mm，腹长 50～55 mm，后翅 39～44 mm。【栖息环境】海拔 1500 m 以下的林荫小溪。【分布】中国台湾特有。【飞行期】全年可见。

[Identification] Male face metallic green. Thorax and abdomen metallic dark green, wings black with bluish white reticulation throughout the wings. Female with coppery green thorax, wings dark brown, with white pseudopterostigma. Abdomen brown. [Measurements] Total length 62-67 mm, abdomen 50-55 mm, hind wing 39-44 mm. [Habitat] Shady montane streams below 1500 m elevation. [Distribution] Endemic to Taiwan of China. [Flight Season] Throughout the year.

台湾单脉色蟌 雄，台湾 | 嘎嘎 摄
Matrona cyanoptera, male from Taiwan | Photo by Gaga

台湾单脉色蟌 雌，台湾 | 嘎嘎 摄
Matrona cyanoptera, female from Taiwan | Photo by Gaga

妈祖单脉色蟌 *Matrona mazu* Yu, Xue & Hämäläinen, 2015

【形态特征】雄性面部金属绿色；胸部和腹部深绿色具金属光泽，后胸具黄色条纹，翅黑色，翅脉在基方1/2处蓝色；腹部第7～10节腹面黄褐色。雌性胸部青铜色，翅深褐色，具白色的伪翅痣；腹部褐色。本种与透顶单脉色蟌相似，但翅更短更宽阔，腹部腹面的浅褐色区域更大。【长度】体长 60～68 mm，腹长 49～51 mm，后翅 38～40 mm。【栖息环境】海拔1500 m以下的山区溪流。【分布】中国海南特有。【飞行期】全年可见。

[Identification] Male face metallic green. Thorax and abdomen metallic dark green, metathorax with yellow stripes, wings black with blue reticulation in basal half. Underside of S7-S10 yellowish brown. Female with coppery green thorax, wings dark brown, with

妈祖单脉色蟌 雌，海南
Matrona mazu, female from Hainan

white pseudopterostigma. Abdomen brown. Similar to *M. basilaris*, but wings proportionally broader and shorter, the underside of abdomen with more extensive pale color. [Measurements] Total length 60-68 mm, abdomen 49-51 mm, hind wing 38-40 mm. [Habitat] Montane streams below 1500 m elevation. [Distribution] Endemic to Hainan of China. [Flight Season] Throughout the year.

妈祖单脉色蟌 雄，海南
Matrona mazu, male from Hainan

黑单脉色蟌 *Matrona nigripectus* Selys, 1879

黑单脉色蟌 雌，云南（德宏）
Matrona nigripectus, female from Yunnan (Dehong)

【形态特征】雄性面部金属绿色；胸部和腹部深绿色具金属光泽，翅黑色，翅脉仅在基方1/3以下深蓝色；腹部第8～10节腹面黑色。雌性胸部青铜色，翅深褐色，翅脉黄褐色，具白色伪翅痣；腹部褐色。本种腹部末端腹面全黑色，容易与透顶单脉色蟌和妈祖单脉色蟌区分。【长度】体长 62～70 mm，腹长 51～57 mm，后翅 37～45 mm。【栖息环境】海拔 2000 m以下的开阔或林荫溪流。【分布】云南；印度、缅甸、泰国、老挝。【飞行期】4—12月。

[Identification] Male face metallic green. Thorax and abdomen metallic dark green, wings black with dark bluish reticulation in the basal one third. Underside of S8-S10 all black. Female with coppery green thorax, wings dark brown with basal area semihyaline and white pseudopterostigma. Abdomen brown. Male of *M. nigripectus* can be distinguished from *M. basilaris* and *M. mazu* by the underside of S8-S10 black. [Measurements] Total length 62-70 mm, abdomen 51-57 mm, hind wing 37-45 mm. [Habitat] Shady or exposed montane streams below 2000 m elevation. [Distribution] Yunnan; India, Myanmar, Thailand, Laos. [Flight Season] April to December.

黑单脉色蟌 雄，云南（临沧）
Matrona nigripectus, male from Yunnan (Lincang)

神女单脉色蟌 *Matrona oreades* Hämäläinen, Yu & Zhang, 2011

神女单脉色蟌 雄, 重庆
Matrona oreades, male from Chongqing

神女单脉色蟌 雄, 重庆
Matrona oreades, male from Chongqing

神女单脉色蟌 雌, 重庆
Matrona oreades, female from Chongqing

【形态特征】雄性复眼褐色和天蓝色，面部金属绿色，上唇淡蓝色；胸部和腹部深绿色具金属光泽，后胸具黄色条纹，翅红棕色，端部色彩加深；腹部第8~10节腹面白色。雌性胸部青铜色，翅红棕色，具白色的伪翅痣；腹部褐色。本种与褐单脉色蟌相似，但翅较窄，雄性面部大面积蓝色，而褐单脉色蟌面部色彩较深。【长度】腹长47~57 mm，后翅 40~48 mm。【栖息环境】海拔 1500 m以下的林荫溪流。【分布】中国特有，分布于甘肃、四川、重庆、湖北。【飞行期】6—9月。

[Identification] Male with eyes brown and sky-blue, face metallic green, labrum pale blue. Thorax and abdomen metallic dark green, metathorax with yellow stripes, wings reddish umber with tips slightly darkened. Underside of S8-S10 white. Female with thorax coppery green, wings reddish umber, with white pseudopterostigma. Abdomen brown. Similar to *M. corephaea*, but wings narrower, male face largely blue, face color darkened in *M. corephaea*. [Measurements] Abdomen 47-57 mm, hind wing 40-48 mm. [Habitat] Shady montane streams below 1500 m elevation. [Distribution] Endemic to China, recorded from Gansu, Sichuan, Chongqing, Hubei. [Flight Season] June to September.

绿色蟌属 Genus *Mnais* Selys, 1853

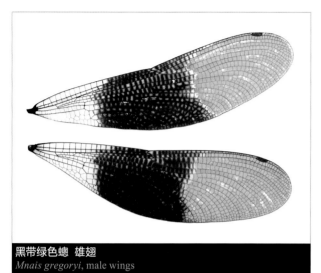

黑带绿色蟌 雄翅
Mnais gregoryi, male wings

本属仅在亚洲东部分布，是体中型的豆娘。由于本属物种的相似性，分类上较困难，是一世界性难题，因此尚未明确本属究竟有多少种。本属雄性多型，通常按翅的色彩分为两型，一类是透翅型，另一类是具色彩的翅型。翅脉在本科中相对较稀疏，基室无横脉，两性具翅痣。雄性腹部末端具白色粉霜。

本属以溪流生境为主，多栖息于狭窄的山区小溪和沟渠。雄性通常停立在溪流中的岩石或者挺水植物上。雌性栖息于溪流附近的森林中，产卵时才会接近水面，它们可以半身潜水产卵。卵产在溪流中的朽木和水生植物的茎干中。

This genus of medium-sized damselflies is confined to eastern Asia. Most species are similar and show considerable variation, the taxonomy of the genus is poorly known and it is unclear how many species there are. Males are polymorphic, all species have a hyaline winged morph and a morph with colored wings. Wing venation rather sparse, median space without crossveins, pterostigma present in both sexes. The apical segments of abdomen in mature males usually with white pruinescence.

Mnais species inhabit narrow montane streams and ditches. Males usually perch on the rocks or emergent plants. Females usually rest in the forest nearby and will approach water for laying eggs, they often oviposit partly submerged. Eggs are inserted into the deadwood or the stalks of aquatic plants.

黑带绿色蟌 雄，黑带型
Mnais gregoryi, colored winged male

安氏绿色蟌 *Mnais andersoni* McLachlan, 1873

【形态特征】雄性多型，透翅型雄性翅稍染褐色，橙翅型雄性翅橙色；胸部和腹部青铜色具金属光泽，后胸具黄条纹，腹部第8~10节覆盖白色粉霜，橙翅型胸部背面随年纪增长逐渐覆盖白色粉霜。雌性翅透明，身体色彩与透翅型雄性相似。【长度】体长 48~50 mm，腹长 39~40 mm，后翅 30~31 mm。【栖息环境】海拔 500~1500 m 森林中的小溪、沟渠和渗流地。【分布】云南（德宏、普洱）；缅甸、泰国、老挝、越南。【飞行期】3—7月。

[Identification] Males polymorphic, the hyaline winged morph with wings slightly tinted brown and the colored winged morph with orange wings. In both morphs thorax and abdomen metallic coppery green, metathorax with yellow stripes and dorsum of abdomen on S8-S10 with white pruinescence, in the orange winged morph the dorsal part of synthorax develops dense white pruinescence as the male gets older. Female with hyaline wings and body color similar to the hyaline winged male morph. [Measurements] Total length 48-50 mm, abdomen 39-40 mm, hind wing 30-31 mm. [Habitat] Streams, ditches and seepages in forested areas at 500-1500 m elevation. [Distribution] Yunnan (Dehong, Pu'er); Myanmar, Thailand, Laos, Vietnam. [Flight Season] March to July.

安氏绿色蟌 雄，透翅型，云南（普洱）
Mnais andersoni, male, hyaline winged morph from Yunnan (Pu'er)

安氏绿色蟌 雌，云南（德宏）
Mnais andersoni, female from Yunnan (Dehong)

安氏绿色蟌 雄，橙翅型，云南（普洱）
Mnais andersoni, male, orange winged morph from Yunnan (Pu'er)

黑带绿色蟌 *Mnais gregoryi* Fraser, 1924

【形态特征】雄性多型,透翅型翅透明,黑带型翅基方1/4透明,中部具1条甚阔的不透明黑带,端方具等大的白色区域,覆盖整个翅端,使翅仅基方透明,翅痣红褐色;腹部第1~3节、第8~10节覆盖白色粉霜。雌性胸部青铜色,腹部大面积青铜色并具光泽;年轻个体翅琥珀色,随年纪增长色彩变浅,翅痣白色。【长度】体长 48~57 mm,腹长 37~45 mm,后翅 36~40 mm。【栖息环境】海拔 2000~3000 m森林中的小溪和沟渠。【分布】中国特有,分布于云南(昆明、丽江、迪庆、红河、大理、德宏)、四川。【飞行期】2—7月。

[Identification] Males polymorphic, the hyaline winged morph wings transparent and the colored winged morph has the basal one fourth of wings hyaline, followed with a broad opaque black band and equally broad opaque white area apically, pterostigma reddish brown. S1-S3 and S8-S10 whitish pruinosed. Female with thorax metallic coppery green, abdomen largely metallic coppery green. Young female with wings amber, the color reducing with age, pterostigma white. [Measurements] Total length 48-57 mm, abdomen 37-45 mm, hind wing 36-40 mm. [Habitat] Montane streams and ditches at 2000-3000 m elevation. [Distribution] Endemic to China, recorded from Yunnan (Kunming, Lijiang, Diqing, Honghe, Dali, Dehong), Sichuan. [Flight Season] February to July.

黑带绿色蟌 雄,透翅型,云南(大理)
Mnais gregoryi, male, hyaline winged morph from Yunnan (Dali)

黑带绿色蟌 雌,云南(大理)
Mnais gregoryi, female from Yunnan (Dali)

黑带绿色蟌 雄，黑带型，云南（大理）
Mnais gregoryi, male, colored winged morph from Yunnan (Dali)

黑带绿色蟌 雄，黑带退化型，云南（大理）
Mnais gregoryi, male, colored winged morph with the black band largely reduced from Yunnan (Dali)

烟翅绿色蟌 *Mnais mneme* Ris, 1916

烟翅绿色蟌 雄，橙翅型，海南
Mnais mneme, male, orange winged morph from Hainan

【形态特征】雄性多型，透翅型胸部和腹部墨绿色具金属光泽，翅透明或稍染烟色；橙翅型胸部覆盖白色粉霜，翅橙色；腹部黑褐色，第8~10节覆盖白色粉霜。雌性身体铜褐色，翅稍染褐色或透明。【长度】体长 48~57 mm，腹长 41~46 mm，后翅 28~35 mm。【栖息环境】海拔 1500 m以下森林中的小溪、沟渠和渗流地。【分布】云南、福建、广西、广东、海南、香港；柬埔寨、老挝、越南。【飞行期】3—7月。

烟翅绿色蟌 雄，透翅型，云南（红河）
Mnais mneme, male, hyaline winged morph from Yunnan (Honghe)

烟翅绿色蟌 雌，广东｜宋睿斌 摄
Mnais mneme, female from Guangdong｜Photo by Ruibin Song

[Identification] Males polymorphic, the hyaline winged morph with thorax and abdomen metallic dark green, wings hyaline or tinted with brown. The orange winged morph with synthorax whitish pruinosed when older, wings orange. Abdomen blackish brown, S8-S10 with white pruinescence. Female with body metallic coppery brown, wings hyaline or tinted with brown. [Measurements] Total length 48-57 mm, abdomen 41-46 mm, hind wing 28-35 mm. [Habitat] Streams, ditches and seepages in the forest below 1500 m elevation. [Distribution] Yunnan, Fujian, Guangxi, Guangdong, Hainan, Hong Kong; Cambodia, Laos, Vietnam. [Flight Season] March to July.

烟翅绿色蟌 雄，橙翅型，云南（红河）
Mnais mneme, male, orange winged morph from Yunnan (Honghe)

黄翅绿色蟌 *Mnais tenuis* Oguma, 1913

【形态特征】雄性多型，透翅型胸部和腹部青铜色具金属光泽，后胸后侧板黄色，翅透明，腹部第8~10节覆盖白色粉霜；橙翅型胸部覆盖白色粉霜，后胸后侧板黄色区域无粉霜，腹部第1~3节、第8~10节覆盖白色粉霜。雌性身体铜褐色，翅稍染褐色。【长度】体长 42~50 mm，腹长 33~42 mm，后翅 27~31 mm。【栖息环境】海拔1500 m以下森林中的小溪、沟渠和渗流地。【分布】浙江、江西、福建、广东、台湾；老挝。【飞行期】2—7月。

[Identification] Males polymorphic, the hyaline winged morph with thorax and abdomen metallic coppery green, metepimeron entirely yellow, wings hyaline, S8-S10 with white pruinescence. The orange winged morph with synthorax whitish pruinosed, metepimeron entirely yellow, S1-S3 and S8-S10 with white pruinescence. Female with body metallic coppery brown, wings slightly tinted with brown. [Measurements] Total length 42-50 mm, abdomen 33-42 mm,

黄翅绿色蟌 雄，透翅型，广东 | 宋黎明 摄
Mnais tenuis, male, hyaline winged morph from Guangdong | Photo by Liming Song

黄翅绿色蟌 雄，橙翅型，广东｜莫善濂 摄
Mnais tenuis, male, orange winged morph from Guangdong｜Photo by Shanlian Mo

黄翅绿色蟌 雌，广东｜宋黎明 摄
Mnais tenuis, female from Guangdong｜Photo by Liming Song

hind wing 27-31 mm. [Habitat] Streams, ditches and seepages in the forest below 1500 m elevation. [Distribution] Zhejiang, Jiangxi, Fujian, Guangdong, Taiwan; Laos. [Flight Season] February to July.

绿色蟌属待定种1 *Mnais* sp. 1

绿色蟌属待定种1 雄，透翅型，云南（西双版纳）
Mnais sp. 1, male, hyaline winged morph from Yunnan (Xishuangbanna)

【形态特征】本种与安氏绿色蟌相似，但体态较细，橙翅型的翅端透明，与后者不同。【长度】体长 44~58 mm，腹长 34~48 mm，后翅 28~35 mm。【栖息环境】海拔 1500 m 以下的林荫小溪。【分布】云南（西双版纳）。【飞行期】3—5月。

绿色蟌属待定种1 雄，橙翅型，云南（西双版纳）
Mnais sp. 1, male, orange winged morph from Yunnan (Xishuangbanna)

绿色蟌属待定种1 雌，云南（西双版纳）
Mnais sp. 1, female from Yunnan (Xishuangbanna)

[Identification] Similar to *M. andersoni* but more slender, orange winged male with the wing tips hyaline, different from that morph of *M. andersoni*. [Measurements] Total length 44-58 mm, abdomen 34-48 mm, hind wing 28-35 mm. [Habitat] Shady streams in forest below 1500 m elevation. [Distribution] Yunnan (Xishuangbanna). [Flight Season] March to May.

绿色蟌属待定种2 *Mnais* sp. 2

【形态特征】本种与黄翅绿色蟌相似，但橙翅型雄性整个腹部覆盖白色粉霜。【长度】体长 47~54 mm，腹长 38~43 mm，后翅 34~36 mm。【栖息环境】海拔 1500 m 以下森林中的开阔溪流。【分布】北京、湖北、安徽。【飞行期】4—7月。

绿色蟌属待定种2 雄，橙翅型，湖北
Mnais sp. 2, male, orange winged morph from Hubei

绿色蟌属待定种2 雄，透翅型，湖北
Mnais sp. 2, male, hyaline winged morph from Hubei

[Identification] Similar to *M. tenuis* but orange winged male with whole abdomen whitish pruinosed. [Measurements] Total length 47-54 mm, abdomen 38-43 mm, hind wing 34-36 mm. [Habitat] Exposed streams in the forest below 1500 m elevation. [Distribution] Beijing, Hubei, Anhui. [Flight Season] April to July.

绿色鳑属待定种2 雄，橙翅型，北京 | 陈炜 摄
Mnais sp. 2, male, orange winged morph from Beijing | Photo by Wei Chen

绿色蟌属待定种3 *Mnais* sp. 3

【形态特征】本种与黄翅绿色蟌相似，但后胸后侧板的黄斑形状不同，橙翅型雄性整个腹部覆盖白色粉霜。【长度】未测量。【栖息环境】海拔 1500 m以下森林中的开阔溪流。【分布】重庆、贵州。【飞行期】4—6月。

[Identification] Similar to *M. tenuis* but yellow spot on metepimeron of different shape, orange winged male with whole abdomen whitish pruinosed. [Measurements] Not measured. [Habitat] Exposed streams in the forest below 1500 m elevation. [Distribution] Chongqing, Guizhou. [Flight Season] April to June.

绿色蟌属待定种3 雄，透翅型，重庆 | 张巍巍 摄
Mnais sp. 3, male, hyaline winged morph from Chongqing | Photo by Weiwei Zhang

绿色蟌属待定种3 雄，橙翅型，重庆 | 张巍巍 摄
Mnais sp. 3, male, orange winged morph from Chongqing | Photo by Weiwei Zhang

绿色蟌属待定种3 雌，重庆 | 张巍巍 摄
Mnais sp. 3, female from Chongqing | Photo by Weiwei Zhang

绿色蟌属待定种4 *Mnais* sp. 4

绿色蟌属待定种4 雄，橙翅型，云南（普洱）
Mnais sp. 4, male, orange winged morph from Yunnan (Pu'er)

【形态特征】橙翅型雄性胸部金属黑绿色，后胸后侧板黄色，翅橙色；腹部黑褐色，第8~10节覆盖白色粉霜。【长度】未测量。【栖息环境】海拔 1200 m森林中的小溪。【分布】云南（普洱）。【飞行期】4—6月。

[Identification] The orange winged male with thorax metallic dark green, metepimeron entirely yellow, orange wings. Abdomen blackish brown, S8-S10 with white pruinescence. [Measurements] Not measured. [Habitat] Streams in the forest at 1200 m elevation. [Distribution] Yunnan (Pu'er). [Flight Season] April to June.

绿色蟌属待定种5 *Mnais* sp. 5

【形态特征】本种与黄翅绿色蟌相似，但橙翅型雄性后胸后侧板具楔状金属斑，透翅型雄性前胸和腹部第1~3节具白色粉霜。【长度】雄性体长 53~54 mm，腹长 43~44 mm，后翅 33~34 mm。【栖息环境】1000 m以下森林中的溪流。【分布】广西。【飞行期】4—6月。

[Identification] Similar to *M. tenuis* but orange winged male with metallic wedge-shaped marking on metepimeron, hyaline winged male with prothorax and S1-S3 whitish pruinosed. [Measurements] Male total

绿色蟌属待定种5 雌，广西
Mnais sp. 5, female from Guangxi

length 53-54 mm, abdomen 43-44 mm, hind wing 33-34 mm. **[Habitat]** Streams in the forest below 1000 m elevation. **[Distribution]** Guangxi. **[Flight Season]** April to June.

绿色螅属待定种5 雄，橙翅型，广西
Mnais sp. 5, male, orange winged morph from Guangxi

绿色螅属待定种5 雄，透翅型，广西
Mnais sp. 5, male, hyaline winged morph from Guangxi

艳色螅属 Genus *Neurobasis* Selys, 1853

本属豆娘主要分布于亚洲的热带地区和新几内亚。全世界已知的13种中有2种在中国分布。雄性通体墨绿色具金属光泽，前翅通常透明（仅有1种例外），有时基方具金属绿色斑，后翅正面具有大面积的金属绿色或者蓝色。雌性体色暗淡，具白色的伪翅痣。翅脉密集，基室具横脉。

本属豆娘栖息于开阔溪流和河流。雄性通常停立在水面的岩石或者挺水植物上。雌性潜水产卵。卵产在溪流中的朽木和水生植物的茎干中。

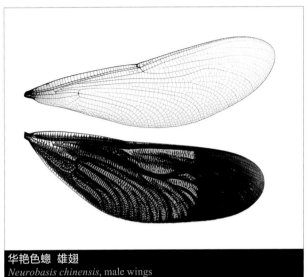

华艳色螅 雄翅
Neurobasis chinensis, male wings

Species of the genus are mainly distributed in tropical Asia and New Guinea. Of the 13 known species two occur in China. The male body metallic dark green, fore wings either entirely hyaline or (in one species) metallic green at base, upperside of hind wings largely metallic green or blue. Female with paler abdomen, wings with white pseudopterostigma. Wing venation dense, median space with crossveins.

Neurobasis species inhabit exposed streams or rivers. Males maintain territories along stretches of water, perching on rocks or emergent plants. Females often oviposit underwater. Eggs are inserted into the deadwood or the stalks of aquatic plants.

华艳色螅 雄 | 宋睿斌 摄
Neurobasis chinensis, male | Photo by Ruibin Song

安氏艳色螅 *Neurobasis anderssoni* Sjöstedt, 1926

【形态特征】一种较粗壮的豆娘，归属于艳色螅属的亚属——中华艳色螅亚属。雄性面部、胸部和腹部青铜色具金属光泽；前翅大面积透明，基方至翅结处半透明并具金属绿色条纹，后翅大面积金属绿色，仅端部透明，背面具青铜色金属光泽。雌性面部和胸部青铜色具金属光泽，前翅基方1/2琥珀色，端部1/2透明，后翅琥珀色，具白色的伪翅痣；腹部基方2节绿色，其余各节褐色，末端具黄色条纹。【长度】腹长 42～56 mm，后翅 30～42 mm。【栖息环境】海拔 500～1000 m的山区溪流。【分布】中国特有，分布于四川、浙江、福建、广西。【飞行期】3—8月。

[Identification] A robust species, placed in a separate subgenus *Sinobasis* Hämäläinen & Orr, 2007. Male with face, thorax and abdomen metallic coppery green. Fore wings largely hyaline, but basal area up to nodus semihyaline with metallic green areas, hind wings largely metallic green, only the tips hyaline, the underside of hind wings with metallic coppery reflections. Female face and thorax metallic coppery green, fore wings with the basal half amber and apical half hyaline, hind wings amber, both wings with narrow white pseudopterostigma. S1-S2 green with rest of abdominal segments brown, the tip with yellow stripes. [Measurements] Abdomen 42-56 mm, hind wing 30-42 mm. [Habitat] Montane streams at 500-1000 m elevation. [Distribution] Endemic to China, recorded from Sichuan, Zhejiang, Fujian, Guangxi. [Flight Season] March to August.

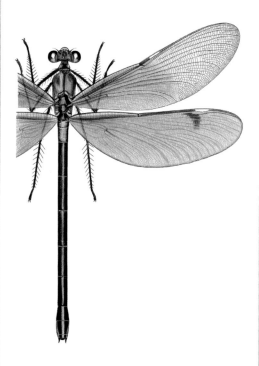

安氏艳色螅 雄 | Albert G. Orr 绘
Neurobasis anderssoni, male | Draw by Albert G. Orr

安氏艳色螅 雌 | Albert G. Orr 绘
Neurobasis anderssoni, female | Draw by Albert G. Orr

华艳色螅 *Neurobasis chinensis* (Linnaeus, 1758)

【形态特征】雄性面部、胸部和腹部铜绿色具金属光泽；前翅透明，后翅正面大面积金属绿色，端部黑色，背面深铜色。雌性身体金属绿色具黄色条纹；前翅透明，后翅琥珀色，翅结处具小白斑，具白色的伪翅痣。【长度】体长 56~60 mm, 腹长 45~48 mm, 后翅 32~35 mm。
【栖息环境】海拔 1500 m以下的溪流和河流。【分布】
贵州、云南、江西、福建、广西、广东、海南、香港、台
湾; 南亚、东南亚。【飞行期】全年可见。

华艳色蟌 雌,广东 | 宋睿斌 摄
Neurobasis chinensis, female from Guangdong | Photo by
Ruibin Song

[Identification] Male face, thorax and abdomen
metallic coppery green. Fore wings hyaline, upper
side of hind wings largely metallic green, with broad
black area at apex, underside of hind wings with dark
coppery reflections. Female body metallic green with
yellow stripes. Fore wings hyaline and hind wings
amber, both wings with white nodal spots and white
pseudopterostigma. [Measurements] Total length
56-60 mm, abdomen 45-48 mm, hind wing 32-35 mm.
[Habitat] Streams and rivers below 1500 m elevation. [Distribution] Guizhou, Yunnan, Jiangxi, Fujian, Guangxi,
Guangdong, Hainan, Hong Kong, Taiwan; South and Southeast Asia. [Flight Season] Throughout the year.

华艳色蟌 雄,云南(西双版纳)
Neurobasis chinensis, male from Yunnan (Xishuangbanna)

华艳色蟌 雌, 云南 (西双版纳)
Neurobasis chinensis, female from Yunnan (Xishuangbanna)

华艳色蟌 雄, 云南 (西双版纳)
Neurobasis chinensis, male from Yunnan (Xishuangbanna)

华艳色蟌 交尾, 云南 (西双版纳)
Neurobasis chinensis, mating pair from Yunnan (Xishuangbanna)

爱色螅属 Genus *Noguchiphaea* Asahina, 1976

本属全世界已知仅2种，分布于东洋界。中国已知1种，仅记录于云南的热带地区。本属豆娘体中型，身体墨绿色具金属光泽，后胸大面积黄色；翅窄而透明，翅结位于基方1/3处，基室无横脉，无翅痣；腹部十分细长或极长。

本属豆娘栖息于森林中的狭窄小溪。它们经常隐秘在丛林深处。雄性喜欢停立在溪流边缘的叶片上。

So far only two species of this Oriental genus have been described. One of which occurs in the tropical part of Yunnan. Species of the genus are medium-sized damselflies, body metallic dark green, metathorax largely yellow. Wings narrow and hyaline, nodus located at basal one third, median space without crossveins, pterostigma absent. Abdomen proportionally long or very long.

Noguchiphaea species inhabit small streamlets in forested mountains. They usually hide in the dense forest. Males prefer to perch on the leaves nearby water.

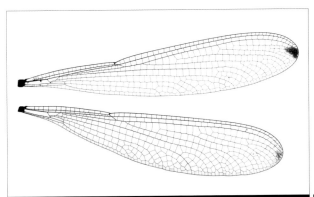

美子爱色螅
Noguchiphaea yoshikoae

美子爱色螅 雄翅
Noguchiphaea yoshikoae, male wings

爱色螅属 雄性肛附器
Genus *Noguchiphaea*, male anal appendages

美子爱色螅 雄
Noguchiphaea yoshikoae, male

美子爱色蟌 *Noguchiphaea yoshikoae* Asahina, 1976

【形态特征】雄性面部、胸部和腹部墨绿色具金属光泽，后胸大面积黄色；翅窄而透明，翅端具小褐斑；腹部第8~10节覆盖白色粉霜。雌性与雄性相似但腹部较短，翅端无褐斑。【长度】体长 49~54 mm，腹长 39~45 mm，后翅 32~34 mm。【栖息环境】海拔 1000~1500 m的林荫小溪。【分布】云南（西双版纳、德宏、红河）；泰国、越南。【飞行期】9—12月。

[Identification] Male face, thorax and abdomen metallic dark green, metathorax largely yellow. Wings narrow and hyaline with small brown spots apically. S8-S10 white pruinosed on dorsum. Female similar to male but the abdomen shorter, wings without brown spots apically. [Measurements] Total length 49-54 mm, abdomen 39-45 mm, hind wing 32-34 mm. [Habitat] Montane shady streams at 1000-1500 m elevation. [Distribution] Yunnan (Xishuangbanna, Dehong, Honghe); Thailand, Vietnam. [Flight Season] September to December.

美子爱色蟌 雌，云南（德宏）
Noguchiphaea yoshikoae, female from Yunnan (Dehong)

美子爱色蟌 雄，云南（红河）
Noguchiphaea yoshikoae, male from Yunnan (Honghe)

褐顶色蟌属 Genus *Psolodesmus* McLachlan, 1870

　　本属为中国台湾地区和日本琉球群岛特有，全世界已知2种。其中有1种分布于中国台湾，但被分成2个亚种，然而它们真正的分类学关系仍未明确。本属是体型较大的豆娘，身体墨绿色具金属光泽；翅部分染有褐色，翅端深褐色，有时雄性翅的亚端方具1个白色宽条纹；翅脉十分密集，基室具横脉，两性均具翅痣。

　　本属豆娘主要栖息于山区溪流。

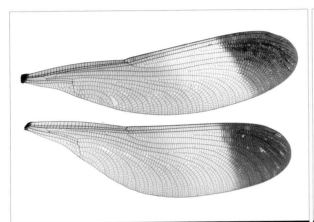

褐顶色蟌指名亚种　雄翅
Psolodesmus mandarinus mandarinus, male wings

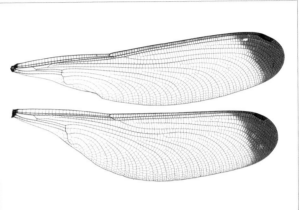

褐顶色蟌南台亚种　雄翅
Psolodesmus mandarinus dorothea, male wings

　　The genus is endemic to Taiwan of China and the Japanese Ryukyu Archipelago, presently two species are recognized. One species is recorded from Taiwan of China but has been divided into two subspecies, although their mutual taxonomic status is somewhat unclear. *Psolodesmus* are moderately large-sized damselflies with metallic dark green body. The wings are partly brownish tinted, wing tips dark brown, in males with or without a broad whitish band in front of the opaque apex. Venation dense, median space with crossveins, pterostigma present in both sexes.

　　Psolodesmus species mainly inhabit montane streams.

褐顶色蟌南台亚种 *Psolodesmus mandarinus dorothea* Williamson, 1904

　　【形态特征】雄性面部、胸部和腹部墨绿色具金属光泽；翅大面积透明并闪烁蓝紫色光泽，翅前缘褐色，翅端具深褐色斑，翅痣深褐色。雌性与雄性相似，翅痣白色。【长度】体长 47~66 mm，腹长 38~54 mm，后翅 34~44 mm。【栖息环境】海拔 1500 m以下的林荫小溪。【分布】中国台湾特有。【飞行期】3—12月。

　　[Identification] Male face, thorax and abdomen metallic dark green. Wings largely hyaline with bluish purple sheen, costal areas narrowly brownish and wing tips narrowly dark brown, pterostigma dark brown. Female similar to male but pterostigma white. [Measurements] Total length 47-66 mm, abdomen 38-54 mm, hind wing 34-

褐顶色蟌南台亚种　交尾，台湾｜嘎嘎 摄
Psolodesmus mandarinus dorothea, mating pair from Taiwan | Photo by Gaga

44 mm. [Habitat] Shady streams in the forest below 1500 m elevation. [Distribution] Endemic to Taiwan of China. [Flight Season] March to December.

褐顶色蟌南台亚种 雄,台湾 | 嘎嘎 摄
Psolodesmus mandarinus dorothea, male from Taiwan | Photo by Gaga

褐顶色蟌南台亚种 雌,台湾 | 嘎嘎 摄
Psolodesmus mandarinus dorothea, female from Taiwan | Photo by Gaga

褐顶色蟌指名亚种 *Psolodesmus mandarinus mandarinus* McLachlan, 1870

褐顶色蟌指名亚种 交尾,台湾 | 嘎嘎 摄
Psolodesmus mandarinus mandarinus, mating pair from Taiwan | Photo by Gaga

【形态特征】雄性面部和胸部墨绿色具金属光泽；翅具蓝紫色光泽，基方3/5浅褐色，翅端深褐色，亚端部具1个白带，翅痣黑褐色；老熟个体翅的背面端部覆盖白色粉霜；腹部深褐色。雌性与雄性相似但翅痣白色。【长度】体长59～66 mm，腹长 47～55 mm，后翅 43～44 mm。【栖息环境】海拔 1500 m以下的林荫小溪。【分布】中国台湾特有。【飞行期】3—12月。

[Identification] Male face, thorax metallic dark green. Wing surface with bluish purple sheen, the basal three fifths pale brownish, wing tip broadly dark brown followed by a sub-apical whitish transverse band, pterostigma blackish brown. In old males the underside of wing tips develops white pruinescence. Abdomen dark brown. Female similar to male but with white pterostigma. [Measurements] Total length 59-66 mm, abdomen 47-55 mm, hind wing 43-44 mm. [Habitat] Shady streams in the forested areas below 1500 m elevation. [Distribution] Endemic to Taiwan of China. [Flight Season] March to December.

褐顶色螅指名亚种 雌,台湾 | 嘎嘎 摄
Psolodesmus mandarinus mandarinus, female from Taiwan | Photo by Gaga

黄细色蟌属 Genus *Vestalaria* May, 1935

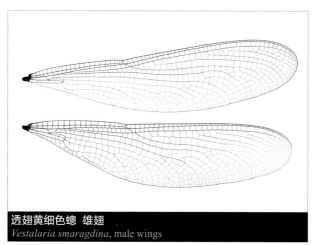

透翅黄细色蟌 雄翅
Vestalaria smaragdina, male wings

本属全世界已知5种，仅分布于东南亚和东亚地区。中国已知4种，分布于华中、华南和西南地区。本属是一类中型豆娘，身体墨绿色具金属光泽，后胸大面积黄色；翅较宽阔，多数种类翅透明，仅有1种稍染褐色；基室无横脉，无翅痣。本属豆娘的雄性可通过肛附器的形状和翅的色彩区分。

本属豆娘栖息于森林中的溪流和沟渠。雄性经常停立在溪流中的挺水植物上。雌性栖息于溪流附近的森林中，产卵时才会接近水面，它们可以半身潜水产卵。卵产在溪流中的朽木和水生植物的茎干中。本属多为晚季节发生，秋季至冬季较常见。在云南南部，它们的飞行期从8月开始，可持续至次年4月。

This genus contains five species, confined to Southeast and East Asia. Four species are recorded from China, found in the Central, South and Southwest regions. Species of the genus are medium-sized damselflies with metallic dark green body and largely yellow metathorax. Wings are moderately broad, hyaline in most species, but one species has wings light brownish throughout. Median space without crossveins, pterostigma absent. Males of different species can be distinguished by the shape of male appendages and color of wings.

Vestalaria species inhabit streams or ditches in forested areas. Males typically perch on plants near streams. Females usually rest in nearby forest and visit water for oviposition, they often oviposit partly submerged. Eggs are inserted into

透翅黄细色蟌 交尾
Vestalaria smaragdina, mating pair

the deadwood or the stalks of aquatic plants. *Vestalaria* species are late season species, and common from autumn to winter. In the south of Yunnan, the flight period starts from August and ends as late as April of the next year.

| 苗黄细色螅 | 透翅黄细色螅 | 褐翅黄细色螅 | 黑角黄细色螅 |
| *Vestalaria miao* | *Vestalaria smaragdina* | *Vestalaria velata* | *Vestalaria venusta* |

黄细色螅属 雄性肛附器
Genus *Vestalaria*, male anal appendages

苗黄细色螅 *Vestalaria miao* (Wilson & Reels, 2001)

【形态特征】雄性面部、胸部和腹部墨绿色具金属光泽，后胸后侧板黄色，翅透明，翅端稍染褐色；腹部第8～10节覆盖白色粉霜，下肛附器甚短。雌性与雄性相似，翅稍染琥珀色。【长度】腹长38～50 mm，后翅36～48 mm。【栖息环境】海拔1000 m以下的山区溪流。【分布】广西、海南、广东；泰国、老挝、越南。【飞行期】4月—次年1月。

[Identification] Male face, thorax and abdomen metallic dark green, metepimeron yellow, wings hyaline with tips narrowly tinted with brown. S8-S10 whitish pruinosed, the inferior appendages very short. Female similar to male, wings slightly tinted with amber. [Measurements] Abdomen 38-50 mm, hind wing 36-48 mm. [Habitat] Montane streams below 1000 m elevation. [Distribution] Guangxi, Hainan, Guangdong; Thailand, Laos, Vietnam. [Flight Season] April to the following January.

 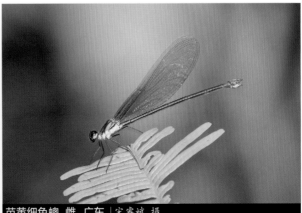

苗黄细色螅 雄，广东 | 宋黎明 摄
Vestalaria miao, male from Guangdong | Photo by Liming Song

苗黄细色螅 雌，广东 | 宋睿斌 摄
Vestalaria miao, female from Guangdong | Photo by Ruibin Song

透翅黄细色蟌 *Vestalaria smaragdina* (Selys, 1879)

透翅黄细色蟌 雌,云南(大理)
Vestalaria smaragdina, female from Yunnan (Dali)

【形态特征】雄性面部、胸部和腹部墨绿色具金属光泽,后胸大面积黄色;翅透明;腹部第8~10节覆盖白色粉霜,上肛附器长,明显长于第10节。雌性与雄性相似,完全成熟后翅稍染褐色,未熟时翅金褐色;产卵瓣具1列尖刺。【长度】体长 53~59 mm,腹长 43~49 mm,后翅 35~39 mm。【栖息环境】海拔 2000 m 以下的开阔溪流和沟渠。【分布】西藏、云南;印度、缅甸、泰国、老挝、越南。【飞行期】8月—次年4月。

[Identification] Male face, thorax and abdomen metallic dark green, metathorax largely yellow. Wings hyaline. S8-S10 whitish pruinosed, superior appendages distinctly longer than S10. Female similar to male, in mature individuals wings slightly brownish tinted, in immatures wings golden brown. The valves of ovipositor bearing a row of sharp spines. [Measurements] Total length 53-59 mm, abdomen 43-49 mm, hind wing 35-39 mm. [Habitat] Exposed streams and ditches below 2000 m elevation. [Distribution] Tibet, Yunnan; India, Myanmar, Thailand, Laos, Vietnam. [Flight Season] August to the following April.

透翅黄细色蟌 雄,云南(德宏)
Vestalaria smaragdina, male from Yunnan (Dehong)

褐翅黄细色蟌 *Vestalaria velata* (Ris, 1912)

褐翅黄细色蟌 雌，广东
Vestalaria velata, female from Guangdong

褐翅黄细色蟌 雄，广东｜吴宏道 摄
Vestalaria velata, male from Guangdong | Photo by Hongdao Wu

褐翅黄细色蟌 雄，广东
Vestalaria velata, male from Guangdong

　　【形态特征】雄性面部、胸部和腹部墨绿色具金属光泽，后胸后侧板黄色；翅褐色，末梢色彩稍加深；腹部第8~10节覆盖白色粉霜，上肛附器长，约为第10节长度的2倍。雌性与雄性相似，翅色彩更深，腹部第9~10节稍带粉霜。【长度】体长 60~65 mm，腹长 45~54 mm，后翅 37~42 mm。【栖息环境】海拔 1000 m 以下的开阔溪流和沟渠。【分布】中国特有，分布于安徽、浙江、福建、广东。【飞行期】7—12月。

　　[Identification] Male face, thorax and abdomen metallic dark green, metepimeron yellow. Wings brown, darkened at the extreme tip. S8-S10 whitish pruinosed, the superior appendages long, about twice as long as S10. Female similar to male, wings slightly darker, S9-S10 slightly pruinosed. **[Measurements]** Total length 60-65 mm, abdomen 45-54 mm, hind wing 37-42 mm. **[Habitat]** Exposed streams and ditches below 1000 m elevation. **[Distribution]** Endemic to China, recorded from Anhui, Zhejiang, Fujian, Guangdong. **[Flight Season]** July to December.

黑角黄细色蟌 *Vestalaria venusta* (Hämäläinen, 2004)

黑角黄细色蟌 雄，重庆
Vestalaria venusta, male from Chongqing

【形态特征】雄性面部、胸部和腹部墨绿色具金属光泽，后胸后侧板黄色，老熟后覆盖蓝白色粉霜；翅透明，翅端稍染褐色；腹部第8~10节覆盖白色粉霜，上肛附器长于第10节。雌性与雄性相似，腹部第9~10节稍微覆盖粉霜。【长度】体长 56~63 mm，腹长 45~51 mm，后翅 37~41 mm。【栖息环境】海拔 1500 m以下的开阔溪流和沟渠。【分布】中国特有，分布于四川、重庆、贵州、安徽、湖北、江西、浙江、福建、广西、广东。【飞行期】7—12月。

黑角黄细色蟌 雌，贵州
Vestalaria venusta, female from Guizhou

[Identification] Male face, thorax and abdomen metallic dark green, metepimeron yellow, gradually pruinosed bluish white with age. Wings hyaline with tips narrowly tinted with brown. S8-S10 whitish pruinosed, superior appendages longer than S10. Female similar to male, S9-S10 slightly pruinosed. [Measurements] Total length 56-63 mm, abdomen 45-51 mm, hind wing 37-41 mm. [Habitat] Exposed streams and ditches below 1500 m elevation. [Distribution] Endemic to China, recorded from Sichuan, Chongqing, Guizhou, Anhui, Hubei, Jiangxi, Zhejiang, Fujian, Guangxi, Guangdong. [Flight Season] July to December.

细色蟌属 Genus *Vestalis* Selys, 1853

本属全球已知16种，主要分布于亚洲的热带地区，尤其在婆罗洲地区种类繁多。在中国仅有1种，多横细色蟌，分布于云南。本属豆娘和黄细色蟌属在身体色彩和翅脉上很相似，但是雄性阳茎的构造十分不同。本属多数种类胸部没有黄色条纹，但中国分布的细色蟌例外，后胸后侧板绿色的楔形金属色斑很容易与黄细色蟌属的种类区分。

本属豆娘以溪流生境为主。多横细色蟌通常栖息于溪流周边的阴暗环境，它们在黄昏时较活跃，雄性会停落到水边的水草上等待雌性。雌性将卵产在溪流中的朽木和水生植物的茎干中。

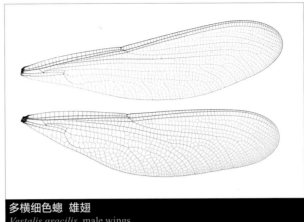

多横细色蟌 雄翅
Vestalis gracilis, male wings

This genus contains 16 species distributed in tropical Asia and is especially diverse in Borneo. Only one species, *V. gracilis*, occurs in China, found in Yunnan. Species of the genus *Vestalis* resemble superficially those of the genus *Vestalaria* in their general color pattern and wing venation, but they have a very different penis structure. Most *Vestalis* species lack yellow markings on the thorax, but the only Chinese species has extensive yellow stripes on the metathorax, however, the metallic green wedge-shaped marking on metepimeron provides an easy distinguishing character from *Vestalaria*.

Vestalis are inhabitants of streams in forested areas. *V. gracilis* usually perch in the shade during the day, they are active after sunset, when males hold territories, perching on marginal and emergent plants as they wait for females. Females insert their eggs into the deadwood or the stalks of aquatic plants.

多横细色蟌 雄
Vestalis gracilis, male

多横细色螅 *Vestalis gracilis* (Rambur, 1842)

【形态特征】雄性面部、胸部和腹部绿色具金属光泽,合胸侧面具较细的黄色条纹;翅透明稍染褐色,末端染有褐色;翅表面闪烁蓝紫色光泽。雌性与雄性相似。【长度】体长 58~66 mm,腹长 48~54 mm,后翅 36~41 mm。【栖息环境】海拔 1000 m以下的林荫小溪。【分布】云南(红河、西双版纳、德宏、临沧);不丹、印度、尼泊尔、缅甸、泰国、老挝、柬埔寨、越南、马来半岛。【飞行期】全年可见。

多横细色螅 雄,云南(德宏)
Vestalis gracilis, male from Yunnan (Dehong)

[Identification] Male face, thorax and abdomen metallic green, sides of synthorax with narrow yellow stripes, Wings hyaline with slight brownish tint, tips of both wings more distinctly brownish tinted. Wing surface with scintillating bluish purple reflections. Female similar to male. [Measurements] Total length 58-66 mm, abdomen 48-54 mm, hind wing 36-41 mm. [Habitat] Shady streams below 1000 m elevation. [Distribution] Yunnan (Honghe, Xishuangbanna, Dehong, Lincang); Bhutan, India, Nepal, Myanmar, Thailand, Laos, Cambodia, Vietnam, Peninsular Malaysia. [Flight Season] Throughout the year.

多横细色蟌 雌，云南（德宏）
Vestalis gracilis, female from Yunnan (Dehong)

3 鼻螅科 Family Chlorocyphidae

鼻螅科是一类较粗壮且面部构造特殊的豆娘。由于其唇基十分突出，所以其面部具有1个较显著的鼻状构造。本科全球已知20属150余种，主要分布于亚洲和非洲的热带地区。中国已知6属20余种。本科豆娘体型小至中型，腹部较短，很多种类翅上具深色斑纹和透明的窗型斑，这些翅窗可以反射出绿色、蓝色、玫瑰红色和铜色等霓虹光泽。

本科豆娘栖息于溪流和河流。雄性经常停落在水边占据领地。许多种类具有蜻蜓中罕见的求偶行为。雄性的足通常具白色胫节，在求偶时向雌性展示。雌性体色灰暗，翅透明。

Chlorocyphids are stout damselflies with an unusual facial structure. The clypeus is protruding, giving the face a snout like appearance. The family contains over 150 species in 20 genera worldwide, mostly distributed in tropical Asia and Africa. Over 20 species in six genera are recorded from China. They are small to medium sized species with a relatively short abdomen, and many species have dark markings and transparent windows on the wings, which may reflect scintillating iridescent greens, blues, magenta and copper.

Species of the family inhabit streams and rivers. Males usually perch near water where they defend territories. Courtship occurs in most species, unusual among odonates. The males often possess white tibiae and display them to the female during courtship. Females are drably colored with hyaline wings.

圣鼻蟌属 Genus *Aristocypha* Laidlaw, 1950

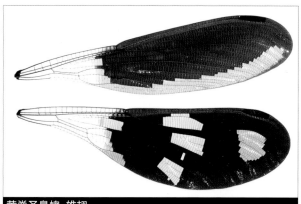

黄脊圣鼻蟌 雄翅
Aristocypha fenestrella, male wings

本属仅分布于亚洲的热带和亚热带地区，全世界已知12种，其中有7种在中国分布。这是一类小型豆娘，在分布于中国的种类中，所有雄性翅具大面积深色斑，后翅具透明的彩虹翅窗；雌性翅透明，翅痣双色，通常为深褐色和白色。

本属豆娘主要栖息于山区溪流。雄性会停栖在光照充足的岩石上或者靠近水面的植物上护卫领地。求偶时会向雌性展示白色的胫节和色彩缤纷的翅。雄性间经常展开争斗，它们面对面飞行，互相炫耀翅，但不展示足的胫节。除了本书所述种类，本属有一个稀有种，簾格圣鼻蟌，仅在中国台湾中部有2个分布点的记录。

The genus is confined to tropical and subtropical Asia, of the 12 known species, seven have definitely been found in China. They are small-sized damselflies. Males of all Chinese species have extensive dark wing markings and transparent, iridescent windows on the hind wing. Female wings hyaline, pterostigma bicolored, usually dark brown and white.

Aristocypha species inhabit montane streams. Males usually perch on sunlit stones or low vegetation guarding territories. Both white tibiae and wing ornamentation are displayed in courtship. Males often contest territory by flying face to face, displaying wing markings but not legs. In addition to species presented here there is a very rare species *A. baibarana* (Matsumura, 1931) known only from two sites in central Taiwan of China.

黄脊圣鼻蟌 雄
Aristocypha fenestrella, male

蓝脊圣鼻螅 *Aristocypha aino* Hämäläinen, Reels & Zhang, 2009

蓝脊圣鼻螅 雄，海南
Aristocypha aino, male from Hainan

【形态特征】雄性面部黑色，头顶和后头各具1对小黄斑；胸部黑色，侧面具黄色条纹，合胸脊具三角形蓝色斑；前翅末端2/3黑色具光泽，后翅端方2/3黑色具光泽，具透明且闪烁蓝紫色光泽的翅窗，前翅翅痣黑色，后翅翅痣蓝色，足的胫节内缘具白色粉霜；腹部黑色。雌性主要黑色具黄色斑纹，翅透明，翅痣黑褐色和白色。【长度】体长29~32 mm，腹长 20~22 mm，后翅 26~30 mm。【栖息环境】海拔 1500 m以下的开阔溪流。【分布】中国海南特有。【飞行期】3—11月。

[Identification] Male face black, vertex and occiput with a pair of small yellow spots. Thorax black with yellow lateral markings, mesothoracic dorsal triangle blue. In fore wings black, iridescent area restricted to the anterior distal two thirds, hind wings with distal two thirds black showing iridescence, with transparent windows shining bluish purple, pterostigma in fore wings black and in hind wings blue, legs with inner surface of tibiae white. Abdomen black. Female largely black with yellow markings, wings hyaline, pterostigma blackish brown and white. [Measurements] Total length 29-32 mm, abdomen 20-22 mm, hind wing 26-30 mm. [Habitat] Exposed streams below 1500 m elevation. [Distribution] Endemic to Hainan of China. [Flight Season] March to November.

蓝脊圣鼻螅 雌，海南｜莫善濂 摄
Aristocypha aino, female from Hainan | Photo by Shanlian Mo

赵氏圣鼻蟌 *Aristocypha chaoi* (Wilson, 2004)

【形态特征】雄性面部黑色，头顶和后头各具1对小蓝斑；胸部黑色具天蓝色条纹，合胸脊具三角形天蓝色斑；前翅透明，翅痣黑色，后翅基方2/3透明，端方1/3黑色并具半透明且闪烁蓝绿色光泽的翅窗，翅痣蓝色；足的胫节内缘具白色粉霜；腹部第2~9节背面大面积蓝色。雌性黑色具黄色斑纹，翅痣褐色和白色。【长度】体长 28~30 mm，腹长 18~20 mm，后翅 23~25 mm。【栖息环境】海拔 1000 m以下的开阔溪流。【分布】贵州、江西、福建、广西、广东；越南。【飞行期】5—9月。

[Identification] Male face black, vertex and occiput with a pair of blue spots. Thorax black with sky-blue stripes, mesothoracic dorsal triangle sky-blue. Fore wings hyaline with black pterostigma, hind wings with basal two thirds hyaline, apical one third black showing iridescence and transparent windows shining bluish green, pterostigma blue. Legs with inner margin of tibiae white. Dorsum of S2-S9 largely blue. Female largely black with yellow markings, pterostigma dark brown and white. [Measurements] Total length 28-30 mm, abdomen 18-20 mm, hind wing 23-25 mm. [Habitat] Exposed streams below 1000 m elevation. [Distribution] Guizhou, Jiangxi, Fujian, Guangxi, Guangdong; Vietnam. [Flight Season] May to September.

赵氏圣鼻蟌 雄，广东
Aristocypha chaoi, male from Guangdong

赵氏圣鼻蟌 雄，广西
Aristocypha chaoi, male from Guangxi

赵氏圣鼻蟌 雌，广东
Aristocypha chaoi, female from Guangdong

西藏圣鼻蟌 *Aristocypha cuneata* (Selys, 1853)

【形态特征】雄性面部黑色，头顶具1对小蓝斑；胸部黑色，侧面具黄色条纹，合胸脊具三角形淡蓝色斑；翅基方1/3透明，端方2/3黑色并具半透明且闪烁蓝紫色光泽的长方形翅窗，最末列的翅窗距翅前缘仅有1~2列翅室，前翅翅痣黑色，后翅翅痣具蓝色条纹；足的胫节内缘具白色粉霜；腹部各节间具甚细的黄色条纹。雌性黑色具更丰富的黄色斑纹。本种与蓝脊圣鼻蟌相似，但后翅最末端的翅窗位于翅痣以内，而蓝脊圣鼻蟌最后1列翅窗位于翅痣下方。【长度】体长 30~33 mm，腹长 19~23 mm，后翅 25~27 mm。【栖息环境】海拔2000 m以下的开阔溪流。【分布】西藏；孟加拉国、印度、尼泊尔、不丹。【飞行期】7—10月。

西藏圣鼻蟌 雄,西藏 | 吴超 摄
Aristocypha cuneata, male from Tibet | Photo by Chao Wu

[Identification] Male face black, vertex with a pair of blue spots. Thorax black with yellow lateral markings, mesothoracic dorsal triangle pale blue. Wings with basal one third hyaline, distal two thirds black showing iridescence with transparent windows shining bluish purple, apical window separated from wing leading margin by only 1-2 cells rows, pterostigma in fore wings black and in hind wings blue. Legs with inner margin of tibiae white. Abdomen black with narrow yellow rings between segments. Female black with more yellow markings. Similar to *A. aino*, but the apical window in hind wing is prior to pterostigma whereas in *A. aino* the window is just below pterostigma. [Measurements] Total length 30-33 mm, abdomen 19-23 mm, hind wing 25-27 mm. [Habitat] Exposed streams below 2000 m elevation. [Distribution] Tibet; Bangladesh, India, Nepal, Bhutan. [Flight Season] July to October.

西藏圣鼻蟌 雄,西藏 | 计云 摄
Aristocypha cuneata, male from Tibet | Photo by Yun Ji

黄脊圣鼻螅 *Aristocypha fenestrella* (Rambur, 1842)

黄脊圣鼻螅 雄，云南（西双版纳）
Aristocypha fenestrella, male from Yunnan (Xishuangbanna)

黄脊圣鼻螅 雌，云南（西双版纳）
Aristocypha fenestrella, female from Yunnan (Xishuangbanna)

【形态特征】雄性面部黑色，头顶和后头各具1对小黄斑；胸部黑色，侧面具黄色条纹，合胸脊具三角形紫色斑；翅基方1/3透明，端方2/3黑色具光泽，前翅黑色区域未到达翅的后缘，后翅具蓝紫色光泽的翅窗，最末列的翅窗距翅前缘有多列翅室，前翅翅痣黑色，后翅翅痣紫色；足的胫节内缘具白色粉霜；腹部黑色。雌性主要黑色具黄色斑纹，翅痣深褐色和白色。【长度】体长 29~34 mm，腹长 20~23 mm，后翅 23~33 mm。【栖息环境】海拔 2000 m以下的开阔溪流。【分布】云南、贵州、广西；缅甸、泰国、柬埔寨、老挝、越南、马来半岛。【飞行期】全年可见。

黄脊圣鼻螅 雄，云南（西双版纳）
Aristocypha fenestrella, male from Yunnan (Xishuangbanna)

[Identification] Male face black, vertex and occiput with a pair of small yellow spots. Thorax black with yellow lateral markings, mesothoracic dorsal triangle violet. Wings with basal one third hyaline, apical two thirds black showing iridescence, in fore wing the black area not extending to the lower border, hind wing with transparent windows shining purple blue, apical window separated from wing leading margin by several cells rows, pterostigma in fore wings black and in hind wings violet. Legs with inner margin of tibiae whitish. Abdomen black. Female black with yellow markings, pterostigma dark brown and white. [Measurements] Total length 29-34 mm, abdomen 20-23 mm, hind wing 23-33 mm. [Habitat] Exposed streams below 2000 m elevation. [Distribution] Yunnan, Guizhou, Guangxi; Myanmar, Thailand, Cambodia, Laos, Vietnam, Peninsular Malaysia. [Flight Season] Throughout the year.

蓝纹圣鼻螅 *Aristocypha iridea* (Selys, 1891)

【形态特征】雄性面部黑色，头顶和后头各具1对黄色斑点；胸部黑色具淡蓝色条纹，合胸脊具紫色和蓝色条纹；前翅端方2/3彩虹绿色，后翅大面积黑色具彩虹光泽和透明且闪烁紫红色光泽的翅窗，最末端的翅窗月牙形，翅痣黑色；足的腿节和胫节内缘白色，腹部黑色具淡蓝色条纹。雌性黑褐色具丰富的黄色斑纹，翅痣深褐色和白色。【长度】体长 29~32 mm，腹长 20~22 mm，后翅 24~28 mm。【栖息环境】海拔 1500 m以下的开阔溪流和沟渠。【分布】云南（德宏）；缅甸、泰国、老挝。【飞行期】4—11月。

蓝纹圣鼻螅 雄，云南（德宏）
Aristocypha iridea, male from Yunnan (Dehong)

蓝纹圣鼻螅 雌，云南（德宏）
Aristocypha iridea, female from Yunnan (Dehong)

[Identification] Male face black, vertex and occiput with a pair of yellow spots. Thorax black with pale blue markings, mesothoracic stripes purple and blue. Fore wings with the apical two thirds iridescent green, hind wings largely black showing iridescence and with transparent windows shining magenta, the outermost window crescent shaped, pterostigma black. Legs with inner margin of tibiae and femora white. Abdomen black with pale blue markings. Female blackish brown with extensive yellow markings, pterostigma dark brown and white. [Measurements] Total length 29-32 mm, abdomen 20-22 mm, hind wing 24-28 mm. [Habitat] Exposed streams and ditches below 1500 m elevation. [Distribution] Yunnan (Dehong); Myanmar, Thailand, Laos. [Flight Season] April to November.

蓝纹圣鼻螅 雄，云南（德宏）
Aristocypha iridea, male from Yunnan (Dehong)

四斑圣鼻蟌 *Aristocypha quadrimaculata* (Selys, 1853)

四斑圣鼻蟌 雄，云南（德宏）
Aristocypha quadrimaculata, male from Yunnan (Dehong)

【形态特征】雄性面部黑色，头顶和后头各具1对甚小的黄斑；胸部黑色，侧面具黄色条纹，合胸脊具狭长的三角形紫色斑；翅基方1/3透明，端方2/3黑色具霓虹绿色，后翅具透明且闪烁蓝紫色光泽的翅窗，最末列翅窗与翅前缘之间有几列翅室，前翅翅痣黑色，后翅翅痣具淡蓝色；足的胫节内缘具白色粉霜；腹部黑色。雌性黑色具黄色条纹。本种与黄脊圣鼻螅相似，但本种体型较小，雄性合胸脊的紫色三角形斑较窄，翅痣淡蓝色且较短。【长度】体长 27~30 mm，腹长 18~21 mm，后翅 22~23 mm。【栖息环境】海拔 500 m以下森林中的溪流。【分布】云南（德宏）；孟加拉国、印度、尼泊尔、不丹、缅甸。【飞行期】全年可见。

[Identification] Male face black, vertex and occiput with a pair of small yellow spots. Thorax black with yellow lateral markings, mesothoracic dorsal triangle violet and narrow. Wings with basal one third hyaline, apical two thirds black showing green iridescence, hind wing with transparent windows shining purple blue, apical window separated from wing leading margin by several cells rows, pterostigma in fore wings black and in hind wings pale blue. Legs with inner margin of tibiae whitish pruinosed. Abdomen black. Female black with yellow markings. Similar to *A. fenestrella*, but differs in smaller size, male mesothoracic triangle narrower, pterostigma pale blue and shorter. [Measurements] Total length 27-30 mm, abdomen 18-21 mm, hind wing 22-23 mm. [Habitat] Streams in forest below 500 m elevation. [Distribution] Yunnan (Dehong); Bangladesh, India, Nepal, Bhutan, Myanmar. [Flight Season] Throughout the year.

四斑圣鼻螅 雌，产卵，云南（德宏）
Aristocypha quadrimaculata, female laying eggs from Yunnan (Dehong)

阳鼻蟌属 Genus *Heliocypha* Fraser, 1949

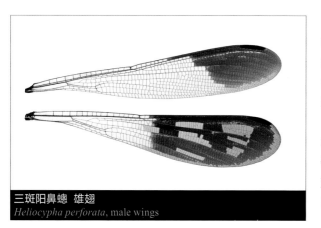

三斑阳鼻蟌 雄翅
Heliocypha perforata, male wings

本属全球已知约10种，分布于东洋界。中国已知2种，主要分布于华南和西南地区。本属豆娘体型较小；雄性后翅末端具黑色区域和翅窗，雌性翅透明，翅痣双色，通常为深褐色和白色。本属豆娘合胸背脊的三角形斑明显短于圣鼻蟌属的多数种类。

本属豆娘主要栖息于低海拔的溪流、沟渠和河流。雄性经常停栖在溪流中的漂浮物上，如朽木和水生植物。雄性求偶时会向雌性展示白色的胫节和色彩缤纷的翅。雄性经常为争夺领地而进行面对面或者并排飞行，以此展示翅和身体。雌性产卵于朽木或水草上。

The genus includes about ten Oriental species. Two species are recorded from China, mainly distributed in the South and Southwest regions. Species of the genus are small-sized. Male hind wings with apical black area and transparent windows, female wings hyaline, pterostigma bicolored, usually dark brown and white. Unlike most *Aristocypha* species the magenta colored dorsal mesothoracic triangle is short.

Species of the genus inhabit lowland streams, ditches and rivers. Males usually perch on emergent deadwood or plants in water. They show their white tibiae and hind wing ornaments to females during courtship. Males often contest territory by flying face to face or side by side displaying wing and body markings. Female oviposit in the deadwood or aquatic plants.

月斑阳鼻蟌 雄
Heliocypha biforata, male

月斑阳鼻蟌 *Heliocypha biforata* (Selys, 1859)

【形态特征】雄性头顶和后头各具1对黄色斑点；胸部黑色，合胸脊具1个三角形紫红色斑，脊两侧具1对紫红色斑点，侧面具淡蓝色条纹；前翅透明，仅末端褐色，后翅端方1/3褐色具闪烁粉红色光泽的翅窗，并在翅的透明区域具翅窗，翅痣黑褐色；足的胫节内缘具白色粉霜；腹部黑色具甚细小的淡蓝色条纹。雌性黑色具黄色细条纹。【长度】体长 25~29 mm，腹长 17~20 mm，后翅 19~23 mm。【栖息环境】海拔 1000 m以下的狭窄小溪和沟渠。【分布】云南、海南；印度、尼泊尔、缅甸、泰国、柬埔寨、老挝、越南、马来半岛。【飞行期】全年可见。

月斑阳鼻蟌 雄,云南(西双版纳)
Heliocypha biforata, male from Yunnan (Xishuangbanna)

月斑阳鼻蟌 雌,云南(西双版纳)
Heliocypha biforata, female from Yunnan (Xishuangbanna)

[Identification] Male vertex and occiput with a pair of yellow spots. Thorax black, dorsally a short magenta, mesothoracic triangle flanked by a pair of pale magenta markings, laterally pale blue markings. Fore wings hyaline with apex narrowly darkened, hind wings with the apical third brown with transparent windows reflecting pinkish magenta light, and a few iridescent windows also in the hyaline area of hind wing, pterostigma blackish brown. Legs with inner margin of tibiae white. Abdomen black with very small pale blue markings. Female black with fine yellow stripes. [Measurements] Total length 25-29 mm, abdomen 17-20 mm, hind wing 19-23 mm. [Habitat] Narrow streams and ditches below 1000 m elevation. [Distribution] Yunnan, Hainan; India, Nepal, Myanmar, Thailand, Cambodia, Laos, Vietnam, Peninsular Malaysia. [Flight Season] Throughout the year.

三斑阳鼻蟌 *Heliocypha perforata* (Percheron, 1835)

【形态特征】雄性头顶和后头各具1对蓝色斑点；胸部黑色，合胸脊具1个较短的三角形紫红色斑，脊两侧具1对较短的蓝斑，合胸侧面具3条宽阔的蓝色条纹；前翅透明，末端1/3褐色，后翅端方1/2黑褐色，具2列闪烁紫色光泽的翅窗，翅痣黑色；足的胫节内缘白色；腹部黑色，第1~9节具蓝色斑点。雌性黑褐色具黄色条纹。【长度】体长 28~31 mm，腹长 17~20 mm，后翅 23~24 mm。【栖息环境】海拔 1000 m以下的溪流、沟渠和河流。【分布】云南、贵州、浙江、福建、广西、广东、海南、香港、台湾；印度、缅甸、泰国、柬埔寨、老挝、越南、马来半岛。【飞行期】全年可见。

[Identification] Male vertex and occiput with a pair of blue spots. Thorax black, dorsally a short, magenta mesothoracic triangle, flanked by a pair of short blue markings, laterally three broad blue stripes. Fore wings hyaline with the apical third brown, hind wings with apical half blackish brown, with two sets of transparent windows shining violet, pterostigma black. Legs with inner margin of tibiae whitish. Abdomen black with blue markings on S1-S9. Female blackish brown with yellow markings. [Measurements] Total length 28-31 mm, abdomen 17-20 mm, hind wing 23-24 mm. [Habitat] Streams, ditches and rivers below 1000 m elevation. [Distribution] Yunnan, Guizhou, Zhejiang, Fujian, Guangxi, Guangdong, Hainan, Hong Kong, Taiwan; India, Myanmar, Thailand, Cambodia, Laos, Vietnam, Peninsular Malaysia. [Flight Season] Throughout the year.

三斑阳鼻蟌 雌，广西
Heliocypha perforata, female from Guangxi

三斑阳鼻蟌 交尾，云南（西双版纳）
Heliocypha perforata, mating pair from Yunnan (Xishuangbanna)

三斑阳鼻蟌 雄，广东
Heliocypha perforata, male from Guangdong

隐鼻螅属 Genus *Heterocypha* Laidlaw, 1950

本属目前仅知1个稀有物种,结合了圣鼻螅属和阳鼻螅属的特征。本属识别特征和栖息环境见种类描述部分。

Only one very poorly known and rare species is known for this genus, which combines characters of the genera *Aristocypha* and *Heliocypha*. For the characters and habitats, see the species account below.

印度隐鼻螅 雄翅
Heterocypha vitrinella, male wings

印度隐鼻螅 *Heterocypha vitrinella* (Fraser, 1935)

【形态特征】雄性头顶和后头各具1对黄色斑点;胸部黑色,合胸脊具1个三角形蓝色斑,侧面具蓝色条纹;前翅末端1/3黑褐色,后翅端方2/5黑褐色具蓝色光泽的1列翅窗和1个位于亚端方的翅窗;足无白色粉霜;腹部黑色具蓝色斑点。雌性黑色具黄色条纹;翅的形状与雄性相似,前翅透明具双色翅痣,后翅末端1/3褐色。【长度】腹长18~20 mm,后翅 21~23 mm。【栖息环境】具静水环境的丛林深处,远离溪流。【分布】云南(德宏);印度。【飞行期】8—11月。

[Identification] Male face black, vertex and occiput with a pair of yellow spots. Thorax black with a short azure blue mesothoracic triangle dorsally, sides of synthorax with azure blue stripes. Fore wings with apical one third blackish brown, hind wings with the apical two fifths blackish brown with a row of vitreous windows and a compact subapical vitreous window, all reflecting iridescent blue ventrally. Legs black with tibiae not pruinosed white. Abdomen black with azure blue markings. Female black with yellow markings. Wings shaped similar to male, fore wings hyaline with bicolored pterostigma, hindwing with apical one third brown. [Measurements] Abdomen 18-20 mm, hind wing 21-23 mm. [Habitat] Deep forests near stagnant water bodies, far away from streams. [Distribution] Yunnan (Dehong); India. [Flight season] August to November.

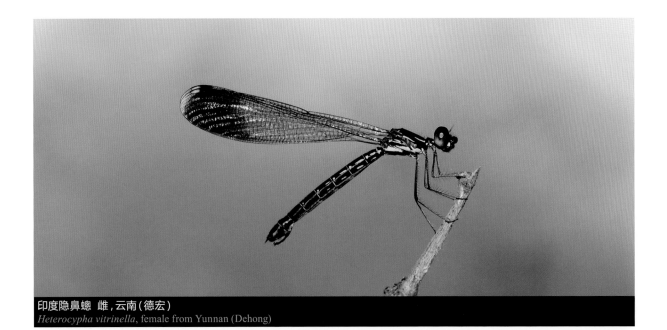

印度隐鼻螅 雌,云南(德宏)
Heterocypha vitrinella, female from Yunnan (Dehong)

印鼻蟌属 Genus *Indocypha* Fraser, 1949

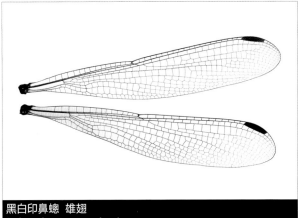

黑白印鼻蟌 雄翅
Indocypha vittata, male wings

本属全世界已知7种，仅分布于东洋界，其中6种在中国分布。除1种特例本属是鼻蟌科体型最大的种类。身体较粗壮，腹部扁平，基方较阔，向端方收缩；翅透明，臀脉与翅后缘的交会处位于第1条结前横脉之前，可以与近似的鼻蟌属相区分。雄性印鼻蟌的翅透明也容易和圣鼻蟌属、阳鼻蟌属种类区分。

本属的多数种类少见或稀有，最珍稀的一种是产自四川的川印鼻蟌，目前仅有1对标本的记录。本属多数种类生活在低海拔河流和溪流，它们栖息于河岸带植被茂盛的丛林中，雄性会在阳光充足时靠近水面，停落在水边的树枝或者叶片上。雌性产卵于水草或者朽木上。

The genus includes seven Oriental species of which six have so far been recorded from China. With one exception they are among the largest species in the family. Body rather robust, with abdomen dorso-ventrally flattened, broader at base and tapering apically. Wings hyaline, the anal vein meets the hind margin of the wing well proximal to the level of the first antenodal, which is a reliable character separating species of this genus from *Rhinocypha*. Males are easily separated from *Aristocypha* and *Heliocypha* by their hyaline wings.

All *Indocypha* species are uncommon or rare, the rarest is *Indocypha svenhedini* (Sjöstedt, 1932) from Sichuan, of which only a single pair has ever been found. Species of the genus inhabit lowland rivers and streams, they usually stay in the shelter of dense marginal vegetation, males approach water during bright sunshine, and perch on the tips of branches or leaves. Females lay their eggs into deadwood and aquatic plants.

黑白印鼻蟌 雄
Indocypha vittata, male

蓝尾印鼻蟌 *Indocypha cyanicauda* Zhang & Hämäläinen, 2018

【形态特征】雄性面部大面积黑色，头顶和后头具小黄斑；胸部黑色具黄条纹，翅稍长于腹部，翅痣黑色，中足和后足胫节内缘具白色条纹；腹部黑色具黄色和淡蓝色斑纹，第1~6节具黄色细条纹，第7~9节背面大面积淡蓝色。身体的黄色条纹随年纪增长变暗，腹部的淡蓝色斑变成蓝白色。雌性黑色具黄色斑纹，翅痣黑色。【长度】体长 31~37 mm，腹长 20~25 mm，后翅 25~28 mm。【栖息环境】海拔 500 m 左右河岸带植被茂盛的河流。【分布】云南（西双版纳、临沧）；越南。【飞行期】10—12月。

[Identification] Male face largely black, vertex and occiput with small yellow spots. Thorax black with yellow markings, wings slightly longer than abdomen, pterostigma black, mid and hind legs with inner face of tibiae white. Abdomen black with yellow and pale blue markings, S1-S6 with yellow stripes, dorsum of S7-S9 largely pale blue. The body yellow markings darkened with age, and the pale blue markings on abdomen turns to bluish white. Female black with yellow markings, pterostigma black. [Measurements] Total length 31-37 mm, abdomen 20-25 mm, hind wing 25-28 mm. [Habitat] Rivers with plenty of fringing vegetation at about 500 m elevation. [Distribution] Yunnan (Xishuangbanna, Lincang); Vietnam. [Flight Season] October to December.

蓝尾印鼻蟌 雄，云南（西双版纳）
Indocypha cyanicauda, male from Yunnan (Xishuangbanna)

蓝尾印鼻蟌 雌，云南（西双版纳）
Indocypha cyanicauda, female from Yunnan (Xishuangbanna)

蓝尾印鼻蟌 雄，云南（西双版纳）
Indocypha cyanicauda, male from Yunnan (Xishuangbanna)

显著印鼻蟌 *Indocypha catopta* Zhang, Hämäläinen & Tong, 2010

【形态特征】雄性面部黑色，上唇、后唇基、头顶和后头具黄色或蓝色斑纹；胸部黑色具黄色斑纹，翅长于腹长，翅痣褐色，足黑色；腹部基方2节黑色具黄色条纹，第3～8节橙黄色，第9～10节主要黑色。雌性与雄性相似。【长度】体长 35～37 mm，腹长 22～24 mm，后翅 28～32 mm。【栖息环境】海拔 1000 m以下森林中的溪流。【分布】中国贵州特有。【飞行期】7—9月。

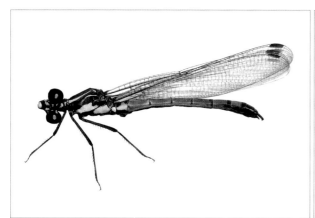
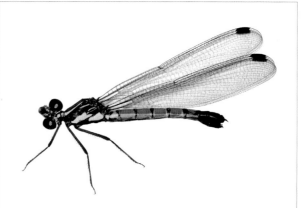

显著印鼻蟌 雄，贵州
Indocypha catopta, male from Guizhou

显著印鼻蟌 雌，贵州
Indocypha catopta, female from Guizhou

[Identification] Male face black, labrum, postclypeus, vertex and occiput with yellow or blue spots. Thorax black with yellow stripes, wings clearly longer than abdomen, pterostigma brown, legs black. Basal two abdominal segments black with yellow markings, S3-S8 orange yellow, S9-S10 largely black. Female similar to male. [Measurements] Total length 35-37 mm, abdomen 22-24 mm, hind wing 28-32 mm. [Habitat] Streams in forest below 1000 m elevation. [Distribution] Endemic to Guizhou of China. [Flight Season] July to September.

卡萨印鼻蟌 *Indocypha katharina* (Needham, 1930)

卡萨印鼻蟌 雌，云南（普洱）
Indocypha katharina, female from Yunnan (Pu'er)

【形态特征】雄性面部黑色，头顶和后头具黄色斑纹；胸部黑色具黄色条纹，翅痣黑色，足的腿节和胫节内缘具白色粉霜；腹部纺锤形，基方3节主要黑色具红褐色斑，第4～8节橙红色，第9～10节主要黑色，有时具甚细的红褐色条纹。雌性黑色具更发达的黄色条纹。【长度】体长 33～39 mm，腹长 22～27 mm，后翅 27～29 mm。【栖息环境】海拔1000 m以下的溪流和河流。【分布】云南（红河、普洱）、贵州、四川、广西、广东；越南。【飞行期】4—7月。

[Identification] Male face black, vertex and occiput with yellow spots. Thorax black with yellow stripes, pterostigma black, legs with inner margin

卡萨印鼻蟌 雄, 云南 (红河)
Indocypha katharina, male from Yunnan (Honghe)

of tibiae and femora white. Abdomen roughly spindle-shaped, basal three segments mainly black with reddish brown markings, S4-S8 dull reddish orange, S9-S10 mainly black, occasionally with reddish brown spots. Female black with more extensive yellow markings. **[Measurements]** Total length 33-39 mm, abdomen 22-27 mm, hind wing 27-29 mm. **[Habitat]** Streams and rivers below 1000 m elevation. **[Distribution]** Yunnan (Honghe, Pu'er), Guizhou, Sichuan, Guangxi, Guangdong; Vietnam. **[Flight Season]** April to July.

红尾印鼻蟌 *Indocypha silbergliedi* Asahina, 1988

红尾印鼻蟌 雄, 云南 (西双版纳)
Indocypha silbergliedi, male from Yunnan (Xishuangbanna)

红尾印鼻螅 雄, 云南 (西双版纳) ｜Adolfo Cordero-Rivera 摄
Indocypha silbergliedi, male from Yunnan (Xishuangbanna) | Photo by Adolfo Cordero-Rivera

【形态特征】雄性面部黑色, 头顶和后头具黄色斑纹; 胸部黑色具黄绿色条纹, 翅痣黑色, 足的胫节内缘具白色粉霜; 腹部第1~6节主要黑色具甚细小的黄色斑纹, 第7~9节橙红色, 第10节黑色具较小的橙色斑。【长度】雄性腹长 21~22.5 mm, 后翅 22~22.5 mm。【栖息环境】海拔 500 m河岸带植被茂盛的河流。【分布】云南 (西双版纳); 泰国、老挝。【飞行期】4—6月。

[Identification] Male face black, vertex and occiput with yellow spots. Thorax black with yellowish green stripes, pterostigma black, legs with inner margin of tibiae white. S1-S6 mainly black with small yellow markings, S7-S9 reddish orange, S10 black with small orange spots. [Measurements] Male abdomen 21-22.5 mm, hind wing 22-22.5 mm. [Habitat] Rivers with plenty of marginal vegetation at 500 m elevation. [Distribution] Yunnan (Xishuangbanna); Thailand, Laos. [Flight Season] April to June.

黑白印鼻螅 *Indocypha vittata* (Selys, 1891)

【形态特征】雄性面部黑色, 后唇基、头顶和后头具丰富的小黄斑; 胸部黑色具黄条纹, 年老后几乎完全黑色, 翅和腹部等长, 翅痣黑色, 足的腿节和胫节内缘白色; 腹部黑色和黄白色或蓝白色, 第3~7节背面大面积黄白色或蓝白色。雌性黑色具黄色斑纹, 翅痣双色。【长度】体长 33~42 mm, 腹长 22~27 mm, 后翅 26~28 mm。【栖息环境】海拔 1000 m以下河岸带植被茂盛的河流。【分布】云南 (西双版纳、普洱); 印度、缅甸、老挝、泰国、越南。【飞行期】4—7月。

[Identification] Male face black, postclypeus, vertex and occiput with small yellow spots. Thorax black with yellow markings, in old males almost black without yellow markings, wings as long as abdomen, pterostigma black, legs with inner margin of tibiae and femora white. Abdomen black and yellowish or bluish white, dorsum of S3-S7 largely yellowish white or bluish white. Female black with yellow markings, pterostigma bicolored. [Measurements] Total length 33-42 mm, abdomen 22-27 mm, hind wing 26-28 mm. [Habitat] Rivers with plenty of fringing vegetation below 1000 m elevation. [Distribution] Yunnan (Xishuangbanna, Pu'er); India, Myanmar, Laos, Thailand, Vietnam. [Flight Season] April to July.

黑白印鼻螅 雄，云南（西双版纳）
Indocypha vittata, male from Yunnan (Xishuangbanna)

黑白印鼻螅 雌，云南（西双版纳）
Indocypha vittata, female from Yunnan (Xishuangbanna)

隼螅属 Genus *Libellago* Selys, 1840

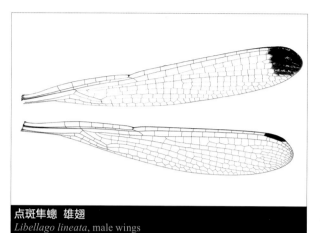

点斑隼螅 雄翅
Libellago lineata, male wings

本属全球已知25种，主要分布于亚洲的热带地区。中国已知仅1种，分布于华南和西南地区。本属是非常短小的豆娘。雄性通常较艳丽，翅明显长于腹部。

本属豆娘主要栖息于低海拔的溪流、沟渠和河流。雄性会停栖在溪流中的漂浮物上，如朽木和水生植物。求偶时会向雌性展示白色的胫节。雄性经常为争夺领地而进行面对面的悬停争斗。雌性产卵于朽木或者水草上。

The genus contains 25 species, mainly distributed in tropical Asia. Only one species is recorded from China, found in the South and Southwest regions. Species of the genus are small-sized damselflies. Males of most species have a colorful body, with wings distinctly longer than the abdomen.

Libellago species inhabit lowland streams, ditches and rivers. Males usually perch over water on emergent deadwood or plants. Males display white tibiae to females during courtship. Males contest territory very often by hovering face to face. Female oviposit in deadwood or aquatic plants.

点斑隼螅 雄
Libellago lineata, male

点斑隼螅 *Libellago lineata* (Burmeister, 1839)

点斑隼螅 雄，海南
Libellago lineata, male from Hainan

点斑隼螅 雌，海南
Libellago lineata, female from Hainan

点斑隼螅 雄，云南（西双版纳）
Libellago lineata, male from Yunnan (Xishuangbanna)

　　【形态特征】雄性面部黑色，头顶和后头具较小的黄色斑纹；胸部黑色具黄色条纹，翅透明，前翅翅端具1个深色斑，翅痣黑色，足胫节内缘具白色粉霜；腹部第1~6节橙色，第7~10节黑色。雌性黑褐色具丰富的黄斑，翅痣灰白色。【长度】体长 20~23 mm，腹长 13~15 mm，后翅 19~22 mm。【栖息环境】海拔 1000 m 以下的溪流和河流。【分布】云南、福建、广西、海南、广东、台湾；南亚、东南亚广布。【飞行期】全年可见。

　　[Identification] Male face black, vertex and occiput with small yellow spots. Thorax black with yellow markings, wings hyaline, fore wings with a dark apical spot, pterostigma black, legs with inner margin of tibiae white. S1-S6 orange, S7-S10 black. Female blackish brown with numerous yellow dots, pterostigma dirty white. [Measurements] Total length 20-23 mm, abdomen 13-15 mm, hind wing 19-22 mm. [Habitat] Streams and rivers below 1000 m elevation. [Distribution] Yunnan, Fujian, Guangxi, Hainan, Guangdong, Taiwan; Widespread in South and Southeast Asia. [Flight Season] Throughout the year.

鼻蟌属 Genus *Rhinocypha* Rambur, 1842

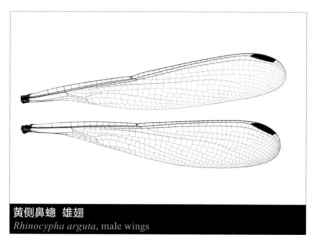

黄侧鼻蟌 雄翅
Rhinocypha arguta, male wings

目前本属是本科种类最多的属，全球已知约有40种，主要分布于亚洲的热带和亚热带地区以及澳新界。但本属应为并系类群，需要重新分类和厘定。中国已知6种，其中最稀有的是西藏的三纹鼻蟌，翅褐红色，腹部黑色具蓝斑。

本属豆娘栖息于森林中的溪流。雄性会停立在阳光充足的岩石、朽木或者水草上占据领地。雌性产卵于朽木或者水草上。

Rhinocypha, in its present composition, is the most speciose genus in the family with some 40 species in tropical and subtropical parts of Asia and Australasia. However, the genus seems to be a paraphyletic group of species in need of reclassification and redefinition. Six species are recorded from China, the rarest is *R. trimaculata* Selys, 1853 which was originally found in Tibet, this species has largely coppery red wings and a black abdomen with bluish markings.

Rhinocypha species inhabit montane streams. Males usually perch on sunlit rocks, deadwood or emergent plants where they maintain territory. Female oviposit into deadwood or aquatic plants.

线纹鼻蟌 雄
Rhinocypha drusilla, male

黄侧鼻蟌 *Rhinocypha arguta* Hämäläinen & Divasiri, 1997

【形态特征】雄性上唇黄色，头顶具小黄斑；胸部黑色，侧面具黄斑，翅透明，足黑色；腹部第2~8节大面积橙色，第9~10节黑色。雌性黑色具甚细小的黄色条纹，翅痣双色。【长度】体长 33~36 mm，腹长 22~24 mm，后翅 25~29 mm。【栖息环境】海拔 1000~1500 m森林中的小溪。【分布】云南（德宏、红河）；泰国、越南。【飞行期】9—12月。

黄侧鼻蟌 雄，云南（德宏）
Rhinocypha arguta, male from Yunnan (Dehong)

黄侧鼻蟌 雌，云南（德宏）
Rhinocypha arguta, female from Yunnan (Dehong)

[Identification] Male labrum yellow, vertex with small yellow spots. Thorax black with yellow lateral markings, wings hyaline, legs black. Dorsal surface of S2-S8 largely orange, S9-S10 black. Female black with small yellow markings, pterostigma bicolored. [Measurements] Total length 33-36 mm, abdomen 22-24 mm, hind wing 25-29 mm. [Habitat] Montane streams at 1000-1500 m elevation. [Distribution] Yunnan (Dehong, Honghe); Thailand, Vietnam. [Flight Season] September to December.

黄侧鼻蟌 雄，云南（红河）
Rhinocypha arguta, male from Yunnan (Honghe)

线纹鼻蟌 *Rhinocypha drusilla* Needham, 1930

线纹鼻蟌 雄，重庆
Rhinocypha drusilla, male from Chongqing

线纹鼻蟌 雌，重庆
Rhinocypha drusilla, female from Chongqing

【形态特征】雄性上唇黄色，额和头顶具小黄斑；胸部黑色，侧面具黄色条纹，翅稍染褐色，后翅端部色彩加深，翅痣双色，足黑色；腹部主要橙色。雌性黑色具黄色斑纹，翅痣双色。【长度】体长 35~38 mm，腹长 24~25 mm，后翅 25~28 mm。【栖息环境】海拔 1500 m 以下森林中的溪流。【分布】中国特有，分布于贵州、重庆、安徽、浙江、福建、广西、广东。【飞行期】7—12月。

[Identification] Male labrum yellow, frons and vertex with small yellow spots. Thorax black with yellow lateral markings, wings slightly tinted with brown, tip of hind wing clearly darkened, pterostigma bicolored, legs black. Abdomen largely orange. Female black with yellow stripes, pterostigma bicolored. [Measurements] Total length 35-38 mm, abdomen 24-25 mm, hind wing 25-28 mm. [Habitat] Streams in forest below 1500 m elevation. [Distribution] Endemic to China, recorded from Guizhou, Chongqing, Anhui, Zhejiang, Fujian, Guangxi, Guangdong. [Flight Season] July to December.

华氏鼻蟌 *Rhinocypha huai* (Zhou & Zhou, 2006)

【形态特征】雄性面部黑色，头顶具小黄斑；胸部黑色，侧面具黄斑，翅透明，翅痣黑褐色，足黑色；腹部各节背面具较大的橙色斑。雌性黑褐色具黄色斑纹。本种和黄侧鼻蟌相似，但雄性腹部第9～10节具橙色斑，雌性黄斑更发达。【长度】体长 30～31 mm，腹长 20～23 mm，后翅 24～27 mm。【栖息环境】海拔 1500 m 以下森林中的溪流。【分布】海南；国外可能分布于越南、老挝。【飞行期】5月—次年1月。

[Identification] Male face black, vertex with small yellow spots. Thorax black with yellow lateral markings, wings hyaline, pterostigma blackish brown, legs black. Dorsal surface of abdomen with large orange spots. Female blackish brown with yellow stripes. Similar to *R. arguta*, but males have orange spots on S9-S10, female with more yellow markings. [Measurements] Total length 30-31 mm, abdomen 20-23 mm, hind wing 24-27 mm. [Habitat] Streams in forest below 1500 m elevation. [Distribution] Hainan; possibly in Vietnam and Laos. [Flight Season] May to the following January.

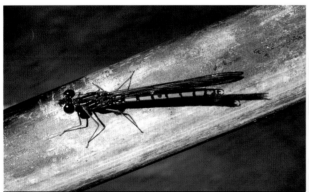

华氏鼻蟌 雄，海南
Rhinocypha huai, male from Hainan

华氏鼻蟌 雌，海南
Rhinocypha huai, female from Hainan

华氏鼻蟌 雄，海南
Rhinocypha huai, male from Hainan

翠顶鼻蟌 *Rhinocypha orea* Hämäläinen & Karube, 2001

【形态特征】雄性身体大面积黑色；胸部侧面具黄斑，前翅透明，后翅更宽阔，正面大面积绿色具霓虹色光泽，末端色彩加深，后翅背面大面积黑色具光泽，末端具1个较大的翠绿色金属斑，中足胫节内缘白色。【长度】体长 29~30 mm，腹长 19~20 mm，后翅 23~24 mm。【栖息环境】海拔 1000 m 以下森林中的阴暗溪流。【分布】广西；越南。【飞行期】5—7月。

[Identification] Male body largely black. Thorax with yellow lateral markings, fore wings hyaline, hind wings broader, upperside largely green showing iridescence, tips darkened, underside largely black with a large metallic green apical spot, mid legs with inner face of tibiae white. [Measurements] Total length 29-30 mm, abdomen 19-20 mm, hind wing 23-24 mm. [Habitat] Shady streams in forest below 1000 m elevation. [Distribution] Guangxi; Vietnam. [Flight Season] May to July.

翠顶鼻蟌 雄，广西
Rhinocypha orea, male from Guangxi

台湾鼻蟌 *Rhinocypha taiwana* Wang & Chang, 2013

【形态特征】雄性面部黑色，头顶具小黄斑；胸部黑色，侧面具黄斑，翅透明，后翅端稍微较深，翅痣双色，足黑色；腹部主要橙色。雌性黑褐色具黄色斑纹，翅色彩与雄性相似。【长度】腹长 22~27 mm，后翅 22~30 mm。【栖息环境】海拔 500 m 以下的林荫小溪。【分布】中国台湾特有。【飞行期】9—12月。

台湾鼻蟌 雄,台湾 | 嘎嘎 摄
Rhinocypha taiwana, male from Taiwan | Photo by Gaga

[Identification] Male face black, vertex with small yellow spots. Thorax black with yellow lateral markings, wings hyaline, hind wing tips slightly darkened, pterostigma bicolored, legs black. Abdomen mainly orange. Female blackish brown with yellow stripes, wings similarly colored to male. [Measurements] Abdomen 22-27 mm, hind wing 22-30 mm. [Habitat] Shady streams in forest below 500 m elevation. [Distribution] Endemic to Taiwan of China. [Flight Season] September to December.

台湾鼻蟌 雄,台湾 | 嘎嘎 摄
Rhinocypha taiwana, male from Taiwan | Photo by Gaga

台湾鼻蟌 雌,台湾 | 嘎嘎 摄
Rhinocypha taiwana, female from Taiwan | Photo by Gaga

4 溪蟌科 Family Euphaeidae

溪蟌科主要分布于亚洲南部和东部，仅有1种分布于西亚至欧洲的东南部。全世界已知9属70余种。中国已知5属30余种，在南方分布广泛。本科多是体中型的豆娘，身体粗壮，足较短；翅无显著的翅柄，翅脉密集，具翅痣。雄性身体通常色彩较暗，翅透明或染有深色；不同种类的雌性通常很相似，身体黑褐色具黄色条纹。

本科豆娘栖息于溪流和河流。雄性具领域行为，通常停落在溪流中的岩石、朽木和水生植物上占据领地。许多种类的雌性可以潜水产卵。多数种类具有很强的飞行能力，比如暗溪蟌属种类可以在河面上长时间巡飞。

Euphaeids are confined to the southern and eastern parts of Asia, the only exception is a western Asiatic species, which occurs also in the extreme south-eastern corner of Europe. The family includes nine genera with over 70 species. More than 30 species in five genera are recorded from China. Species of the family are medium-sized damselflies with a rather stout body and short legs. The bases of the wings are not stalked, venation is rather dense and a pterostigma is present. Males are usually rather dark colored, with either hyaline or opaque wings. Females of different species are often rather similarly colored and marked, with the body blackish brown bearing yellow stripes.

Species of the family inhabit streams and rivers. Males usually perch on rocks, deadwood or emergent plants while maintaining a territory. Females of many species oviposit underwater. Many species are strong-flying damselflies, notably species of genus *Dysphaea*, whose males often patrol along the rivers for long period.

华丽暗溪螅 雄（左）
Dysphaea gloriosa, male (left)

透顶溪螅 雄（右）
Euphaea masoni, male (right)

宽带溪螅 雄
Euphaea ornata, male

异翅溪蟌属 Genus *Anisopleura* Selys, 1853

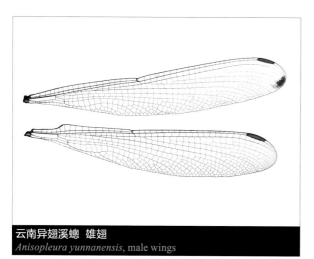

云南异翅溪蟌 雄翅
Anisopleura yunnanensis, male wings

本属全世界已知11种，分布于东洋界。中国已知5种，在南方分布较广泛。本属豆娘体中型，身体具有发达的黄色条纹；翅大面积透明，雄性前翅末端具小褐斑，后翅的前缘脉在亚基方不同程度的向前弯曲。本属雄性可以通过肛附器的形状区分。

本属豆娘主要栖息于具有一定海拔高度的山区溪流。有些种类可以在高山上生活，例如云南异翅溪蟌可以生活在 2500 m 以上的高山溪流。雄性停落在溪流中的岩石、朽木和水生植物上占据领地。雌性潜水产卵。

The genus contains 11 species, distributed entirely in the Oriental realm. Five species are recorded from China, widespread in the southern part. Species of the genus are medium-sized, bodies with extensive yellow markings. The wings are largely hyaline, in males the fore wing with small brown apical spots, the costal vein of the hind wing is more or less expanded and kinked anteriorly near the base. Males of the genus can be distinguished by the shape of their anal appendages.

Anisopleura species inhabit montane streams at moderate altitudes. Some species, such as *A. yunnanensis* can live in high montane streams above 2500 m. Males usually perch on rocks, deadwood or emerging plants while maintaining a territory. Females oviposit underwater.

云南异翅溪蟌 交尾
Anisopleura yunnanensis, mating pair

蓝斑异翅溪螅
Anisopleura furcata

斧尾异翅溪螅
Anisopleura pelecyphora

庆元异翅溪螅
Anisopleura qingyuanensis

三彩异翅溪螅
Anisopleura subplatystyla

云南异翅溪螅
Anisopleura yunnanensis

异翅溪螅属 雄性肛附器
Genus *Anisopleura*, male anal appendages

蓝斑异翅溪螅 *Anisopleura furcata* Selys, 1891

【形态特征】雄性面部下方大部分蓝色，头部大面积黑色，头顶具1对小黄斑；胸部黑色具宽阔的黄色条纹，后翅前缘脉在亚基方仅稍微弯曲；腹部黑色，侧面具黄色条纹，第9～10节具粉霜。【长度】体长 45～50 mm，腹长

蓝斑异翅溪螅 雄，泰国│Matti Hämäläinen 摄
Anisopleura furcata, male from Thailand │ Photo by Matti Hämäläinen

35～39 mm，后翅 27～31 mm。【栖息环境】森林中的溪流。【分布】云南；缅甸、泰国。【飞行期】泰国北部5—11月。

[Identification] Male face largely blue at lower part, head otherwise mainly black, vertex with a pair of small yellow spots. Thorax black with broad yellow stripes, costal border of hind wing only moderately angulated near base. Abdomen black with yellow lateral stripes, S9-S10 pruinosed. [Measurements] Total length 45-50 mm, abdomen 35-39 mm, hind wing 27-31 mm. [Habitat] Streams in forest. [Distribution] Yunnan; Myanmar, Thailand. [Flight Season] May to November in the north of Thailand.

斧尾异翅溪蟌 *Anisopleura pelecyphora* Zhang, Hämäläinen & Cai, 2014

斧尾异翅溪蟌 雄，云南（临沧）
Anisopleura pelecyphora, male from Yunnan (Lincang)

斧尾异翅溪螅 雄，云南（临沧）
Anisopleura pelecyphora, male from Yunnan (Lincang)

【形态特征】雄性面部黑色具蓝斑；胸部黑色具宽阔的黄色条纹，后翅前缘脉在亚基方稍呈角度弯曲；腹部黑色具黄色条纹，第9～10节具粉霜。雌性与雄性相似，但腹部较短。【长度】体长 46～49 mm，腹长 35～38 mm，后翅 29～30 mm。【栖息环境】海拔 800～1500 m的山区溪流。【分布】中国云南（普洱、临沧）特有。【飞行期】9—11月。

[Identification] Male face black with blue spots. Thorax black with broad yellow stripes, Costa of hind wing only moderately angulated near base. Abdomen black with yellow stripes, S9-S10 pruinosed. Female similar to male but abdomen shorter. [Measurements] Total length 46-49 mm, abdomen 35-38 mm, hind wing 29-30 mm. [Habitat] Montane streams at 800-1500 m elevation. [Distribution] Endemic to Yunnan (Pu'er, Lincang) of China. [Flight Season] September to November.

庆元异翅溪螅 *Anisopleura qingyuanensis* Zhou, 1982

【形态特征】雄性面部黑色具蓝斑；胸部黑色具淡蓝色和黄色条纹，后翅前缘脉在亚基方稍微弯曲；腹部黑色具黄色条纹，第9～10节具粉霜。雌性与雄性相似，但腹部较短。两性在完全成熟以前条纹为蓝色。【长度】体长41～47 mm，腹长 30～36 mm，后翅 29～30 mm。【栖息环境】海拔 2000 m以下森林中的溪流。【分布】云南、贵州、湖北、湖南、江西、浙江、福建、广西、广东；老挝、越南。【飞行期】6—10月。

[Identification] Male face black with blue spots. Thorax black with pale blue and yellow stripes, Costa of hind wing only moderately angulated near base. Abdomen black with yellow stripes, S9-S10 pruinosed. Female similar to male, but abdomen shorter. The body maculation is blue in both sexes before fully maturity is reached. [Measurements] Total length 41-47 mm, abdomen 30-36 mm, hind wing 29-30 mm. [Habitat] Montane streams below 2000 m elevation. [Distribution] Yunnan, Guizhou, Hubei, Hunan, Jiangxi, Zhejiang, Fujian, Guangxi, Guangdong; Laos, Vietnam. [Flight Season] June to October.

庆元异翅溪螅 雄，贵州
Anisopleura qingyuanensis, male from Guizhou

庆元异翅溪蟌 雄 , 贵州
Anisopleura qingyuanensis, male from Guizhou

庆元异翅溪蟌 雌 , 贵州
Anisopleura qingyuanensis, female from Guizhou

三彩异翅溪蟌 *Anisopleura subplatystyla* Fraser, 1927

【形态特征】雄性面部黑色具淡蓝色斑点；胸部黑色，具淡蓝色的肩前条纹，合胸侧面具2条宽阔的黄色条纹，后翅前缘脉在亚基方显著弯曲；腹部黑色具黄色条纹，第9～10节具粉霜。【长度】雄性体长 46 mm，腹长 37 mm，后翅 29 mm。【栖息环境】海拔 1000 m森林中的溪流。【分布】云南（红河）；印度、尼泊尔、不丹、泰国、越南。【飞行期】5—11月。

[Identification] Male face black with pale blue spots. Thorax black with pale blue antehumeral stripes and two broad lateral yellow stripes, Costa of hind wing distinctly angulated near base. Abdomen black with yellow stripes, S9-S10 pruinosed. [Measurements] Male total length 46 mm, abdomen 37 mm, hind wing 29 mm. [Habitat] Streams in forest at 1000 m elevation. [Distribution] Yunnan (Honghe); India, Nepal, Bhutan, Thailand, Vietnam. [Flight Season] May to November.

三彩异翅溪蟌 雄，云南（红河）
Anisopleura subplatystyla, male from Yunnan (Honghe)

云南异翅溪蟌 *Anisopleura yunnanensis* Zhu & Zhou, 1999

【形态特征】雄性面部黑色具淡蓝色斑点；胸部黑色，具圆弧形肩前条纹，合胸侧面具2条宽阔的黄色条纹，后翅前缘脉在亚基方呈明显的角状；腹部黑色具黄色条纹，第9～10节具粉霜。雌性与雄性相似，但腹部较短。雌性未熟时翅基方染有琥珀色。【长度】体长 39～50 mm，腹长 28～39 mm，后翅 27～30 mm。【栖息环境】海拔1000～2500 m森林中的溪流。【分布】云南，广西；可能分布于越南。【飞行期】6—11月。

云南异翅溪螅 雄,云南(大理)
Anisopleura yunnanensis, male from Yunnan (Dali)

云南异翅溪螅 雌,云南(大理)
Anisopleura yunnanensis, female from Yunnan (Dali)

[Identification] Male face black with pale blue spots. Thorax black with arc-shaped antehumeral stripes and two broad lateral yellow stripes, Costa of hind wing distinctly angulated near base. Abdomen black with yellow stripes, S9-S10 pruinosed. Female similar to male, but abdomen shorter. Immature female with amber infusion at the wing bases. [Measurements] Total length 39-50 mm, abdomen 28-39 mm, hind wing 27-30 mm. [Habitat] Streams in forest at 1000-2500 m elevation. [Distribution] Yunnan, Guangxi; possibly Vietnam. [Flight Season] June to November.

云南异翅溪螅 雄,云南(德宏)
Anisopleura yunnanensis, male from Yunnan (Dehong)

尾溪蟌属 Genus *Bayadera* Selys, 1853

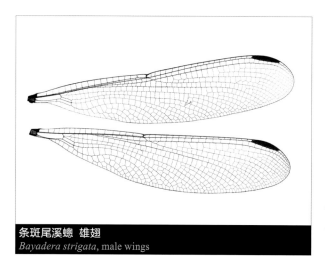

条斑尾溪蟌 雄翅
Bayadera strigata, male wings

本属全世界已知17种，主要分布于亚洲的热带和亚热带地区。中国已知12种，在南方广泛分布。本属是一类体中型、体色较暗的豆娘，多数种类翅透明，少数种类翅端具褐斑；雌性身体通常暗色具黄色条纹。本属雄性可通过肛附器的形状并结合身体色彩进行区分。

本属豆娘栖息于具有一定海拔高度的山区溪流。有些种类可以在高山生活，例如条斑尾溪蟌可以生活在2000 m以上的高山环境。雄性停落在溪流周边的枝头上占据领地，有时翅半张开。雌雄连结产卵。本属有集群产卵的习性，产卵位置通常在水边的灌木上、陡峭的土坡上或者水中的朽木上。

The genus contains 17 species, mainly distributed in tropical and subtropical Asia. 12 species are recorded from China, widespread in the southern part of the country. They are medium-sized, dark bodied damselflies, wings usually hyaline but sometimes with brown tips. Females body usually dark with yellow markings. Males of the genus can be distinguished by the shape of appendages combined with details of body markings.

条斑尾溪蟌 雄
Bayadera strigata, male

二齿尾溪螅

Bayadera bidentata

巨齿尾溪螅

Bayadera melanopteryx

条斑尾溪螅

Bayadera strigata

透翅尾溪螅

Bayadera hyalina

锯突尾溪螅

Bayadera serrata

褐翅尾溪螅

Bayadera nephelopennis

墨端尾溪螅

Bayadera hatvan

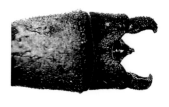

大陆尾溪螅

Bayadera continentalis

短尾尾溪螅

Bayadera brevicauda

尾溪螅属 雄性肛附器

Genus *Bayadera*, male anal appendages

巨齿尾溪螅 集群产卵
Bayadera melanopteryx, laying eggs in group.

Bayadera species inhabit montane streams at moderate altitudes. Some species, such as *Bayadera strigata* can live in high montane above 2000 m. Males holding territories by perching on the tips of branches, sometimes with wings partly open. Oviposition takes place in tandem. Pairs of *Bayadera* usually lay eggs in large groups at one site, eggs are laid into the branches of small bushes, muddy slopes of the stream bank or deadwood in the water.

二齿尾溪螅 *Bayadera bidentata* Needham, 1930

二齿尾溪螅 连结产卵，广东 | 莫善濂 摄
Bayadera bidentata, laying eggs into plants in tandem from Guangdong | Photo by Shanlian Mo

【形态特征】雄性面部黑色，上唇和面部侧面蓝色；胸部黑色，侧面具蓝灰色粉霜，翅稍染褐色，末梢具小白斑；腹部黑色。雌性黑色具黄色条纹。广西的雌性与其他地区的身体黄色条纹有差异。【长度】体长 41~53 mm，腹长 30~40 mm，后翅 28~31 mm。【栖息环境】海拔 2500 m 以下森林中的溪流。【分布】四川、贵州、湖北、浙江、福建、广西、广东；越南。【飞行期】5—8月。

[Identification] Male face black, labrum and sides of face blue. Thorax black with bluish grey pruinescence laterally, wings slightly tinted with brown, with the extreme wing tip narrowly whitish. Abdomen black.

Female black with yellow markings. Yellow body markings of females from Guangxi differ from those from other parts of China. **[Measurements]** Total length 41-53 mm, abdomen 30-40 mm, hind wing 28-31 mm. **[Habitat]** Streams in forest below 2500 m elevation. **[Distribution]** Sichuang, Guizhou, Hubei, Zhejiang, Fujian, Guangxi, Guangdong; Vietnam. **[Flight Season]** May to August.

二齿尾溪螅 雄, 广西
Bayadera bidentata, male from Guangxi

二齿尾溪螅 雌, 广西
Bayadera bidentata, female from Guangxi

短尾尾溪螅 *Bayadera brevicauda* Fraser, 1928

【形态特征】雄性面部黑色，上唇淡蓝色；胸部黑色，侧面具黄色条纹，翅透明；腹部黑色，第1~2节、第8~10节具蓝白色粉霜。雌性黑褐色具黄色条纹。老熟雄性胸部稍染粉霜。【长度】体长 42~45 mm，腹长 31~35 mm，后翅 27~31 mm。【栖息环境】海拔 1500 m 以下森林中的溪流。【分布】中国台湾特有。【飞行期】3—11月。

[Identification] Male face black, labrum pale blue. Thorax black with yellow lateral stripes, wings hyaline. Abdomen black, S1-S2 and S8-S10 with bluish white pruinescence. Female blackish brown with yellow markings. Thorax of aged male slightly pruinosed. [Measurements] Total length 42-45 mm, abdomen 31-35 mm, hind wing 27-31 mm. [Habitat] Streams in forest below 1500 m elevation. [Distribution] Endemic to Taiwan of China. [Flight Season] March to November.

短尾尾溪螅 雄，台湾 | 嘎嘎 摄
Bayadera brevicauda, male from Taiwan | Photo by Gaga

短尾尾溪螅 雌，台湾 | 嘎嘎 摄
Bayadera brevicauda, female from Taiwan | Photo by Gaga

大陆尾溪螅 *Bayadera continentalis* Asahina, 1973

大陆尾溪螅 雌，广西
Bayadera continentalis, female from Guangxi

【形态特征】雄性面部黑色，上唇淡蓝色；胸部黑色具黄色条纹和蓝灰色粉霜，翅稍染褐色；腹部黑色，第1~2节、第8~10节具蓝白色粉霜。雌性黑色具黄色条纹。【长度】广西个体体长 38~43 mm，腹长 29~33 mm，后翅 29~30 mm。【栖息环境】海拔 1000~2500 m 森林中的溪流。【分布】贵州、浙江、福建、广西、广东；越南。【飞行期】4—6月。

[Identification] Male face black, labrum pale blue. Thorax black with yellow stripes and bluish grey pruinescence, wings slightly tinted with brown. Abdomen black, S1-S2 and S8-S10 with bluish white pruinescence. Female black with yellow markings. [Measurements] Individuals from Guangxi with total length 38-43 mm, abdomen 29-33 mm, hind wing 29-30 mm. [Habitat] Streams in forest at 1000-2500 m elevation. [Distribution] Guizhou, Zhejiang, Fujian, Guangxi, Guangdong; Vietnam. [Flight Season] April to June.

大陆尾溪螅 雄，广西
Bayadera continentalis, male from Guangxi

墨端尾溪螅 *Bayadera hatvan* Hämäläinen & Kompier, 2015

墨端尾溪螅 雌，云南（普洱）
Bayadera hatvan, female from Yunnan (Pu'er)

【形态特征】雄性面部黑色，上唇和面部侧面淡蓝色；胸部黑色具发达的黄色条纹，翅端部具褐斑；腹部黑色，第1~8节具甚细的黄色条纹。雌性身体黑色具橙黄色条纹，翅端无褐斑。【长度】体长 42~50 mm，腹长 31~38 mm，后翅 30~32 mm。【栖息环境】海拔 500~1500 m 森林中的溪流。【分布】云南（普洱）；越南。【飞行期】5—7月。

[Identification] Male face black, labrum and sides of face pale blue. Thorax black with strong yellow markings, wings hyaline with brown tips. Abdomen black, S1-S8 with narrow yellow markings. Female body black with orange yellow stripes, wings without brown tips. [Measurements] Total length 42-50 mm, abdomen 31-38 mm, hind wing 30-32 mm. [Habitat] Streams in forest at 500-1500 m elevation. [Distribution] Yunnan (Pu'er); Vietnam. [Flight Season] May to July.

墨端尾溪螅 雄，云南（普洱）
Bayadera hatvan, male from Yunnan (Pu'er)

透翅尾溪蟌 *Bayadera hyalina* Selys, 1879

透翅尾溪蟌 雄，云南（德宏）
Bayadera hyalina, male from Yunnan (Dehong)

【形态特征】雄性面部黑色，上唇淡蓝色；胸部黑色具甚细的黄白色条纹，翅透明；腹部黑色。雌性黑色具橙黄色条纹。贵州和四川的雄性个体有时翅端具褐斑。【长度】体长 44~54 mm，腹长 33~42 mm，后翅 30~33 mm。【栖息环境】海拔 1000~2000 m森林中的溪流。【分布】四川、贵州、云南；印度、泰国、老挝、越南。【飞行期】6—11月。

[Identification] Male face black, labrum pale blue. Thorax black with fine yellowish white stripes, wings hyaline. Abdomen black. Female black with orange yellow markings. Males from Sichuan and Guizhou occasionally with brown wing tips. **[Measurements]** Total length 44-54 mm, abdomen 33-42 mm, hind wing 30-33 mm. **[Habitat]** Streams in forest at 1000-2000 m elevation. **[Distribution]** Sichuan, Guizhou, Yunnan; India, Thailand, Laos, Vietnam. **[Flight Season]** June to November.

透翅尾溪螅 雄,贵州
Bayadera hyalina, male from Guizhou

透翅尾溪螅 雌,云南(红河)
Bayadera hyalina, female from Yunnan (Honghe)

科氏尾溪螅 *Bayadera kirbyi* Wilson & Reels, 2001

科氏尾溪螅 雄,云南(红河)
Bayadera kirbyi, male from Yunnan (Honghe)

【形态特征】雄性面部黑色,上唇和面部侧面淡蓝黄色;胸部黑色具甚细的灰白色条纹,翅稍染褐色,有时端部具黑褐色斑;腹部黑色。【长度】雄性体长 47~53 mm,腹长 36~41 mm,后翅 27~34 mm。【栖息环境】海拔 1500 m以下森林中的溪流。【分布】中国特有,分布于云南(红河)、海南。【飞行期】6—10月。

[Identification] Male face black, labrum and sides of face pale bluish yellow. Thorax black with fine greyish white stripes, wings slightly tinted with pale brown and wing tips sometimes brown. Abdomen black. [Measurements] Male total length 47-53 mm, abdomen 36-41 mm, hind wing 27-34 mm. [Habitat] Streams in forest below 1500 m elevation. [Distribution] Endemic to China, recorded from Yunnan (Honghe), Hainan. [Flight Season] June to October.

科氏尾溪螅 雄,云南(红河)
Bayadera kirbyi, male from Yunnan (Honghe)

巨齿尾溪螅 *Bayadera melanopteryx* Ris, 1912

【形态特征】雄性面部黑色，上唇淡蓝色；胸部黑色具蓝灰色粉霜，翅端具甚大的褐色斑，斑的大小和形状在不同分布地的差异较大；腹部黑色。雌性黑褐色具黄色条纹，翅的色彩与雄性相似。【长度】体长 44~51 mm，腹长 34~40 mm，后翅 28~30 mm。【栖息环境】海拔 500~2500 m森林中的溪流。【分布】广泛分布于西北、华中、华南和西南地区；越南。【飞行期】6—9月。

巨齿尾溪螅 雄，贵州
Bayadera melanopteryx, male from Guizhou

巨齿尾溪螅 雄，广西
Bayadera melanopteryx, male from Guangxi

巨齿尾溪螅 雌，贵州
Bayadera melanopteryx, female from Guizhou

巨齿尾溪螅 连结产卵，贵州
Bayadera melanopteryx, laying eggs in tandem from Guizhou

[Identification] Male face black, labrum pale blue. Thorax black with bluish grey pruinescence laterally, wings with apical brown patches which are variable in size and shape according to distribution. Abdomen black. Female blackish brown with yellow markings, wing color similar to male. [Measurements] Total length 44-51 mm, abdomen 34-40 mm, hind wing 28-30 mm. [Habitat] Streams in forest at 500-2500 m elevation. [Distribution] Widespread in the Northwest, Central, South and Southwest regions; Vietnam. [Flight Season] June to September.

褐翅尾溪螅 *Bayadera nephelopennis* Davies & Yang, 1996

【形态特征】雄性面部黑色，上唇和面部侧面淡蓝色；胸部黑色，侧面具蓝灰色粉霜，翅染有黄褐色；腹部黑色。雌性黑色具黄色条纹。【长度】体长 50~56 mm，腹长 38~44 mm，后翅 34~38 mm。【栖息环境】海拔 500~2000 m森林中的溪流。【分布】四川；越南。【飞行期】6—9月。

[Identification] Male face black, labrum and sides of face pale blue. Thorax black with bluish grey pruinescence laterally, wings tinted with yellowish brown. Abdomen black. Female black with yellow markings. [Measurements] Total length 50-56 mm, abdomen 38-44 mm, hind wing 34-38 mm. [Habitat] Streams in forest at 500-2000 m elevation. [Distribution] Sichuan; Vietnam. [Flight Season] June to September.

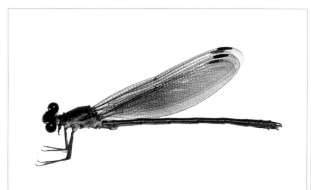

褐翅尾溪螅 雄，四川
Bayadera nephelopennis, male from Sichuan

褐翅尾溪螅 雌，四川
Bayadera nephelopennis, female from Sichuan

锯突尾溪螅 *Bayadera serrata* Davies & Yang, 1996

【形态特征】雄性面部大面积黄色，头顶和后头黑色；胸部黑色具发达的蓝黄色条纹，翅透明；腹部黑色，第1~9节具黄色条纹。雌性与雄性相似。【长度】体长 54~58 mm，腹长 40~46 mm，后翅 36~39 mm。【栖息环境】海拔 500~2000 m的林荫小溪。【分布】云南（大理、普洱、西双版纳）；泰国、老挝、越南。【飞行期】5—8月。

锯突尾溪螅 雄，未熟，云南（普洱）
Bayadera serrata, immature male from Yunnan (Pu'er)

锯突尾溪螅 雌，云南（普洱）
Bayadera serrata, female from Yunnan (Pu'er)

锯突尾溪螅 雄，云南（普洱）
Bayadera serrata, male from Yunnan (Pu'er)

[Identification] Male face largely yellow, vertex and occiput black. Thorax black with strongly developed bluish yellow markings, wings hyaline. Abdomen black, S1-S9 with yellow markings. Female similar to male. [Measurements] Total length 54-58 mm, abdomen 40-46 mm, hind wing 36-39 mm. [Habitat] Streams in forest at 500-2000 m elevation. [Distribution] Yunnan (Dali, Pu'er, Xishuangbanna); Thailand, Laos, Vietnam. [Flight Season] May to August.

条斑尾溪螅 *Bayadera strigata* Davies & Yang, 1996

【形态特征】雄性面部黑色，上唇和面部侧面淡蓝色；胸部黑色具发达的黄色条纹，翅透明；腹部黑色，第1~7节具细小的黄色斑纹。雌性黑色具橙黄色斑纹。【长度】体长 50~54 mm，腹长 37~42 mm，后翅31~36 mm。【栖息环境】海拔 1500~2500 m森林中的溪流。【分布】中国云南（大理、保山）特有。【飞行期】6—9月。

[Identification] Male face black, labrum and sides of face pale blue. Thorax black with strong yellow markings, wings hyaline. Abdomen black, S1-S7 with yellow markings. Female black with orange yellow markings. [Measurements] Total length 50-54 mm, abdomen 37-

条斑尾溪螅 雌雄连结，云南（大理）
Bayadera strigata, pair in tandem from Yunnan (Dali)

42 mm, hind wing 31-36 mm. [Habitat] Streams in forest at 1500-2500 m elevation. [Distribution] Endemic to Yunnan (Dali, Baoshan) of China. [Flight Season] June to September.

条斑尾溪蟌 雄，云南（大理）
Bayadera strigata, male from Yunnan (Dali)

条斑尾溪蟌 雌，云南（大理）
Bayadera strigata, female from Yunnan (Dali)

隐溪螅属 Genus *Cryptophaea* Hämäläinen, 2003

本属仅分布于亚洲的热带和亚热带地区，全世界已知的3种都在中国分布，但仅局限在云南和广西。本属是体中型、腹部甚细长的豆娘；翅透明而狭长，基方具柄。

本属豆娘主要栖息于山区的林荫小溪，喜欢阴暗的环境。有时也会停在阴暗溪流中阳光透射的区域。雄性通常会停落在水边悬挂的枝头上占据领地。

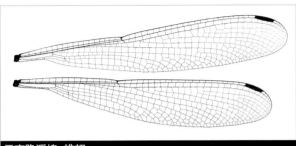

云南隐溪螅 雄翅
Cryptophaea yunnanensis, male wings

The genus is confined to tropical and subtropical Asia, the three known species have all been recorded from China, but are restricted to Yunnan and Guangxi. They are medium-sized damselflies with a slim and proportionally long abdomen. Wings hyaline, narrow and stalked at base.

越南隐溪螅 雄
Cryptophaea vietnamensis, male

Cryptophaea species prefer shady montane streams, and usually perch in shade. Sometimes they show up in sunny spots in shaded stream sites. Males usually perch on the tips of overhanging branches maintaining territories.

优雅隐溪螅
Cryptophaea saukra

越南隐溪螅
Cryptophaea vietnamensis

云南隐溪螅
Cryptophaea yunnanensis

隐溪螅属 雄性肛附器
Genus *Cryptophaea*, male anal appendages

优雅隐溪螅 *Cryptophaea saukra* Hämäläinen, 2003

优雅隐溪螅 雄，云南（普洱）
Cryptophaea saukra, male from Yunnan (Pu'er)

【形态特征】雄性面部大面积黄色；胸部黑色具黄色条纹，后胸具白色粉霜；腹部黑色，第1～8节具甚细的黄色条纹，第9～10节覆盖粉霜，第10节背面具1个短刺突。雌性腹部较短，未熟时胸部条纹为蓝色，成熟后为黄色和蓝白色。【长度】体长 44～58 mm，腹长 35～50 mm，后翅 32～34 mm。【栖息环境】海拔 1000 m以下的林荫小溪。【分布】云南（普洱）；泰国。【飞行期】5—7月。

[Identification] Male face mainly yellow. Thorax black with yellow markings, metathorax whitish pruinosed. Abdomen black, S1-S8 with narrow yellow stripes, S9-S10 pruinosed, S10 with a short spine dorsally. Female abdomen shorter, the thoracic stripes blue when immature, yellow and bluish white when fully mature. [Measurements] Total length 44-58 mm, abdomen 35-50 mm, hind wing 32-34 mm. [Habitat] Shady streams below 1000 m elevation. [Distribution] Yunnan (Pu'er); Thailand. [Flight Season] May to July.

优雅隐溪螅 雌，云南（普洱）
Cryptophaea saukra, female from Yunnan (Pu'er)

越南隐溪螅 *Cryptophaea vietnamensis* (Van Tol & Rozendaal, 1995)

【形态特征】雄性面部黑色具大面积绿黄色斑；胸部黑色具蓝绿色和蓝色条纹，翅稍染褐色；腹部黑色，第1~4节侧缘具甚小的黄色斑点，第9~10节稍微覆盖粉霜。雌性胸部背面具红褐色斑，胸部侧面和腹部具黄色条纹。【长度】体长 41~53 mm，腹长 32~45 mm，后翅 29~31 mm。【栖息环境】海拔 1000 m以下的林荫小溪。【分布】云南（红河）、广西；老挝、越南。【飞行期】4—10月。

[Identification] Male face black with extensive greenish yellow markings. Thorax black with bluish green and blue markings, wings slightly tinted with brown. Abdomen black, S1-S4 with small yellow spots laterally, S9-S10 slightly pruinosed. Female with broad reddish brown spots on dorsum of thorax, sides of thorax and abdomen with yellow stripes. [Measurements] Total length 41-53 mm, abdomen 32-45 mm, hind wing 29-31 mm. [Habitat] Shady streams below 1000 m elevation. [Distribution] Yunnan (Honghe), Guangxi; Laos, Vietnam. [Flight Season] April to October.

越南隐溪螅 雄，广西
Cryptophaea vietnamensis, male from Guangxi

越南隐溪螅 雌，云南（红河）
Cryptophaea vietnamensis, female from Yunnan (Honghe)

云南隐溪螅 *Cryptophaea yunnanensis* (Davies & Yang, 1996)

【形态特征】雄性面部主要黑色，上唇黄色；胸部黑色具淡蓝色条纹；腹部黑色，第1~8节侧缘具甚细的淡蓝色条纹，第9~10节稍微覆盖粉霜。雌性黑色，未成熟时胸部条纹为蓝色，成熟后为黄褐色。【长度】体长 42~53 mm，腹长 33~43 mm，后翅 30~31 mm。【栖息环境】海拔 1500 m以下的林荫小溪。【分布】中国云南（西双版纳、普洱）特有。【飞行期】5—7月。

云南隐溪螅 雌，云南（西双版纳）
Cryptophaea yunnanensis, female from Yunnan (Xishuangbanna)

[Identification] Male face mainly black, labrum yellow. Thorax black with pale blue markings. Abdomen black, S1-S8 with narrow pale blue stripes laterally, S9-S10 slightly pruinosed. Female black, the thoracic stripes blue when immature, yellowish brown when fully mature.[Measurements] Total length 42-53 mm, abdomen 33-43 mm, hind wing 30-31 mm. [Habitat] Shady streams below 1500 m elevation. [Distribution] Endemic to Yunnan (Xishuangbanna, Pu'er) of China. [Flight Season] May to July.

云南隐溪螅 雄，云南（西双版纳）
Cryptophaea yunnanensis, male from Yunnan (Xishuangbanna)

暗溪螅属 Genus *Dysphaea* Selys, 1853

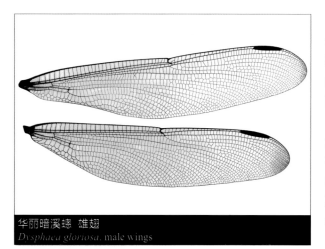

华丽暗溪螅 雄翅
Dysphaea gloriosa, male wings

本属全世界已知9种，分布于亚洲的热带和亚热带地区。中国已知3种，分布于华南和西南地区。本属是本科中体型最大且十分粗壮的豆娘，它们身体色彩较深，雄性翅上具有深色斑纹或者完全深色。本属中国已知的雄性可以通过翅的色彩区分。

本属豆娘主要栖息于低地的河流，但它们的种群数量明显少于同生境下的溪螅属豆娘。雄性通常会停落在河流中的岩石或者水生植物上，并会经常巡逻，它们的飞行能力很强，有时可以连续巡飞较长时间。交尾结束后雌雄连结飞行至合适地点，之后雄性与雌性分离，雌性沿水生植物潜入水下产卵，雄性护卫。

The genus includes nine described species confined to tropical and subtropical Asia. Three species are recorded from China, distributed in the South and Southwest regions. The stout bodied *Dysphaea* species are among the largest damselflies in their family. Their body is dark, the wings of males are usually tinted with dark spots, or are entirely dark. Males of the Chinese species can be distinguished by the color pattern of the wings.

Dysphaea species prefer lowland rivers, but they are always less abundant than the co-occurring *Euphaea* species. Males usually perch on the rocks or emergent plants but are strong fliers and will sometimes patrol along stretches of the river. Pairs in tandem can be seen flying in search of oviposition sites after mating. Once located, the female uncouples and descends alone under the water surface to lay eggs on submerged plants, the male remaining guarding nearby.

黑斑暗溪螅 交尾
Dysphaea basitincta, mating pair

黑斑暗溪螆 *Dysphaea basitincta* Martin, 1904

【形态特征】雄性通体黑色，腹部侧缘具甚小的黄斑；翅基方具宽阔的黑斑，翅端具略小的黑斑，中央稍染褐色。雌性黑褐色具黄色条纹，翅稍染褐色。【长度】体长 54~57 mm，腹长 40~44 mm，后翅 38~41 mm。【栖息环境】海拔 500 m 以下的河流和宽阔溪流。【分布】云南（红河）、广西、海南；越南。【飞行期】4—7月。

[Identification] Male body black throughout, abdomen with tiny yellow spots laterally. wings with bases broadly and tips more narrowly black, the intervening hyaline section with slight brownish tint. Female blackish brown with yellow markings, wings slightly tinted with brown. [Measurements] Total length 54-57 mm, abdomen 40-44 mm, hind wing 38-41 mm. [Habitat] Rivers and wide streams below 500 m elevation. [Distribution] Yunnan (Honghe), Guangxi, Hainan; Vietnam. [Flight Season] April to July.

黑斑暗溪螆 雄，广西
Dysphaea basitincta, male from Guangxi

华丽暗溪螅 *Dysphaea gloriosa* Fraser, 1938

华丽暗溪螅 雄,云南(西双版纳)
Dysphaea gloriosa, male from Yunnan (Xishuangbanna)

【形态特征】雄性面部主要黑色;胸部黑色具黄色条纹,翅完全琥珀色,翅端色彩稍微加深;腹部黑色具小黄斑。雌性黑褐色具更发达的黄色条纹,翅稍染淡褐色。【长度】体长 47~51 mm,腹长 34~39 mm,后翅 32~36 mm(根据云南标本)。【栖息环境】海拔 500 m以下的河流。【分布】云南(西双版纳)、海南;印度、不丹、泰国、柬埔寨、老挝、越南。【飞行期】4—10月。

[Identification] Male face mainly black. Thorax black with yellow markings, wings amber throughout, tips slightly darker. Abdomen with small yellow spots. Female blackish brown with more extensive yellow markings, wings slightly tinted with pale brown. [Measurements] Total length 47-51 mm, abdomen 34-39 mm, hind wing 32-36 mm (based on Yunnan specimens). [Habitat] Rivers below 500 m elevation. [Distribution] Yunnan (Xishuangbanna), Hainan; India, Bhutan, Thailand, Cambodia, Laos, Vietnam. [Flight Season] April to October.

华丽暗溪螅 交尾，云南（西双版纳）
Dysphaea gloriosa, mating pair from Yunnan (Xishuangbanna)

华丽暗溪螅 雌雄连结，海南 ｜莫善濂 摄
Dysphaea gloriosa, pair in tandem from Hainan ｜ Photo by Shanlian Mo

浩淼暗溪蟌 *Dysphaea haomiao* Hämäläinen, 2012

【形态特征】雄性整个身体包括翅黑色；腹部具甚小的黄斑。雌性黑褐色具黄色条纹，翅稍染褐色。云南红河的雄性个体翅的色彩有差异，有些色彩较淡，为深褐色。【长度】体长 50～56 mm，腹长 37～41 mm，后翅 37～38 mm。【栖息环境】海拔 500 m 以下的宽阔河流。【分布】云南（红河）、贵州、广西；越南。【飞行期】4—7月。

浩淼暗溪蟌 雄，云南（红河）
Dysphaea haomiao, male from Yunnan (Honghe)

[Identification] Male black throughout including wings. Abdomen with tiny yellow spots laterally. Female blackish brown with yellow markings, wings slightly tinted with brown. Color of wings in males from Honghe of Yunnan is variable, some males have less darkened wings, which are dark brown. [Measurements] Total length 50-56 mm, abdomen 37-41 mm, hind wing 37-38 mm. [Habitat] Rivers below 500 m elevation. [Distribution] Yunnan (Honghe), Guizhou, Guangxi; Vietnam. [Flight Season] April to July.

浩淼暗溪蟌 雄，云南（红河）
Dysphaea haomiao, male from Yunnan (Honghe)

浩淼暗溪蟌 雌，云南（红河）
Dysphaea haomiao, female from Yunnan (Honghe)

溪蟌属 Genus *Euphaea* Selys, 1840

本属全世界已知约30种，在亚洲的热带和亚热带地区分布广泛。中国已知8种，主要分布在华南和西南地区。这些体中型的豆娘十分粗壮，身体色彩较深并具黄色或棕色条纹；雄性翅上具深色斑纹，并具彩虹色泽，有些翅完全深色；雄性的腹部第10节具1个锥形突起，可以与近似的暗溪蟌属区分。本属中国已知的雄性可以通过翅的色彩区分。

本属豆娘栖息于河流和山区溪流。雄性通常会停落在河流中的岩石或者水生植物上占据领地，时而短暂巡飞。交尾结束后雌雄连结飞行寻找合适的地点产卵。雌性潜水产卵。

透顶溪蟌 雄翅
Euphaea masoni, male wings

The genus contains about 30 species, widely distributed in tropical and subtropical Asia. Eight species are recorded from China, mainly found in the South and Southwest regions. These rather robustly built, medium-sized damselflies have a dark body with yellow or brownish stripes. Wings of males are usually marked and may be brilliantly iridescent. Some species have entirely dark, matt wings. In males the last abdominal segment has a distinct pyramidal prominence, a character which easily separates *Euphaea* from *Dysphaea* species. Males of Chinese *Euphaea* species can be distinguished by the differences in wing pattern.

褐翅溪蟌 雄（左）和方带溪蟌 雄（右）
Euphaea opaca, male (left) and *Euphaea decorate*, male (right)

Euphaea species inhabit rivers and montane streams. Males usually perch on the rocks or emergent plants guarding territory, sometimes patrol for short time. After mating some species fly in tandem for a suitable site for laying eggs. Females oviposit underwater.

透顶溪蟌
Euphaea masoni

浩淼暗溪蟌
Dysphaea haomiao

溪蟌属与暗溪蟌属 雄性腹部末端比较
Comparison of the male abdominal tip between genera *Euphaea* and *Dysphaea*

方带溪蟌 *Euphaea decorata* Hagen, 1853

方带溪蟌 雌,广东
Euphaea decorata, female from Guangdong

方带溪蟌 雄，广东
Euphaea decorata, male from Guangdong

【形态特征】雄性面部黑色；胸部黑色具甚细的褐色条纹，前翅透明，后翅亚端部具1个甚大的黑带；腹部黑色。雌性黑褐色具黄色条纹，翅前缘染有褐色，在后翅中更显著。【长度】体长 37～42 mm，腹长 28～32 mm，后翅 25～27 mm。【栖息环境】海拔 1500 m以下的山区溪流。【分布】华南和西南地区广布；越南。【飞行期】4—11月。

[Identification] Male face black. Thorax black with narrow brownish stripes, fore wings hyaline, hind wings with a subapical dark opaque band. Abdomen black. Female blackish brown with yellow stripes. Wings slightly tinted with pale brown in costal areas, more extensively in hind wing. [Measurements] Total length 37-42 mm, abdomen 28-32 mm, hind wing 25-27 mm. [Habitat] Montane streams below 1500 m elevation. [Distribution] Widespread in South and Southwest regions; Vietnam. [Flight Season] April to November.

台湾溪蟌 *Euphaea formosa* Hagen, 1869

台湾溪蟌 雌,台湾 | 嘎嘎 摄
Euphaea formosa, female from Taiwan | Photo by Gaga

【形态特征】雄性面部黑色;胸部黑色具红褐色条纹,前翅透明,后翅具1个甚大的深铜色带,色带的宽度大于翅长的1/2,使翅仅在基方1/3处和端部透明;腹部第1~5节暗红色,第6~10节黑色。雌性黑色具黄色条纹,翅具褐色斑,在后翅更显著。【长度】体长 39~52 mm,腹长 32~40 mm,后翅 31~34 mm。【栖息环境】海拔1500 m以下的山区溪流。【分布】中国台湾特有。【飞行期】2—11月。

[Identification] Male face black. Thorax black with reddish brown stripes, fore wings hyaline and hind wings with a broad opaque band with dark coppery sheen. This band covers more than half of the wing length, leaving only the basal third and wing tip hyaline. S1-S5 dull red, S6-S10 black. Female black with yellow markings, wings with brown markings, much more extensive in hind wing. [Measurements] Total length 39-52 mm, abdomen 32-40 mm, hind wing 31-34 mm. [Habitat] Montane streams below 1500 m elevation. [Distribution] Endemic to Taiwan of China. [Flight Season] February to November.

台湾溪蟌 雄,台湾 | 嘎嘎 摄
Euphaea formosa, male from Taiwan | Photo by Gaga

绿翅溪螅 *Euphaea guerini* Rambur, 1842

绿翅溪螅 雄, 广西
Euphaea guerini, male from Guangxi

【形态特征】雄性身体主要黑色; 胸部侧面具较暗的黄色条纹, 前翅黑色, 基方具甚小的透明区域, 端部透明, 后翅全黑色, 翅正面基方的2/3处具闪烁光泽的蓝绿色或苹果绿色斑, 此色斑在翅背面反射深蓝色或者紫色光泽; 腹部黑色。【长度】雄性体长 53 mm, 腹长 42 mm, 后翅 31 mm。【栖息环境】海拔 1000 m以下的山区溪流。【分布】广西; 柬埔寨、老挝、越南。【飞行期】4—9月。

[Identification] Male body mainly black. Thorax with obscure yellow stripes laterally, fore wings black with a small hyaline area at base, tips more broadly hyaline, hind wings wholly black the upperside with a brilliant iridescent turquoise or apple green patch in the basal two thirds; this area on the underside reflects brilliant iridescent ultramarine or purple. Abdomen black. [Measurements] Male total length 53 mm, abdomen 42 mm, hind wing 31 mm. [Habitat] Montane streams below 1000 m elevation. [Distribution] Guangxi; Cambodia, Laos, Vietnam. [Flight Season] April to September.

绿翅溪螅 雄, 广西
Euphaea guerini, male from Guangxi

透顶溪蟌 *Euphaea masoni* Selys, 1879

【形态特征】雄性身体主要为黑色；胸部侧面具较暗的黄色条纹，前翅在中央具1条宽阔的黑色带，宽度为翅长的1/2，后翅黑色，基方和端部具甚小的透明区域。雌性黑色具黄色条纹，翅透明。【长度】体长 45～48 mm，腹长 35～38 mm，后翅 28～30 mm。【栖息环境】海拔 1000 m以下的河流和宽阔溪流。【分布】云南（西双版纳、普洱）；印度、缅甸、泰国、柬埔寨、老挝、越南。【飞行期】全年可见。

[Identification] Male body mainly black. Thorax with obscure yellow stripe laterally, fore wings with a broad black band in the middle, covering half of the wing length, hind wings black, with extreme base and wing tips narrowly hyaline. Female black with yellow stripes, wings hyaline. [Measurements] Total length 45-48 mm, abdomen 35-38 mm, hind wing 28-30 mm. [Habitat] Rivers and wide streams below 1000 m elevation. [Distribution] Yunnan (Xishuangbanna, Pu'er); India, Myanmar, Thailand, Cambodia, Laos, Vietnam. [Flight Season] Throughout the year.

透顶溪蟌 交尾，云南（西双版纳）
Euphaea masoni, mating pair from Yunnan (Xishuangbanna)

透顶溪蟌 雄,云南(西双版纳)
Euphaea masoni, male from Yunnan (Xishuangbanna)

透顶溪蟌 雌,云南(西双版纳)
Euphaea masoni, female from Yunnan (Xishuangbanna)

黄翅溪蟌 *Euphaea ochracea* Selys, 1859

黄翅溪蟌 雄，云南（西双版纳）
Euphaea ochracea, male from Yunnan (Xishuangbanna)

黄翅溪蟌 交尾，云南（西双版纳）
Euphaea ochracea, mating pair from Yunnan (Xishuangbanna)

黄翅溪蟌 雌，云南（西双版纳）
Euphaea ochracea, female from Yunnan (Xishuangbanna)

【形态特征】雄性面部黑色；胸部黑色具红褐色圆圈形条纹，翅染有红褐色，前翅的红褐色区域略超过翅的 1/2，后翅则伸达翅痣处；腹部黑色，第1～6节侧缘具红褐色条纹。雌性黑色具黄色条纹，翅透明，基方稍染褐色。【长度】体长 41～46 mm，腹长 31～35 mm，后翅 26～30 mm。【栖息环境】海拔 1500 m 以下的山区溪流。【分布】云南广布；南亚、东南亚广布。【飞行期】5—12月。

[Identification] Male face black. Thorax black with reddish brown, partly-formed loop markings, wings largely with deep reddish brown tint, in the fore wing this color extending over the basal half, in hind wing as far as the pterostigma. Abdomen black with reddish brown lateral stripes on S1-S6. Female black bodied with yellow markings, wings largely clear with bases slightly tinted with pale brown. [Measurements] Total length 41-46 mm, abdomen 31-35 mm, hind wing 26-30 mm. [Habitat] Montane streams below 1500 m elevation. [Distribution] Widespread in Yunnan; Widespread in South and Southeast Asia. [Flight Season] May to December.

褐翅溪螅 *Euphaea opaca* Selys, 1853

【形态特征】雄性身体主要黑色；胸部和腹部具甚细的褐色条纹；翅深褐色。雌性黑色具黄色条纹，翅透明，前缘基方染褐色。【长度】体长 55~60 mm，腹长 40~46 mm，后翅 37~40 mm。【栖息环境】海拔 500 m 以下的溪流和河流。【分布】中国特有，分布于安徽、湖北、浙江、福建、广东、香港。【飞行期】4—8月。

[Identification] Male body mainly black. Thorax and abdomen with fine pale brownish stripes. Wings entirely dark brown. Female with black body and extensive yellow markings, wings hyaline, costal areas pale brownish tinted.

褐翅溪螅 雄，广东
Euphaea opaca, male from Guangdong

褐翅溪蟌 交尾
Euphaea opaca, mating pair

褐翅溪蟌 雌雄连结
Euphaea opaca, pair in tandem

褐翅溪蟌 雄性护卫雌性
Euphaea opaca, male guarding female

褐翅溪蟌 雌性潜水产卵
Euphaea opaca, female laying eggs under water

[Measurements] Total length 55-60 mm, abdomen 40-46 mm, hind wing 37-40 mm. [Habitat] Streams and rivers below 500 m elevation. [Distribution] Endemic to China, recorded from Anhui, Hubei, Zhejiang, Fujian, Guangdong, Hong Kong. [Flight Season] April to August.

宽带溪蟌 *Euphaea ornata* (Campion, 1924)

宽带溪蟌 雌，海南
Euphaea ornata, female from Hainan

【形态特征】雄性面部黑色；胸部黑色具黄褐色条纹，前翅透明，基方稍染褐色，后翅中央显著加阔，具1条红褐色宽带，端方透明；腹部黑色，第1~6节侧缘具黄褐色条纹。雌性黑色具黄色条纹，翅透明，基方稍染褐色。【长度】体长 39~47 mm，腹长 29~37 mm，后翅 27~29 mm。【栖息环境】海拔 1500 m以下的山区溪流。【分布】中国海南特有。【飞行期】3—11月。

[Identification] Male face black. Thorax black with yellowish brown stripes, fore wings hyaline with bases slightly tinted with brown, hind wings distinctly broadened in the middle, with a median opaque reddish brown band, tips broadly hyaline. Abdomen black, S1-S6 with yellowish brown stripes laterally. Female black with yellow markings, wings hyaline with bases slightly tinted with pale brown. [Measurements] Total length 39-47 mm, abdomen 29-37 mm, hind wing 27-29 mm. [Habitat] Montane streams below 1500 m elevation. [Distribution] Endemic to Hainan of China. [Flight Season] March to November.

宽带溪蟌 雄,海南
Euphaea ornata, male from Hainan

宽带溪蟌 交尾,海南
Euphaea ornata, mating pair from Hainan

华丽溪蟌 *Euphaea superba* Kimmins, 1936

【形态特征】雄性面部黑色，上唇具黄斑；胸部黑色具红褐色条纹，翅深褐色，翅脉红褐色；腹部第1~6节红褐色，第7~10节黑色。雌性黑褐色具黄色条纹，翅稍染褐色。本种与褐翅溪蟌相似，但翅脉红褐色，胸部具更发达的红褐色条纹，这些条纹并不随年纪增长而褪去。【长度】体长 51~56 mm，腹长 37~44 mm，后翅 35~37 mm。【栖息环境】海拔 1000 m 以下的河流和宽阔溪流。【分布】贵州、广西；越南。【飞行期】4—7月。

[Identification] Male face black, labrum with yellow spots. Thorax black with reddish brown stripes, wings dark brown with distinctive reddish brown venation. S1-S6 dark reddish, S7-S10 black. Female blackish brown with yellow markings, wings slightly tinted with brown. Similar to *E. opaca*, but venation distinctive reddish brown, the thoracic reddish brown stripes more developed and not reduced when aged. [Measurements] Total length 51-56 mm, abdomen 37-44 mm, hind wing 35-37 mm. [Habitat] Rivers and wide streams below 1000 m elevation. [Distribution] Guizhou, Guangxi; Vietnam. [Flight Season] April to July.

华丽溪蟌 雄，贵州
Euphaea superba, male from Guizhou

华丽溪蟌 雌雄连结，贵州 | 莫善濂 摄
Euphaea superba, pair in tandem from Guizhou | Photo by Shanlian Mo

华丽溪蟌 雄，贵州 | 莫善濂 摄
Euphaea superba, male from Guizhou | Photo by Shanlian Mo

溪螅属待定种 *Euphaea* sp.

溪螅属待定种 雄，云南（红河）
Euphaea sp, male from Yunnan (Honghe)

　　【形态特征】雄性面部黑色；胸部黑色，侧面具黄色条纹，前翅中央黑色，基方和端方透明，后翅黑色，背面中央具1个甚大的霓虹紫色或蓝色斑，具霓虹色彩；腹部黑色。雌性黑褐色具黄色条纹，翅透明，基方稍染褐色。本种仅后翅背面具霓虹色斑，第9节腹面缺少一簇紧毛，与绿翅溪螅不同。【长度】体长 43~49 mm，腹长 33~38 mm，后翅 28~32 mm。【栖息环境】海拔 500 m 以下的宽阔溪流。【分布】云南（红河）、广西。【飞行期】4—7月。

[Identification] Male face black. Thorax black with yellow markings laterally, fore wings black with hyaline bases and tips, underside of hind wings black with a median iridescent purple or blue patch. Abdomen black. Female blackish brown with yellow markings, wings hyaline with bases slightly tinted with brown. The median iridescent purple patch only appears in the underside of hind wings in this species and it lacks a tuft of setae under S9, thus differing from *E. guerini*. [Measurements] Total length 43-49 mm, abdomen 33-38 mm, hind wing 28-32 mm. [Habitat] Wide streams below 500 m elevation. [Distribution] Yunnan (Honghe), Guangxi. [Flight Season] April to July.

溪螅属待定种 雌，云南（红河）
Euphaea sp, female from Yunnan (Honghe)

5 大溪螁科 Family Philogangidae

关于大溪螁是否是蜻蜓中最古老的类群学界一直存在争议。本科仅包含1属，即大溪螁属，主要分布于东洋界。它们是体型粗壮且巨大的豆娘，栖息于茂盛森林中的溪流。大溪螁多数时间停歇在树干或树枝上，腹部稍微翘起，翅展开。欲交配的雄性停落在水面附近。交尾在树上完成，持续较长时间。交尾结束后雄性护卫雌性，雌性将卵产在漂于水面的树枝上。

Members of the family Philogangidae are arguably among the most ancient of dragonflies. The family includes a single genus, *Philoganga*, mainly distributed in the Oriental region. They are stout bodied and large-sized damselflies and inhabit streams in dense forests. Most of the time individuals perch on tree trunks or branches with the abdomen slightly raised and the wings spread. Males perch near the water when they intend to mate. Mating also takes place on trees, and is of long duration. After mating females oviposit on the branches overhanging water while guarded by the male.

大溪螅 雄
Philoganga vetusta, male

壮大溪螅指名亚种 雄
Philoganga robusta robusta, male

大溪蟌属 Genus *Philoganga* Kirby, 1890

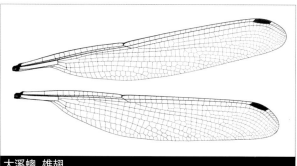

大溪蟌 雄翅
Philoganga vetusta, male wings

本属全球已知4种，中国已知2种，广布于南方各地。这些神秘的昆虫翅窄而长，基方具较长的翅柄。雄性的上肛附器较长但下肛附器甚短，雌性的产卵器十分发达。

The genus contains four species, two of them recorded from China, widely distributed in the south. They are elusive insects with narrow, long wings and a long stalk at the wing base. Male superior appendages long but the inferiors reduced, the female ovipositor well developed.

大溪蟌 雄
Philoganga vetusta, male

壮大溪螅指名亚种 *Philoganga robusta robusta* Navás, 1936

壮大溪螅指名亚种 雄，广东
Philoganga robusta robusta, male from Guangdong

壮大溪螅指名亚种 雌，广东 | 宋黎明 摄
Philoganga robusta robusta, female from Guangdong | Photo by Liming Song

壮大溪螅指名亚种 雄，广西
Philoganga robusta robusta, male from Guangxi

壮大溪螅指名亚种 雌，广西
Philoganga robusta robusta, female from Guangxi

【形态特征】雄性复眼深蓝色，面部黑色，上唇具黄斑；胸部黑色，合胸脊两侧具1对甚细的黄色条纹，肩前条纹显著，合胸侧面具2条宽阔的黄色条纹；腹部黑色具黄色斑纹。雌性与雄性相似但腹部更粗壮。【长度】体长65～77 mm，腹长 46～57 mm，后翅 49～59 mm。【栖息环境】海拔 1500 m以下森林中的溪流。【分布】四川、贵州、云南、河南、湖北、湖南、江西、浙江、福建、广西、广东、海南；越南。【飞行期】4—8月。

[Identification] Male eyes dark blue, face black, labrum with yellow spots. Thorax black with a pair of linear yellow stripes besides dorsal carina and long antehumeral stripes, sides with two broad yellow stripes. Abdomen black with yellow markings. Female similar to male, but with more robust abdomen. [Measurements] Total length 65-77 mm, abdomen 46-57 mm, hind wing 49-59 mm. [Habitat] Streams in forest below 1500 m elevation. [Distribution] Sichuan, Guizhou, Yunnan, Henan, Hubei, Hunan, Jiangxi, Zhejiang, Fujian, Guangxi, Guangdong, Hainan; Vietnam. [Flight Season] April to August.

大溪螅 *Philoganga vetusta* Ris, 1912

【形态特征】雄性复眼上方黑色下方深蓝色，上唇
具黄斑。胸部黑色，合胸脊两侧具1对甚细的黄色条纹，
肩前条纹有时完整，有时仅为下半段，有时缺失，合胸
侧面具2条宽阔的黄色条纹。腹部橙色和黑色。雌性更
粗壮，腹部黑色，各节具黄绿色斑纹，第9节背面具1个
甚大的T形斑，与雌性的壮大溪螅不同。本种身体色彩
变异较大。【长度】体长 57~73 mm，腹长 42~56 mm，
后翅 45~55 mm。【栖息环境】海拔 1500 m以下森林
中的溪流。【分布】云南（红河）、四川、贵州、湖南、江
西、浙江、福建、广西、广东、海南、香港；老挝、越南。
【飞行期】3—8月。

大溪螅 雌，海南
Philoganga vetusta, female from Hainan

[Identification] Male eyes black above and
dark blue below, labrum with yellow spots. Thorax black with a pair of linear yellow stripes besides dorsal carina,
antehumeral stripes complete, half or absent, sides with two broad yellow stripes. Abdomen orange and black. Female
stouter, abdomen balck with yellowish green stripes, S9 with a large T-mark dorsally, a different character from female

大溪螅 雄，海南
Philoganga vetusta, male from Hainan

of *P. robusta*. Body maculation of this species is variable. **[Measurements]** Total length 57-73 mm, abdomen 42-56 mm, hind wing 45-55 mm. **[Habitat]** Streams in forest below 1500 m elevation. **[Distribution]** Yunnan (Honghe), Sichuan, Guizhou, Hunan, Jiangxi, Zhejiang, Fujian, Guangxi, Guangdong, Hainan, Hong Kong; Laos, Vietnam. **[Flight Season]** March to August.

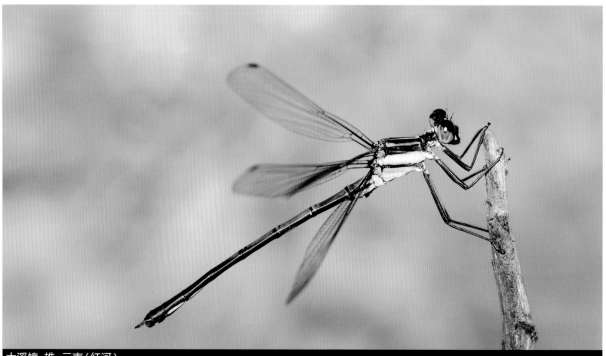

大溪螁 雄,云南(红河)
Philoganga vetusta, male from Yunnan (Honghe)

大溪螁 雌,云南(红河)
Philoganga vetusta, female from Yunnan (Honghe)

大溪螅 雄, 广东 | 宋黎明 摄
Philoganga vetusta, male from Guangdong | Photo by Liming Song

大溪螅 雌, 广东 | 宋黎明 摄
Philoganga vetusta, female from Guangdong | Photo by Liming Song

6 ▷ 黑山蟌科 Family Philosinidae

　　本科仅包括2属豆娘，共计12种，仅分布于东洋界。中国已知2属3种，主要分布在华南和西南地区。本科豆娘体型粗壮，翅具明显的翅柄，停歇时翅展开，它们的身体通常具有非常鲜艳的色彩或身披浓密的白色粉霜。

　　本科豆娘栖息于较低海拔的河流和森林中的溪流。它们通常躲在河岸带茂盛的树林中。雄性会停落在水面附近的遮蔽环境中占据领地。

This family contains two genera 12 species worldwide, confined to Oriental region. A total of three in both genera are recorded from China, distributed in the South and Southwest regions. Members of the family are stout species, with wings stalked at base, perching with spread wings. They are usually colorful and heavily whitish pruinosed.

Species of the family inhabit lowland rivers and streams in forest. They usually hide in the dense bordering forest. Males perch near water with shelter for territory.

海南鲨山螅, 雄 | 莫善濂 摄
Rhinagrion hainanense, male
| Photo by Shanlian Mo

覆雪黑山螅, 雄 | 宋睿斌 摄
Philosina alba, male | Photo
by Ruibin Song

黑山螅属 Genus *Philosina* Ris, 1917

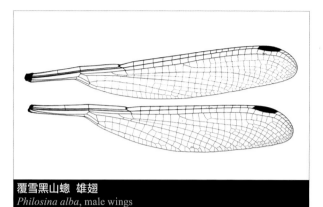

覆雪黑山螅 雄翅
Philosina alba, male wings

本属全球已知2种，分布于中国、老挝和越南。本属豆娘体型较大且粗壮，雄性身体被浓厚的白色粉霜覆盖，很容易与鲨山螅属豆娘区分。

本属豆娘栖息于较低海拔的河流和溪流。它们具有很强的飞行能力，红尾黑山螅可以像蜓一样悬停飞行，而覆雪黑山螅可以在密林里快速穿梭。雄性会在水边占据领地，通常停在树枝上，雌性在河岸陡坡的潮湿土壤上产卵，雄性护卫。

The genus contains only two species, distributed in China, Laos and Vietnam. They are moderately large sized and robust damselflies. The male body bears dense whitish pruinescence, which becomes an outstanding character for separating them from *Rhinagrion* species.

Philosina species inhabit rivers and streams in lowland. Both fly strongly. *P. buchi* can hover like many aeshnids and *P. alba* can fly fast through the forest. Males hold territory at the water margins, usually perching on branches. Females were observed laying eggs into the mud of the steep bank while males guarded them.

红尾黑山螅 雄
Philosina buchi, male

覆雪黑山螅 *Philosina alba* Wilson, 1999

【形态特征】雄性复眼褐色和绿色，面部黑色具黄绿色斑纹；胸部黑色并稍微覆盖粉霜，具黄色的肩前条纹，侧面具2条黄色条纹；腹部覆盖白色粉霜。雌性黑色具黄色条纹。【长度】体长 48~52 mm，腹长 35~40 mm，后翅 33~36 mm。【栖息环境】海拔 1000 m 以下森林中的溪流。【分布】广东、海南；老挝、越南。【飞行期】4—7月。

[Identification] Male eyes brown and green, face black with yellowish green markings. Thorax black and slightly pruinosed, with yellow antehumeral stripes, sides with two yellow stripes. Abdomen with dense whitish pruinescence. Female black with yellow markings. [Measurements] Total length 48-52 mm, abdomen 35-40 mm, hind wing 33-36 mm. [Habitat] Streams in forest below 1000 m elevation. [Distribution] Guangdong, Hainan; Laos, Vietnam. [Flight Season] April to July.

覆雪黑山螅 雄，广东 | 宋睿斌 摄
Philosina alba, male from Guangdong | Photo by Ruibin Song

覆雪黑山螅 交尾，海南 | 莫善濂 摄
Philosina alba, mating pair from Hainan | Photo by Shanlian Mo

覆雪黑山螅 雄，广东 | 宋睿斌 摄
Philosina alba, female from Guangdong | Photo by Ruibin Song

红尾黑山螅 *Philosina buchi* Ris, 1917

【形态特征】雄性复眼蓝黑色，面部黑色覆盖白色粉霜；胸部黑色，黄色的肩前条纹和侧面的条纹被白色粉霜覆盖，翅端具小褐斑；腹部覆盖白色粉霜，第7~9节红色。雌性黑色具黄色条纹。【长度】体长 58~60 mm，腹长 44~45 mm，后翅 37~38 mm。【栖息环境】海拔 1000 m 以下的溪流和河流。【分布】四川、贵州、福建、广西、广东；越南。【飞行期】4—8月。

[Identification] Male eyes bluish black, face black with white pruinescence. Thorax black, the whitish pruinescence covering the yellow antehumeral stripes and lateral stripes, wings with small brown tips. Abdomen with whitish pruinescence, S7-S9 red. Female black with yellow markings. [Measurements] Total length 58-60 mm, abdomen 44-45 mm, hind wing 37-38 mm. [Habitat] Streams and rivers below 1000 m elevation. [Distribution] Sichuan, Guizhou, Fujian, Guangxi, Guangdong; Vietnam. [Flight Season] April to August.

红尾黑山螅 雄，广西
Philosina buchi, male from Guangxi

红尾黑山螅 雄，广西
Philosina buchi, male from Guangxi

红尾黑山螅 雌，广西
Philosina buchi, female from Guangxi

红尾黑山螅 交尾，广东 | 宋睿斌 摄
Philosina buchi, mating pair from Guangdong | Photo by Ruibin Song

鲨山螅属 Genus *Rhinagrion* Calvert, 1913

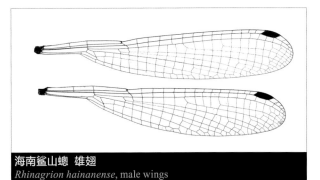

海南鲨山螅 雄翅
Rhinagrion hainanense, male wings

本属全世界已知10种,主要分布于亚洲的热带地区。中国已知仅1种,海南鲨山螅。同黑山螅相比,鲨山螅体型更小,但身体具有更斑驳的色彩。

鲨山螅属豆娘栖息于低海拔的溪流和河流,喜欢在密林中穿梭。雌性将卵产在石壁的苔藓上。

This genus includes ten species distributed in tropical Asia. Only one species is recorded from China, *Rhinagrion hainanense* Wilson & Reels, 2001. Compared with *Philosina*, *Rhinagrion* species are smaller, but the body with more complex markings.

Rhinagrion species inhabit streams and rivers in lowland, prefer to fly in forest. Females oviposit in the moss on rocks.

海南鲨山螅 雌
Rhinagrion hainanense, female

海南鲨山螅 *Rhinagrion hainanense* Wilson & Reels, 2001

【形态特征】雄性复眼蓝黑色,面部黑色;胸部黑色,具黄色的肩前条纹,合胸侧面具2条黄色条纹;腹部黄褐色,第1~7节具白色斑纹,第8~10节背面黑色具白色斑点,腹面红色。雌性身体红褐色具蓝灰色和黑色条纹。【长度】腹长 32~34 mm,后翅 24~26 mm。【栖息环境】海拔 500 m 以下的林荫小溪和宽阔河流河岸带植被茂盛的河段。【分布】贵州、广西、海南;泰国、柬埔寨、老挝、越南。【飞行期】4—8月。

[Identification] Male eyes bluish black, face black. Thorax black with yellow antehumeral stripes, sides with two yellow stripes. Abdomen yellowish brown, S1-S7 with white markings, S8-S10 black dorsally with white spots, sternites red. Female basically reddish brown with bluish grey and black markings. [Measurements] Abdomen 32-34 mm, hind wing 24-26 mm. [Habitat] Shady streams and large rivers with plenty of bordering vegetation below 500 m elevation. [Distribution] Guizhou, Guangxi, Hainan; Thailand, Cambodia, Laos, Vietnam. [Flight Season] April to August.

海南鲨山螅 雄, 海南 | 莫善濂 摄
Rhinagrion hainanense, male from Hainan | Photo by Shanlian Mo

海南鲨山螅 雌, 海南 | 莫善濂 摄
Rhinagrion hainanense, female from Hainan | Photo by Shanlian Mo

7 色蟌总科待定科 **Family "Calopterygoidea *incertae sedis*"**

　　根据最新分子分类系统，中国有6属30余种豆娘的分类地位尚未明确。此处暂将它们放在"色蟌总科待定科"中进行描述。

　　Based on the new molecular system of classification, over 30 species in six genera are considered to be of "incertae sedis". They were here listed as "Calopterygoidea *incertae sedis*".

黑缅山螅 雄
Burmargiolestes melanothorax, male

宾黑古山螅 雄
Priscagrion pinheyi, male

野蟌属 Genus *Agriomorpha* May, 1933

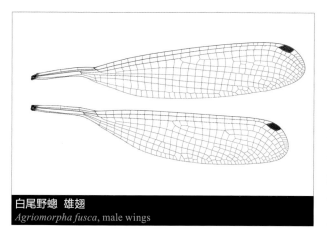

白尾野蟌 雄翅
Agriomorpha fusca, male wings

本属全球已知仅2种，分布于亚洲的热带和亚热带区域，这2种豆娘都分布于中国的华南和西南地区。本属是体中型、体色灰暗的豆娘，身体具黄色和白色斑纹；翅透明，具有2条结前横脉，IR3的起点位于翅结以外；停歇时翅合并。

本属豆娘栖息于茂盛森林中，喜欢狭窄的林荫溪流、渗流地和具有细小滴流的石壁。它们通常在这些潮湿阴暗的森林中活动，雄性会在水边的植物叶片或枝条上停立占据领地，雄性之间会展开激烈的面对面争斗。交尾持续时间较长。雌性产卵于泥土或者朽木中。

This genus contains two species, distributed in tropical and subtropical Asia, both found in South and Southwest China. The genus includes medium-sized and dark species with bodies with yellow and white stripes. Wings hyaline with two antenodals and IR3 begins beyond the nodus. They perch with wings closed.

Agriomorpha species inhabit dense forests, preferring narrow and shady streams, seepages and precipices with trickles. They usually keep to damp shadowy places in forest, males usually perching on leaves or branches defending territory, over which they can fight fiercely face to face. Mating behavior is protracted and females oviposit into mud or deadwood.

白尾野蟌 交尾 | 宋睿斌 摄
Agriomorpha fusca, mating pair | Photo by Ruibin Song

白尾野螅　两雄面对面争斗
Agriomorpha fusca, males fighting face to face

白尾野螅　雌，产卵 │宋睿斌 摄
Agriomorpha fusca, female laying eggs │ Photo by Ruibin Song

白尾野螅 *Agriomorpha fusca* May, 1933

白尾野螅 雄, 海南
Agriomorpha fusca, male from Hainan

白尾野螅 雌, 海南
Agriomorpha fusca, female from Hainan

白尾野螅 雌, 云南 (红河)
Agriomorpha fusca, female from Yunnan (Honghe)

【形态特征】雄性面部橙黄色，头顶和后头黑色；胸部黑色具甚细的白色条纹，足黄褐色，腹部黑色，第2～7节具白斑，华南地区个体第8～10节白色，云南个体第9～10节白色，下肛附器缺失。雌性与雄性相似，但腹部更粗壮。【长度】体长46～50 mm，腹长38～42 mm，后翅29～31 mm。【栖息环境】海拔1500 m以下森林中的狭窄林荫溪流和渗流地。【分布】云南（红河）、福建、广西、海南、广东、香港；越南。【飞行期】3—9月。

[Identification] Male face orange yellow, vertex and occiput black. Thorax black with fine white stripes, legs yellowish brown. Abdomen black, S2-S7 with white spots, individuals from South China with S8-S10 white, those from Yunnan with S9-S10 white, inferior appendages absent. Female similar to male but abdomen more robust. [Measurements] Total length 46-50 mm, abdomen 38-42 mm, hind wing 29-31 mm. [Habitat] Shady narrow streams and seepages in forest below 1500 m elevation. [Distribution] Yunnan (Honghe), Fujian, Guangxi, Hainan, Guangdong, Hong Kong; Vietnam. [Flight Season] March to September.

白尾野螈 雄，云南（红河）
Agriomorpha fusca, male from Yunnan (Honghe)

兴隆野螅 *Agriomorpha xinglongensis* (Wilson & Reels, 2001)

【形态特征】雄性面部黑色，额具白色条纹；胸部黑色具白色条纹，足黄褐色；腹部黑色，第2~7节侧面具白斑，第8~10节白色，下肛附器约为上肛附器长度的1/3。雌性与雄性相似，体色更灰暗。本种与白尾野螅十分相似，但具明显的下肛附器。【长度】体长 39~42 mm，腹长 33~36 mm，后翅 23~27 mm。【栖息环境】海拔 1000 m以下森林中的渗流石壁和小型瀑布。【分布】中国海南特有。【飞行期】4—6月。

[Identification] Male face black, frons with white markings. Thorax black with white stripes, legs yellowish brown. Abdomen black, S2-S7 with white spots laterally, S8-S10 white, the inferior appendages one third length of the superiors. Female similar to male but color darker. Similar to *A. fusca* but can be easily distinguished by the presence of the inferior appendages. [Measurements] Total length 39-42 mm, abdomen 33-36 mm, hind wing 23-27 mm. [Habitat] Precipices with trickles and small waterfalls in forest below 1000 m elevation. [Distribution] Endemic to Hainan of China. [Flight Season] April to June.

兴隆野螅 雄，海南
Agriomorpha xinglongensis, male from Hainan

野螅属待定种 *Agriomorpha* sp.

野螅属待定种 雄,云南(德宏)
Agriomorpha sp., male from Yunnan (Dehong)

【形态特征】雄性面部淡蓝色,头顶和后头黑色;胸部几乎黑色,足黄色;腹部黑色,第9节背面具白斑。雌性与雄性相似,但胸部具黄色条纹。【长度】体长 38~43 mm,腹长 31~36 mm,后翅 28~29 mm。【栖息环境】海拔 1000~1500 m森林中的渗流地。【分布】云南(德宏)。【飞行期】5—6月。

[Identification] Male face pale blue, vertex and occiput black. Thorax almost black, legs yellow. Abdomen black, S9 with a white spot dorsally. Female similar to male, but thorax with yellow stripes. [Measurements] Total length 38-43 mm, abdomen 31-36 mm, hind wing 28-29 mm. [Habitat] Seepages in forest at 1000-1500 m elevation. [Distribution] Yunnan (Dehong). [Flight Season] May to June.

野螅属待定种 雌,云南(德宏)
Agriomorpha sp., female from Yunnan (Dehong)

野蟌属待定种 交尾，云南（德宏）
Agriomorpha sp, mating pair from Yunnan (Dehong)

缅山螅属 Genus *Burmargiolestes* Kennedy, 1925

本属全球已知2种，分布于亚洲的热带地区。中国已知1种，仅分布于云南。本属豆娘体中型，身体深色具稀疏的黄色和白色斑纹；翅透明，具2条结前横脉；停歇时翅合并。缅山螅腹部末端3节大面积黑色仅有甚细的白色条纹，而野螅属的腹部末端2~3节主要白色。

本属豆娘栖息于热带雨林中，喜欢狭窄的林荫溪流和渗流地，通常都在这些潮湿阴暗的森林中活动，雄性会在水边的植物叶片或枝条上停立占据领地，雌性产卵于泥土或者朽木中。

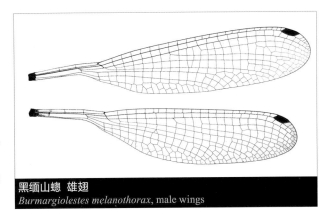

黑缅山螅 雄翅
Burmargiolestes melanothorax, male wings

The genus contains two species confined to tropical Asia. A single species is recorded from China but only found in Yunnan. Species of the genus are medium-sized, dark species, their bodies with sparse yellow and white stripes. Wings hyaline with two antenodals. They perch with wings closed. *Burmargiolestes* species have the last three abdominal segments largely black with fine white stripes, whereas *Agriomorpha* species have the last two or three abdominal segments mainly white.

Burmargiolestes species live in tropical rain forests, favouring narrow and shady streams and seepages. Individuals usually keep to shaded moist places and males perch on leaves or branches guarding territory, while the female oviposits into mud or deadwood.

黑缅山螅 雄
Burmargiolestes melanothorax, male

黑缅山螅 *Burmargiolestes melanothorax* (Selys, 1891)

【形态特征】雄性面部橙黄色，头顶和后头黑色；胸部黑色具白色条纹，足浅褐色；腹部黑色，第3~7节侧面具灰褐色斑，第7~9节背面后缘具白色细纹。雌性与雄性相似。【长度】体长 37~44 mm，腹长 31~36 mm，后翅 24~26 mm。【栖息环境】海拔 1500 m以下森林中的渗流地。【分布】云南（西双版纳、普洱）；缅甸、泰国、老挝、越南。【飞行期】4—9月。

[Identification] Male face orange yellow, vertex and occiput black. Thorax black with white stripes, legs pale brown. Abdomen black, S3-S7 with grey brown spots laterally, S7-S9 with dorsal fine white stripes to the posterior margin. Female similar to male. [Measurements] Total length 37-44 mm, abdomen 31-36 mm, hind wing 24-26 mm. [Habitat] Seepages in forest below 1500 m elevation. [Distribution] Yunnan (Xishuangbanna, Pu'er); Myanmar, Thailand, Laos, Vietnam. [Flight Season] April to September.

黑缅山螅 雄，云南（普洱）
Burmargiolestes melanothorax, male from Yunnan (Pu'er)

黑缅山螅 雌，云南（西双版纳）
Burmargiolestes melanothorax, female from Yunnan (Xishuangbanna)

黑缅山螅 交尾，云南（普洱）
Burmargiolestes melanothorax, mating pair from Yunnan (Pu'er)

凸尾山蟌属 Genus *Mesopodagrion* McLachlan, 1896

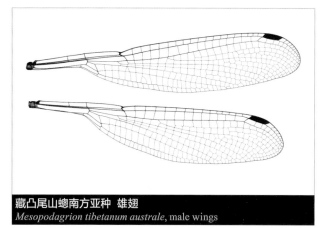

藏凸尾山蟌南方亚种 雄翅
Mesopodagrion tibetanum australe, male wings

本属全球已知2种，分布于中国、缅甸和中南半岛。本属豆娘体中型，身体黑色具黄色斑纹；翅透明，具有2条结前横脉，IR3的起点位于翅结以外，停歇时翅张开；本属雄性的显著特征是第10腹节背板后缘中央呈锥形突起，下肛附器甚短。

本属豆娘主要生活在高山环境，喜欢狭窄溪流和小型水潭。雄性会在水边的植物叶片或枝条上停立占据领地，雌性产卵于枯枝或者岩石的苔藓中。

The genus contains two species distributed in China, Myanmar and Indochina. They are medium-sized damselflies, body black with yellow markings. Wings hyaline with two antenodals and IR3 begins beyond the nodus, they perch with wings spread. Males of the genus possess a prominent pyramidal structure on the dorsum of S10, inferior appendages very short.

Mesopodagrion species mainly live in high mountains, freguenting narrow streams and small pools. Males usually perch on leaves or branches near water where they defend territory. Females oviposit into deadwood or moss on rocks.

藏凸尾山蟌南方亚种 交尾
Mesopodagrion tibetanum australe, mating pair

藏凸尾山蟌南方亚种
Mesopodagrion tibetanum australe

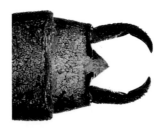

藏凸尾山蟌指名亚种
Mesopodagrion tibetanum tibetanum

雅州凸尾山蟌
Mesopodagrion yachowense

凸尾山蟌属待定种
Mesopodagrion sp.

凸尾山蟌属 雄性肛附器
Genus *Mesopodagrion*, male anal appendages

藏凸尾山蟌南方亚种 *Mesopodagrion tibetanum australe* Yu & Bu, 2009

【形态特征】雄性上唇淡黄色，前唇基淡蓝色；胸部蓝黑色具金属光泽，具较长的肩前条纹，合胸侧面具黄色条纹，翅痣黄褐色；腹部蓝黑色具金属光泽，第1~8节具黄斑，第9~10节覆盖白色粉霜。雌性黑色具黄斑。【长度】体长 38~46 mm，腹长 29~36 mm，后翅 28~32 mm。【栖息环境】海拔 2000~3000 m森林中的溪流、沟渠和小型湿地。【分布】四川、云南（大理、丽江、普洱、红河）；缅甸、越南。【飞行期】5—8月。

[Identification] Male labrum pale yellow, anteclypeus pale blue. Thorax metallic bluish black, with long antehumeral stripes, sides with yellow stripes, pterostigma yellowish brown. Abdomen metallic bluish black, S1-S8 with yellow spots laterally,

藏凸尾山蟌南方亚种 雌，云南（大理）
Mesopodagrion tibetanum australe, female from Yunnan (Dali)

S9-S10 whitish pruinosed. Female black with yellow spots. [Measurements] Total length 38-46 mm, abdomen 29-36 mm, hind wing 28-32 mm. [Habitat] Streams, ditches and small wetlands in forest at 2000-3000 m elevation. [Distribution] Sichuan, Yunnan (Dali, Lijiang, Pu'er, Honghe); Myanmar, Vietnam. [Flight Season] May to August.

藏凸尾山螅南方亚种 雄, 云南(大理)
Mesopodagrion tibetanum australe, male from Yunnan (Dali)

藏凸尾山螅指名亚种 *Mesopodagrion tibetanum tibetanum* McLachlan, 1896

【形态特征】雄性面部黑色具淡蓝色斑；胸部黑色具淡蓝色条纹，肩前条纹和合胸侧面条纹甚阔，翅痣褐色；腹部黑色，第1~8节侧面具淡蓝色斑，第9~10节覆盖白色粉霜。本亚种翅痣黑褐色，而南方亚种翅痣黄褐色。两亚种的身体色彩和肛附器构造也都不同。【长度】腹长 35~38 mm，后翅 27~31 mm。【栖息环境】海拔 2000~3000 m 森林中的溪流、沟渠和小型湿地。【分布】四川、云南；缅甸。【飞行期】5—8月。

[Identification] Male face black with pale blue markings. Thorax black with pale blue stripes, antehumeral stripes and lateral stripes broad, pterostigma brown. Abdomen black, S1-S8 with pale blue spots laterally, S9-S10 whitish pruinosed. The pterostigma of this subspecies is blackish brown but yellowish brown in *M. tibetanum australe*. The body color and male anal appendages also differ in the two subspecies. [Measurements] Abdomen 35-38 mm, hind wing 27-31 mm. [Habitat] Streams, ditches and small wetlands in forest at 2000-3000 m elevation. [Distribution] Sichuan, Yunnan; Myanmar. [Flight Season] May to August.

藏凸尾山螅指名亚种 雄，云南（红河）
Mesopodagrion tibetanum tibetanum, male from Yunnan (Honghe)

雅州凸尾山螅 *Mesopodagrion yachowense* Chao, 1953

【形态特征】雄性面部褐色；胸部黑色具黄色条纹，翅痣深褐色；腹部黑色具黄斑。雌性与雄性相似。【长度】体长 42~46 mm，腹长 33~36 mm，后翅 29~33 mm。【栖息环境】海拔 1000~2000 m森林中的沟渠和小型水潭。【分布】中国特有，分布于陕西、甘肃、四川、安徽、河南、湖北、湖南、浙江、江西。【飞行期】6—8月。

[Identification] Male face brown. Thorax black with yellow markings, pterostigma dark brown. Abdomen black with yellow spots. Female similar to male. [Measurements] Total length 42-46 mm, abdomen 33-36 mm, hind wing

29-33 mm. [Habitat] Ditches and small pools in forest at 1000-2000 m elevation. [Distribution] Endemic to China, recorded from Shaanxi, Gansu, Sichuan, Anhui, Henan, Hubei, Hunan, Zhejiang, Jiangxi. [Flight Season] June to August.

雅州凸尾山螅 雄,湖北 | 莫善濂 摄
Mesopodagrion yachowense, male from Hubei | Photo by Shanlian Mo

雅州凸尾山螅 雌,湖北 | 莫善濂 摄
Mesopodagrion yachowense, female from Hubei | Photo by Shanlian Mo

凸尾山螅属待定种 *Mesopodagrion* sp.

【形态特征】雄性面部褐色,上唇淡黄色;胸部黑色具黄色条纹;腹部黑色具黄斑。本种翅痣黄色,与雅州凸尾山螅不同。【长度】雄性体长 44 mm,腹长 35 mm,后翅 29 mm。【栖息环境】海拔 1000～2000 m森林中的沟渠和小型水潭。【分布】广东。【飞行期】6—8月。

[Identification] Male face brown, labrum pale yellow. Thorax black with yellow markings. Abdomen black with yellow spots. Similar to *M. yachowense* but pterostigma yellow. [Measurements] Male total length 44 mm, abdomen 35 mm, hind wing 29 mm. [Habitat] Ditches and small pools in forest at 1000-2000 m elevation. [Distribution] Guangdong. [Flight Season] June to August.

凸尾山螅属待定种 雄,广东 | 莫善濂 摄
Mesopodagrion sp., male from Guangdong | Photo by Shanlian Mo

古山蟌属 Genus *Priscagrion* Zhou & Wilson, 2001

　　本属全球已知2种，分布于中国南部和越南。本属豆娘体中型，身体具有鲜艳的蓝绿色条纹；翅透明，具有3条结前横脉，IR3的起点位于翅结以外，停歇时翅展开。雄性的下肛附器与上肛附器近等长。

　　本属豆娘栖息于森林中的小型瀑布和具有细小滴流的石壁。雄性停落在水边的叶片或枝条上。

The genus contains two species distributed in the south of China and Vietnam. These are medium-sized damselflies, body with bright bluish green stripes. Wings hyaline with three antenodals, IR3 begins beyond the nodus. They perch with wings spread. In males the inferior appendages almost as long as the superiors.

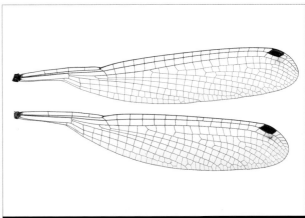

古山蟌属待定种　雄翅
Priscagrion sp., male wings

克氏古山蟌　雄
Priscagrion kiautai, male

Priscagrion species inhabit small waterfalls, slopes or precipice with trickles in forests. Males usually perch on leaves or branches near water.

克氏古山螅
Priscagrion kiautai

宾黑古山螅
Priscagrion pinheyi

古山螅属 雄性肛附器
Genus *Priscagrion*, male anal appendages

克氏古山螅 *Priscagrion kiautai* Zhou & Wilson, 2001

克氏古山螅 雄，重庆
Priscagrion kiautai, male from Chongqing

【形态特征】雄性面部大面积黄色，头顶和后头黑色；胸部黑色具绿黄色的肩前条纹，侧面具2条黄色条纹，翅端具小褐斑；腹部黑色具白色斑，第8~10节蓝色。雌性黑色具黄斑，翅端无褐斑。【长度】体长 54~55 mm，腹长 43~44 mm，后翅 34~35 mm。【栖息环境】海拔 1000~2000 m森林中的滴水陡坡和石壁。【分布】四川、重庆、贵州；越南。【飞行期】6—8月。

[Identification] Male face largely yellow, vertex and occiput black. Thorax black with greenish yellow antehumeral stripes, sides with two yellow stripes, wings with small brown apical spots. Abdomen black with white spots, S8-S10 blue. Female black with yellow markings, wings without brown tips. [Measurements] Total length 54-55 mm, abdomen 43-44 mm, hind wing 34-35 mm. [Habitat] Slopes and precipices with trickles in forest at 1000-2000 m elevation. [Distribution] Sichuan, Chongqing, Guizhou; Vietnam. [Flight Season] June to August.

宾黑古山螅 *Priscagrion pinheyi* Zhou & Wilson, 2001

【形态特征】雄性面部大面积黄绿色，头顶和后头黑色；胸部黑色具绿色的肩前条纹，侧面具2条黄绿色条纹，翅端无褐斑；腹部黑色具小黄斑，第8~10节蓝色。【长度】雄性体长 56 mm，腹长 46 mm，后翅 36 mm。【栖息环境】海拔 1000~1500 m森林中的滴水陡坡和石壁。【分布】中国广西特有。【飞行期】6—8月。

[Identification] Male face largely yellowish green, vertex and occiput black. Thorax black with green antehumeral stripes, sides with two yellowish green stripes, wings without brown apical spots. Abdomen black with small yellow spots, S8-S10 blue. [Measurements] Male total length 56 mm, abdomen 46 mm, hind wing 36 mm. [Habitat] Slopes and precipices with trickles in forest at 1000-1500 m elevation. [Distribution] Endemic to Guangxi of China. [Flight Season] June to August.

宾黑古山螅 雄，广西
Priscagrion pinheyi, male from Guangxi

古山螅属待定种 *Priscagrion* sp.

【形态特征】雄性复眼褐色，面部黄绿色，头顶和后头黑色；胸部黑色具绿色的肩前条纹，侧面具2条蓝色条纹，翅端无褐斑；腹部黑色具白色斑，第8～10节蓝色。【长度】雄性体长 55 mm，腹长 45 mm，后翅 33 mm。【栖息环境】海拔 1000～1500 m森林中的滴水陡坡和石壁。【分布】贵州。【飞行期】6—8月。

[Identification] Male eyes brown, face yellowish green, vertex and occiput black. Thorax black with green antehumeral stripes, sides with two blue stripes, wings without brown apical spots. Abdomen black with white spots, S8-S10 blue. [Measurements] Male total length 55 mm, abdomen 45 mm, hind wing 33 mm. [Habitat] Slopes and precipices with trickles in forest at 1000-1500 m elevation. [Distribution] Guizhou. [Flight Season] June to August.

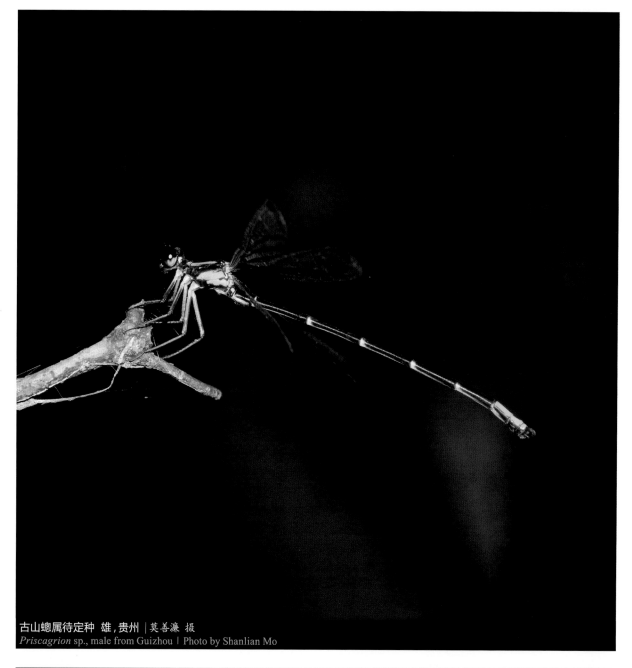

古山螅属待定种 雄,贵州 | 莫善濂 摄
Priscagrion sp., male from Guizhou | Photo by Shanlian Mo

扇山螅属 Genus *Rhipidolestes* Ris, 1912

褐顶扇山螅 交尾 ｜宋睿斌 摄
Rhipidolestes truncatidens, mating pair ｜ Photo by Ruibin Song

　　本属全球已知20余种，多数种类分布于中国和日本，少数记录于缅甸、泰国、老挝和越南。中国已知16种，南方广布。本属多数种类分布狭窄，仅在一两座山脉分布。这些豆娘体中型，体色灰暗，具稀疏的黄色或红色条纹；翅透明，具有2条结前横脉，IR 3的起点位于翅结以内；停歇时双翅展开；雄性腹部第9节背面具1个锥状突起，肛附器构造复杂，是区分种类的重要依据。

　　本属豆娘栖息于潮湿阴暗的森林，喜欢十分狭窄的林荫溪流、渗流地和具有渗流的陡坡和崖壁。雄性经常停落在地面低处的枝条或者叶片上。雌性在朽木和泥土中产卵。

This genus contains over 20 species, mostly distributed in China and Japan, a few others recorded from Myanmar, Thailand, Laos and Vietnam. 16 species are recorded from China, widespread in the southern part. Most species have a small range occurring on just one or two mountains. They are medium-sized, dark damselflies, with scattered yellow or red markings. Wings hyaline with two antenodals and IR3 begins before the nodus. They perch with wings spread. Males have a spine on the dorsum of S9, the structure of anal appendages are complex and facilitate the identification of the species.

Rhipidolestes species inhabit shaded moist places in forests, frequenting narrow, shady streams, seepages and slopes or precipices with trickles. Males usually perch on leaves or branches low above ground. Females lay eggs into mud or deadwood.

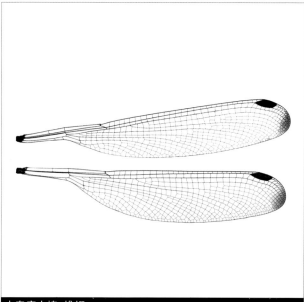

水鬼扇山螅 雄翅
Rhipidolestes nectans, male wings

扇山螅属待定种4
Rhipidolestes sp. 4

扇山螅属 雄性肛附器
Genus *Rhipidolestes,* male anal appendages

扇山螅属待定种1
Rhipidolestes sp. 1

扇山螅属待定种2
Rhipidolestes sp. 2

扇山螅属待定种3
Rhipidolestes sp. 3

扇山螅属 雄性肛附器
Genus *Rhipidolestes,* male anal appendages

棘扇山螅
Rhipidolestes aculeatus

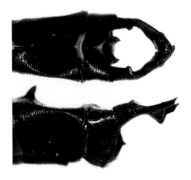

艾伦扇山螅
Rhipidolestes alleni

珍妮扇山螅
Rhipidolestes janetae

褐带扇山螅
Rhipidolestes fascia

愉快扇山螅
Rhipidolestes jucundus

李氏扇山螅
Rhipidolestes lii

水鬼扇山螅
Rhipidolestes nectans

黄白扇山螅
Rhipidolestes owadai

褐顶扇山螅
Rhipidolestes truncatidens

扇山螅属 雄性肛附器
Genus *Rhipidolestes*, male anal appendages

棘扇山蟌 *Rhipidolestes aculeatus* Ris, 1912

【形态特征】雄性面部黑色，上唇黄色；胸部和腹部黑色具淡黄色斑纹，翅痣和足红色。雌性与雄性相似。【长度】体长 37~48 mm。【栖息环境】海拔 2000 m 以下森林中的林荫溪流。【分布】中国台湾；日本。【飞行期】2—8月。

[Identification] Male face black, labrum yellow. Thorax and abdomen black with pale yellow markings, pterostigma and legs red. Female similar to male. [Measurements] Total length 37-48 mm. [Habitat] Shady streams in forest below 2000 m elevation. [Distribution] Taiwan of China; Japan. [Flight Season] February to August.

棘扇山蟌 交尾, 台湾 | 嘎嘎 摄
Rhipidolestes aculeatus, mating pair from Taiwan | Photo by Gaga

艾伦扇山蟌 *Rhipidolestes alleni* Wilson, 2000

【形态特征】雄性面部大面积黄色，头顶和后头黑色；胸部黑色具黄色条纹，翅端具褐斑，翅痣黄色具黑褐色边缘，足褐色；腹部黑色。雌性与雄性相似，翅端无褐斑。【长度】体长 47~52 mm，腹长 39~43 mm，后翅 30~34 mm。【栖息环境】海拔 1000 m 以下森林中的渗流地和具有细小滴流的陡峭石壁。【分布】中国广西特有。【飞行期】5—7月。

[Identification] Male face largely yellow, vertex and occiput black. Thorax black with yellow stripes. Wings with brown apical spots, pterostigma yellow with the margin blackish brown, legs brown. Abdomen black. Female similar to male, wings without brown apical spots. **[Measurements]** Total length 47-52 mm, abdomen 39-43 mm, hind wing 30-34 mm. **[Habitat]** Seepages and precipices with trickles in forest below 1000 m elevation. **[Distribution]** Endemic to Guangxi of China. **[Flight Season]** May to July.

艾伦扇山螅　雄，广西
Rhipidolestes alleni, male from Guangxi

艾伦扇山螅　雌，广西
Rhipidolestes alleni, female from Guangxi

黄蓝扇山螅 *Rhipidolestes cyanoflavus* Wilson, 2000

黄蓝扇山螅 雄, 海南 | Graham Reels 摄
Rhipidolestes cyanoflavus, male from Hainan | Photo by Graham Reels

【形态特征】雄性面部蓝色; 胸部黑色具蓝黄色斑纹, 足和翅痣蓝白色; 腹部黑色。雌性与雄性相似。【长度】腹长 37~48 mm, 后翅 28~37 mm。【栖息环境】海拔 1500 m 以下森林中的渗流地和狭窄溪流。【分布】中国特有, 分布于海南、广东。【飞行期】4—6月。

[Identification] Male face blue. Thorax black with bluish yellow stripes, legs and pterostigma bluish white. Abdomen black. Female similar to male. [Measurements] Abdomen 37-48 mm, hind wing 28-37 mm. [Habitat] Seepages and narrow streams in forest below 1500 m elevation. [Distribution] Endemic to China, recorded from Hainan, Guangdong. [Flight Season] April to June.

褐带扇山螅 *Rhipidolestes fascia* Zhou, 2003

【形态特征】雄性面部黑色; 胸部黑色具黄色条纹, 翅染琥珀色, 中央具1条黑褐色宽带, 翅痣黄色, 足黄褐色; 腹部黑色。【长度】雄性体长 42 mm, 腹长 32 mm, 后翅 37 mm。【栖息环境】1500 m 以下森林中的渗流地和狭窄溪流。【分布】中国特有, 分布于贵州、重庆。【飞行期】5—7月。

[Identification] Male face black. Thorax black with yellow stripes, wings tinted with amber, with a broad median blackish brown band, pterostigma yellow, legs yellowish brown. Abdomen black. [Measurements] Male total length 42 mm, abdomen 32 mm, hind wing 37 mm. **[Habitat]** Seepages and narrow streams in forest below 1500 m elevation. **[Distribution]** Endemic to China, recorded from Guizhou, Chongqing. **[Flight Season]** May to July.

褐带扇山螅 雄，重庆
Rhipidolestes fascia, male from Chongqing

珍妮扇山螅 *Rhipidolestes janetae* Wilson, 1997

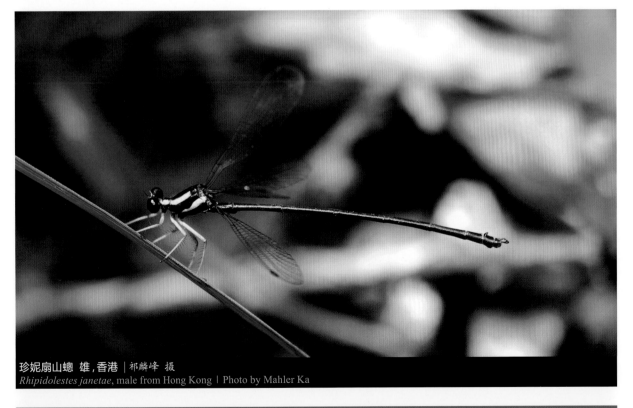

珍妮扇山螅 雄，香港 |祁麟峰 摄
Rhipidolestes janetae, male from Hong Kong | Photo by Mahler Ka

【形态特征】雄性面部黄色，头顶和后头黑色；胸部黑色具黄色条纹，翅痣黑褐色，足黄色；腹部黑色。雌性与雄性相似。【长度】体长 38～48 mm，腹长 31～40 mm，后翅 25～27 mm。【栖息环境】海拔 1000 m 以下森林中的渗流地和具有细小滴流的陡峭石壁。【分布】中国特有，分布于广东、香港。【飞行期】4—6月。

[Identification] Male face yellow, vertex and occiput black. Thorax black with yellow stripes, pterostigma blackish brown, legs yellow. Abdomen black. Female similar to male. [Measurements] Total length 38-48 mm, abdomen 31-40 mm, hind wing 25-27 mm. [Habitat] Seepages and precipices with trickles in forest below 1000 m elevation. [Distribution] Endemic to China, recorded from Guangdong, Hong Kong. [Flight Season] April to June.

珍妮扇山蟌 雌，香港 | 祁麟峰 摄
Rhipidolestes janetae, female from Hong Kong | Photo by Mahler Ka

愉快扇山螅 *Rhipidolestes jucundus* Lieftinck, 1948

愉快扇山螅 雄,贵州 | 莫善濂 摄
Rhipidolestes jucundus, male from Guizhou | Photo by Shanlian Mo

【形态特征】雄性面部黄褐色；胸部黑色具黄色条纹, 翅痣红色, 足红褐色；腹部黑色。【长度】雄性体长 52 mm, 腹长 42 mm, 后翅 33 mm。【栖息环境】海拔 1000~1500 m森林中具有细小渗流的石壁。【分布】中国特有, 分布于福建、贵州。【飞行期】5—7月。

[Identification] Male face yellowish brown. Thorax black with yellow stripes, pterostigma red, legs reddish brown. Abdomen black. [Measurements] Male total length 52 mm, abdomen 42 mm, hind wing 33 mm. [Habitat] Precipices with trickles in forest at 1000-1500 m elevation. [Distribution] Endemic to China, recorded from Fujian, Guizhou. [Flight Season] May to July.

李氏扇山螅 *Rhipidolestes lii* Zhou, 2003

【形态特征】雄性面部黄褐色, 头顶和后头黑色；胸部黑色具黄色条纹, 翅端具褐斑, 翅痣黑色, 足红色；腹部黑色。雌性与雄性相似但翅端无褐斑。【长度】体长 45~48 mm, 腹长 36~39 mm, 后翅 29~31 mm。【栖息环境】海拔 1000~1500 m森林中的渗流地和具有细小滴流的陡峭石壁。【分布】中国贵州特有。【飞行期】5—7月。

[Identification] Male face yellowish brown, vertex and occiput black. Thorax black with yellow stripes, wings with brown tips, pterostigma black, legs red. Abdomen black. Female similar to male but wings without brown tips.

[Measurements] Total length 45-48 mm, abdomen 36-39 mm, hind wing 29-31 mm. [Habitat] Seepages and precipices with trickles in forest at 1000-1500 m elevation. [Distribution] Endemic to Guizhou of China. [Flight Season] May to July.

李氏扇山螅 雄,贵州 | 莫善濂 摄
Rhipidolestes lii, male from Guizhou | Photo by Shanlian Mo

李氏扇山螅 雌,贵州 | 莫善濂 摄
Rhipidolestes lii, female from Guizhou | Photo by Shanlian Mo

李氏扇山螅 雄,贵州 | 莫善濂 摄
Rhipidolestes lii, male from Guizhou | Photo by Shanlian Mo

水鬼扇山螅 *Rhipidolestes nectans* (Needham, 1929)

【形态特征】雄性面部黑色具白色粉霜；胸部黑色，侧面具浓密的粉霜，翅端部具褐斑，翅痣褐色，足黑褐色；腹部黑色，第8~10节具粉霜。雌性黑色具黄色条纹，翅透明，翅痣淡黄色。【长度】体长 45~54 mm，腹长 36~43 mm，后翅 30~34 mm。【栖息环境】海拔1000 m以下森林中的渗流地和狭窄溪流。【分布】中国浙江特有。【飞行期】4—7月。

水鬼扇山螅 雌, 浙江 | 莫善濂 摄
Rhipidolestes nectans, female from Zhejiang | Photo by Shanlian Mo

[Identification] Male face black with white pruinescence. Thorax black, sides heavily pruinosed, wings with brown apical spots, pterostigma brown, legs blackish brown. Abdomen black, S8-S10 pruinosed. Female black with yellow stripes, wings hyaline, pterostigma pale yellow. [Measurements] Total length 45-54 mm, abdomen 36-43 mm, hind wing 30-34 mm. [Habitat] Seepages and narrow streams in forest below 1000 m elevation. [Distribution] Endemic to Zhejiang of China. [Flight Season] April to July.

水鬼扇山螅 雄, 浙江 | 莫善濂 摄
Rhipidolestes nectans, male from Zhejiang | Photo by Shanlian Mo

黄白扇山蟌 *Rhipidolestes owadai* Asahina, 1997

黄白扇山蟌 雄，云南（红河）
Rhipidolestes owadai, male from Yunnan (Honghe)

【形态特征】雄性面部黄白色，头顶和后头黑色；胸部黑色具黄白色条纹，翅端具褐斑，翅痣黄白色具黑色边缘，足黄褐色；腹部黑色。【长度】雄性体长 53 mm，腹长 43 mm，后翅 35 mm。【栖息环境】海拔 1000～1500 m 森林中的渗流地和具有细小滴流的陡峭石壁。【分布】云南（红河）、广西；老挝、越南。【飞行期】4—6月。

[Identification] Male face yellowish white, vertex and occiput black. Thorax black with yellowish white stripes. Wings with brown apical spots, pterostigma yellowish white with the margin black, legs yellowish brown. Abdomen black. [Measurements] Male total length 53 mm, abdomen 43 mm, hind wing 35 mm. [Habitat] Seepages and precipices with trickles in forest at 1000-1500 m elevation. [Distribution] Yunnan (Honghe), Guangxi; Laos, Vietnam. [Flight Season] April to June.

褐顶扇山螅 *Rhipidolestes truncatidens* Schmidt, 1931

【形态特征】雄性面部黑色具黄色斑纹；胸部黑色具黄色条纹，侧面稍微覆盖白色粉霜，翅端部具褐斑，翅痣淡粉红色，足黑色；腹部黑色，第8～10节具粉霜。雌性黑色具黄色条纹，腹部黑色。【长度】体长 49～56 mm，腹长 39～46 mm，后翅 33～36 mm。【栖息环境】海拔 2000 m 以下森林中的渗流地和具有细小滴流的陡峭石壁。【分布】中国特有，分布于福建、广东。【飞行期】3—7月。

[Identification] Male face black with yellow markings. Thorax black with yellow stripes, sides slightly whitish pruinosed, wings with brown apical spots, pterostigma pink, legs black. Abdomen black, S8-S10 pruinosed. Female black with yellow stripes, abdomen black. [Measurements] Total length 49-56 mm, abdomen 39-46 mm, hind wing 33-36 mm. [Habitat] Seepages and precipices with trickles in forest below 2000 m elevation. [Distribution] Endemic to China, recorded from Fujian, Guangdong. [Flight Season] March to July.

褐顶扇山螅 雄，广东｜宋睿斌 摄
Rhipidolestes truncatidens, male from Guangdong | Photo by Ruibin Song

褐顶扇山螅 雌，广东｜宋睿斌 摄
Rhipidolestes truncatidens, female from Guangdong | Photo by Ruibin Song

褐顶扇山螅 雄，广东
Rhipidolestes truncatidens, male from Guangdong

扇山螅属待定种1 *Rhipidolestes* sp. 1

【形态特征】雄性面部黑色；胸部黑色具红褐色条纹，翅端具小褐斑，翅痣红色，足红褐色；腹部黑色。雌性面部和胸部具黄色斑纹，翅痣淡黄色，足黄褐色。【长度】体长 48~64 mm，腹长 38~52 mm，后翅 32~40 mm。【栖息环境】海拔 1000 m 以下森林中具有细小滴流的陡峭石壁。【分布】湖北。【飞行期】5—7月。

[Identification] Male face black. Thorax black with reddish brown stripes, wings with small brown apical spots, pterostigma red, legs reddish brown. Abdomen black. Female face and thorax with yellow markings, pterostigma pale yellow, legs yellowish brown. [Measurements] Total length 48-64 mm, abdomen 38-52 mm, hind wing 32-40 mm. [Habitat] Precipices with trickles in forest below 1000 m elevation. [Distribution] Hubei. [Flight Season] May to July.

扇山螅属待定种1 雄，湖北
Rhipidolestes sp. 1, male from Hubei

扇山螅属待定种2 *Rhipidolestes* sp. 2

【形态特征】雄性面部红褐色，头顶和后头黑色；胸部黑色具黄色条纹，翅端具小褐斑，翅痣黄色，足红色；腹部黑色。雌性黑色具黄色条纹，翅痣淡黄色，足黄褐色。【长度】体长 51~57 mm，腹长 41~47 mm，后翅 34~35 mm。【栖息环境】海拔 1500~2000 m森林中具有细小渗流的石壁。【分布】浙江。【飞行期】6—8月。

扇山螅属待定种2 雄，浙江 |莫善濂 摄
Rhipidolestes sp. 2, male from Zhejiang | Photo by Shanlian Mo

扇山螅属待定种2 雌，浙江 |莫善濂 摄
Rhipidolestes sp. 2, female from Zhejiang | Photo by Shanlian Mo

[Identification] Male face reddish brown, vertex and occiput black. Thorax black with yellow stripes, wings with small brown apical spots, pterostigma yellow, legs red. Abdomen black. Female black with yellow stripes, pterostigma pale yellow, legs yellowish brown. [Measurements] Total length 51-57 mm, abdomen 41-47 mm, hind wing 34-35 mm. [Habitat] Precipices with trickles in forest at 1500-2000 m elevation. [Distribution] Zhejiang. [Flight Season] June to August.

扇山螅属待定种3 *Rhipidolestes* sp. 3

扇山螅属待定种3 雌，广东 |莫善濂 摄
Rhipidolestes sp. 3, female from Guangdong | Photo by Shanlian Mo

【形态特征】雄性面部红褐色，头顶和后头黑色；胸部黑色具红色条纹，翅端具小褐斑，翅痣和足红色；腹部黑色。雌性黑色具黄褐色条纹，足黄褐色。【长度】体长 39~47 mm，腹长 31~38 mm，后翅 25~29 mm。【栖息环境】海拔 1000 m 以下森林中的渗流地。【分布】广东。【飞行期】4—6月。

[Identification] Male face reddish brown, vertex and occiput black. Thorax black with red stripes, wings with small brown apical spots, pterostigma and legs red. Abdomen black. Female black with yellowish brown stripes, legs yellowish brown. [Measurements] Total length 39-47 mm, abdomen 31-38 mm, hind wing 25-29 mm. [Habitat] Seepages in forest below 1000 m elevation. [Distribution] Guangdong. [Flight Season] April to June.

扇山蟌属待定种3 雄，广东 | 宋睿斌 摄
Rhipidolestes sp. 3, male from Guangdong | Photo by Ruibin Song

扇山螅属待定种3 雌,广东 | 宋睿斌 摄
Rhipidolestes sp. 3, female from Guangdong | Photo by Ruibin Song

扇山螅属待定种3 雄,广东 | 宋睿斌 摄
Rhipidolestes sp. 3, male from Guangdong | Photo by Ruibin Song

扇山螅属待定种4 *Rhipidolestes* sp. 4

【形态特征】雄性面部黑色，上唇黄色；胸部黑色具黄色条纹，翅端具褐斑，翅痣淡褐色具黑褐色边缘，足淡褐色；腹部黑色。【长度】雄性体长 54 mm，腹长 44 mm，后翅 32 mm。【栖息环境】海拔 1000 m以下森林中的渗流地和具有细小滴流的陡峭石壁。【分布】广西。【飞行期】5—7月。

[Identification] Male face black, labrum yellow. Thorax black with yellow stripes. Wings with brown apical spots, pterostigma pale brown with the margin blackish brown, legs pale brown. Abdomen black. [Measurements] Male total length 54 mm, abdomen 44 mm, hind wing 32 mm. [Habitat] Seepages and precipices with trickles in forest below 1000 m elevation. [Distribution] Guangxi. [Flight Season] May to July.

扇山螅属待定种4 雄，广西
Rhipidolestes sp. 4, male from Guangxi

华山螅属 Genus *Sinocnemis* Wilson & Zhou, 2000

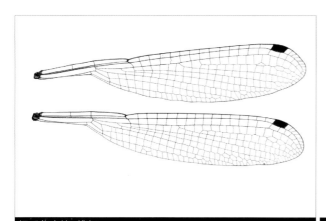

杨氏华山螅
Sinocnemis yangbingi

杨氏华山螅 雄翅
Sinocnemis yangbingi, male wings

华山螅属 雄性肛附器
Genus *Sinocnemis*, male anal appendages

本属中国特有，已知2种，分布于中部至西南地区。本属豆娘体中型，身体具有黄色和蓝色条纹；翅透明，具有2条结前横脉，IR3的起点位于亚翅结下方；停歇时翅展开；雄性的下肛附器与上肛附器近等长。

本属豆娘主要生活在森林中的溪流，它们经常停落在水边的叶片上。

This genus is endemic to China and includes two species found from the Central to Southwest regions. They are medium-sized damselflies, body with yellow and blue stripes. Wings hyaline with two antenodals and IR3 beginning at the level of the subnodus. They perch with wings spread. Male inferior appendages almost the same length as the superiors.

Sinocnemis inhabit streams in forest, they usually perch on leaves by the water's edge.

杨氏华山螅 雄
Sinocnemis yangbingi, male

杨氏华山螅 *Sinocnemis yangbingi* Wilson & Zhou, 2000

【形态特征】雄性面部黑色，上唇淡蓝色；胸部黑色具黄绿色的肩前条纹，合胸侧面具2条黄色条纹；腹部黑色具淡黄色斑，第8~10节背面蓝色。雌性黑色具黄色条纹，腹部第8~10节具灰白色斑。【长度】体长 40~47 mm，腹长 30~38 mm，后翅 28~30 mm。【栖息环境】海拔 1000~2000 m森林中的溪流。【分布】中国特有，分布于甘肃、河南、湖北、湖南、四川、贵州。【飞行期】4—8月。

[Identification] Male face black, labrum pale blue. Thorax black with yellowish green antehumeral stripes, sides with two yellow stripes. Abdomen black with pale yellow markings, S8-S10 blue dorsally. Female black with yellow stripes, S8-S10 with greyish white spots. [Measurements] Total length 40-47 mm, abdomen 30-38 mm, hind wing 28-30 mm. [Habitat] Streams in forest at 1000-2000 m elevation. [Distribution] Endemic to China, recorded from Gansu, Henan, Hubei, Hunan, Sichuan, Guizhou. [Flight Season] April to August.

杨氏华山螅 雌雄连结，湖北
Sinocnemis yangbingi, pair in tandem from Hubei

杨氏华山螅 雄, 湖北
Sinocnemis yangbingi, male from Hubei

杨氏华山螅 雌, 湖北
Sinocnemis yangbingi, female from Hubei

8 拟丝螅科 Family Pseudolestidae

　　本科全世界仅1种，即丽拟丝螅。这是一种体型较小的豆娘，为中国海南特有。这种豆娘广泛分布于海南全岛茂盛森林中的溪流，多栖息于光线较暗的环境。雄性在小溪边缘占据领地，通常停立于植物的叶片或枝条顶端。雄性常会展开激烈的争斗，面对面飞行，腹部翘起，后翅向下伸，有时会有多只雄性参与争斗。交尾时停落在溪流边缘，交尾结束后雌性先停歇一段时间，雄性护卫于其身旁，然后再飞到溪边寻找合适的产卵地点。雌性通常在泥土和朽木上产卵。

　　The family contains just one species, *Pseudolestes mirabilis* Kirby, 1900, a small-sized species endemic to Hainan of China. This damselfly is widespread in the well-forested streams of the island, they typically perch in shade. Males usually maintain territories and perch on leaves or the tips of branches at stream margins. They fight fiercely, flying face to face, curving up their tails, with the hind wings held downwards. Sometimes several males fight together. When mating they perch near the stream. After mating, the female first perches for a short period with the male guarding, then the female flies to the stream to find a suitable site for laying eggs. Females lay eggs into mud and deadwood in streams.

丽拟丝螅 雄

Pseudolestes mirabilis, male

丽拟丝螅 雄性飞行
Pseudolestes mirabilis, male in flight

丽拟丝螅 雄性飞行
Pseudolestes mirabilis, male in flight

丽拟丝螅 三雄争斗
Pseudolestes mirabilis, three males fighting

丽拟丝螅 两雄面对面飞行争斗
Pseudolestes mirabilis, two males fighting by flying face to face

丽拟丝螅 雄性护卫停歇中的雌性
Pseudolestes mirabilis, male guarding the perching female

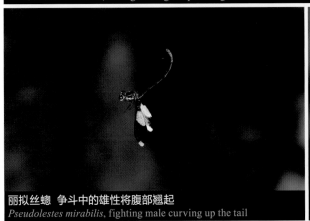

丽拟丝螅 争斗中的雄性将腹部翘起
Pseudolestes mirabilis, fighting male curving up the tail

丽拟丝螅 交尾
Pseudolestes mirabilis, mating pair

丽拟丝螅 雌性产卵于朽木中
Pseudolestes mirabilis, female laying eggs into deadwood

拟丝螅属 Genus *Pseudolestes* Kirby, 1900

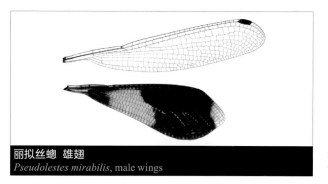

丽拟丝螅 雄翅
Pseudolestes mirabilis, male wings

本属最显著的特征是后翅长度仅为前翅的3/4，雄性的后翅具金色、银色和黑色斑，雌性的翅则具黑色、褐色和白色斑。

The outstanding character of this species is its short, falcate hind wings, which are about three fourths length of the fore wings with metallic gold, silver and black markings in males, female hind wings are tinted with black, brown and white markings.

丽拟丝螅 雄
Pseudolestes mirabilis, male

丽拟丝螅 *Pseudolestes mirabilis* Kirby, 1900

丽拟丝螅 雄，海南
Pseudolestes mirabilis, male from Hainan

丽拟丝螅 雌，海南
Pseudolestes mirabilis, female from Hainan

　　【形态特征】雄性面部主要蓝色；胸部黑色，侧面具黄色细条纹，前翅透明，后翅正面黑色，中央和端部具金色斑，背面中央和亚端方具银白色"鳞片"；腹部黑色，第1节侧面具黄色条纹。雌性前翅透明，后翅琥珀色，亚端方具1个较大的深褐色斑和白色的小端斑。【长度】体长 37~40 mm，腹长 27~31 mm，前翅 28~29 mm，后翅 21~22 mm。【栖息环境】海拔 1500 m 以下森林中的溪流、渗流地和小型瀑布。【分布】中国海南特有。【飞行期】3—10月。

丽拟丝螅 雄，海南
Pseudolestes mirabilis, male from Hainan

[Identification] Male face mainly blue. Thorax black with lateral fine yellow stripes, fore wings hyaline, upperside of hind wings black with median and apical opaque golden spots, underside with luminous silvery white bundled wax fibres centrally and subapically. Abdomen black, S1 with lateral yellow stripe. Female with hyaline fore wings, hind wings amber, with a large subapical dark brown spot and small white spot apically. [Measurements] Total length 37-40 mm, abdomen 27-31 mm, fore wings 28-29 mm, hind wings 21-22 mm. [Habitat] Streams, seepages and small waterfalls in forests below 1500 m elevation. [Distribution] Endemic to Hainan of China. [Flight Season] March to October.

丽拟丝螅 雌，海南
Pseudolestes mirabilis, female from Hainan

9 ▸ 丝螅科 Family Lestidae

　　本科全球已知9属150余种，世界性分布。中国已知4属20余种，全国广布。本科豆娘体型从小型至大型不等，多数是中小型豆娘。翅通常透明，具较长的翅柄和翅痣，结前横脉通常2条，弓脉位于第2条结前横脉以下。有些种类停歇时翅半张开。

　　本科豆娘栖息于水草丛生的湿地和山区溪流。雄性通常会停歇在挺水植物的茎干上，或者吊挂在具有林荫遮蔽的水潭上方。许多种类以雌雄连结的方式产卵。

This family contains nine genera with over 150 species, distributed all over the world. Over 20 species in four genera are recorded from China where they are widespread throughout the country. They rang from small to large in size, with most species small to medium sized. Usually they have hyaline wings, with long stalk at the base, long pterostigma and two antenodals, arc located below the second antenodal. Some species perch with wings half spread.

Species of the family inhabit standing water with plenty of emergent plants or montane streams. Males usually perch on the stalks of emergent plants or hang from the branches above the pools in shady forests. Many species oviposit in tandem.

蕾尾丝螅 连结产卵，宋睿斌 摄
Lestes nodalis, laying eggs in tandem | Photo by Ruibin Song

印丝螅属 Genus *Indolestes* Fraser, 1922

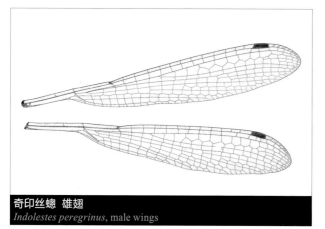

奇印丝螅 雄翅
Indolestes peregrinus, male wings

本属全球已知30余种，分布于亚洲和大洋洲。中国已知5种，分布于华中、华南和西南地区。本属豆娘体中型，身体蓝色或者褐色具丰富的斑纹；翅透明，后翅的四边室长于前翅；停歇时翅合并。

本属豆娘栖息于水草茂盛的静水环境，包括水稻田。它们的种群数量明显小于丝螅属种类。它们通过雌雄连结的方式将卵插入距离水面有一定高度的挺水植物茎干内或水面上方的树枝中。

The genus contains over 30 species, distributed in Asia and Oceania. Five species are recorded from China, found in the Central, South and Southwest regions. They are medium-sized damselflies, with blue or brown body strongly marked. Wings hyaline, the quadrangle in the hind wings longer than in the fore wings. They perch with wings closed.

Indolestes species inhabit standing water with plenty of emergent plants, including paddy fields. They are less abundant than *Lestes* species. They oviposit in tandem, their eggs being inserted into the stalks of emergent plants well above water level or into branches above water.

黄面印丝螅 雄
Indolestes assamicus, male

黄面印丝螈 *Indolestes assamicus* **Fraser, 1930**

黄面印丝螈 雄，云南（保山）
Indolestes assamicus, male from Yunnan (Baoshan)

【形态特征】雄性面部黑褐色，上唇和前唇基淡蓝色；胸部淡蓝色，背脊上具1条黑色条纹，翅黄褐色；腹部淡蓝色，第1～8节具黑色的背中条纹。雌性身体淡褐色具黑色条纹。【长度】体长 39～44 mm，腹长 33～36 mm，后翅 19～21 mm。【栖息环境】海拔 2000～3000 m的静水环境包括水稻田。【分布】云南（大理、保山、丽江）；印度。【飞行期】10月—次年6月。

[Identification] Male face dark brown, labrum and anteclypeus pale blue. Thorax pale blue, dorsal carina with a broad black stripe, wings tinted with yellowish brown. Abdomen pale blue, S1-S8 with black stripes along the dorsal

黄面印丝螅 雌,云南(保山)
Indolestes assamicus, female from Yunnan (Baoshan)

carina. Female pale brown with black stripes. [Measurements] Total length 39-44 mm, abdomen 33-36 mm, hind wing 19-21 mm. [Habitat] Standing water including paddy fields from 2000-3000 m elevation. [Distribution] Yunnan (Dali, Baoshan, Lijiang); India. [Flight Season] October to the following June.

蓝印丝螅 *Indolestes cyaneus* (Selys, 1862)

【形态特征】雄性面部黑褐色,上唇和前唇基淡蓝色;胸部黑色具淡蓝色条纹,翅稍染褐色;腹部黑褐色,第1～6节和第10节具蓝斑。雌性黑褐色具黄色条纹。【长度】体长 47～50 mm,腹长 37～41 mm,后翅 25～30 mm。

蓝印丝螅 雄,西藏│吴超 摄
Indolestes cyaneus, male from Tibet │ Photo by Chao Wu

蓝印丝螅 雌,西藏│张巍巍 摄
Indolestes cyaneus, female from Tibet │ Photo by Weiwei Zhang

【栖息环境】海拔 1000～2500 m的湿地。【分布】分布于西藏、湖北、贵州、台湾；不丹、印度。【飞行期】12月—次年9月。

[Identification] Male face dark brown, labrum and anteclypeus pale blue. Thorax black with pale blue stripes, wings slightly tinted with brown. Abdomen dark brown, S1-S6 and S10 with blue markings. Female dark brown with yellow markings. [Measurements] Total length 47-50 mm, abdomen 37-41 mm, hind wing 25-30 mm. [Habitat] Standing water from 1000-2500 m elevation. [Distribution] Tibet, Hubei, Guizhou, Taiwan; Bhutan, India. [Flight Season] December to the following September.

蓝印丝螅 连结产卵, 西藏 | 吴超 摄
Indolestes cyaneus, laying eggs in tandem from Tibet | Photo by Chao Wu

奇印丝螅 *Indolestes peregrinus* (Ris, 1916)

【形态特征】雄性面部黑褐色，上唇淡蓝色；胸部蓝色，合胸脊具1条黑色条纹，翅透明；腹部蓝色具黑褐色斑。雌性微染淡蓝色具黑褐色斑纹。【长度】体长 34～36 mm，腹长 27～29 mm，后翅 19～20 mm。【栖息环境】海拔 2500 m以下的静水环境，包括水稻田。【分布】云南、贵州、安徽、江苏、江西、浙江、福建、广东、台湾；朝鲜半岛、日本。【飞行期】全年可见。

[Identification] Male face dark brown, labrum pale blue. Thorax blue with a broad black stripe along dorsal carina, wings hyaline. Abdomen blue with dark brown markings. Female slightly tinted with pale blue with dark brown

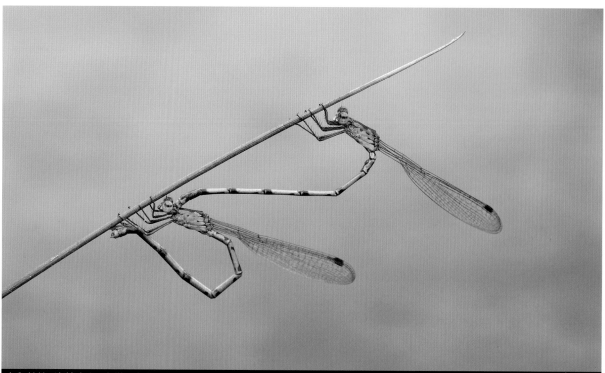

奇印丝螅 连结产卵,安徽 | 秦彧 摄
Indolestes peregrinus, laying eggs in tandem from Anhui | Photo by Yu Qin

奇印丝螅 雄,云南(德宏)
Indolestes peregrinus, male from Yunnan (Dehong)

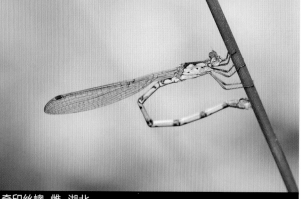

奇印丝螅 雌,湖北
Indolestes peregrinus, female from Hubei

markings. **[Measurements]** Total length 34-36 mm, abdomen 27-29 mm, hind wing 19-20 mm. **[Habitat]** Standing water including paddy field below 2500 m elevation. **[Distribution]** Yunnan, Guizhou, Anhui, Jiangsu, Jiangxi, Zhejiang, Fujian, Guangdong, Taiwan; Korean peninsula, Japan. **[Flight Season]** Throughout the year.

丝螅属 Genus *Lestes* Leach, 1815

　　本属全球已知80余种,世界性分布。中国已知10余种,全国广布。本属多数种类身体具有绿色的金属光泽;前翅和后翅四边室的形状相同,而印丝螅属和黄丝螅属后翅的四边室更长。

　　本属豆娘通常栖息于静水环境,包括小型水潭和水草茂盛的沼泽以及湖泊和水库的边缘。雄性通常停歇在挺水植物上,翅半张开。雌雄连结产卵,卵被插入水面上方挺水植物的茎干中。

This genus contains over 80 species, distributed all over the world. More than ten species are recorded from China. Most damselflies of the genus have a metallic green body. The shape of quadrangle is similar in both wings, but in *Indolestes* and *Sympecma* quadrangle is longer in hind wings.

Lestes species usually frequent standing water habitats, including small pools and swamps with plenty of emergent plants as well as the margin of lakes and reservoirs. Males typically perch on emergent plants with their wings half spread. They oviposit in tandem, eggs being inserted into the stalks of emergent plants well above water level.

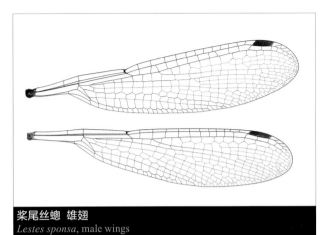

桨尾丝螅 雄翅
Lestes sponsa, male wings

足尾丝螅
Lestes dryas

桨尾丝螅
Lestes sponsa

日本丝螅
Lestes japonicus

蓝绿丝螅
Lestes temporalis

丝螅属 雄性肛附器
Genus *Lestes*, male anal appendages

桨尾丝螅 雄
Lestes sponsa, male

整齐丝螅 *Lestes concinnus* Hagen, 1862

【形态特征】雄性身体主要浅褐色；翅透明；腹部具深褐色条纹，上肛附器白色末端黑色。雌性与雄性相似。【长度】雄性腹长 28~32 mm，后翅 19~22 mm。【栖息环境】森林中的杂草池塘。【分布】海南、台湾；南亚、东南亚、澳新界。【飞行期】全年可见。

[Identification] Male mainly pale brown. Wings hyaline. Abdomen with dark brown markings, superior appendages white with black tips. Female similar to male. [Measurements] Male abdomen 28-32 mm, hind wing 19-22 mm. [Habitat] Grassy ponds in forest. [Distribution] Hainan, Taiwan; South and Southeast Asia, Australasia. [Flight Season] Throughout the year.

整齐丝螅 雄，柬埔寨 | Oleg E. Kosterin 摄
Lestes concinnus, male from Cambodia | Photo by Oleg E. Kosterin

整齐丝螅 雌，柬埔寨 | Oleg E. Kosterin 摄
Lestes concinnus, female from Cambodia | Photo by Oleg E. Kosterin

足尾丝螅 *Lestes dryas* Kirby, 1890

【形态特征】雄性复眼深蓝色，面部金属绿色；胸部墨绿色具金属光泽，侧面覆盖蓝灰色粉霜；腹部墨绿色具金属光泽，第1~3节、第8~10节覆盖白色粉霜。雌性身体青铜色具黄色条纹；产卵管超出第10节末端。【长度】体长38~41 mm，腹长 29~32 mm，后翅 22~25 mm。【栖息环境】海拔 1000 m以下挺水植物茂盛的静水环境。【分布】黑龙江、吉林、辽宁、内蒙古、河北；广布于欧亚大陆北部从爱尔兰至中国和日本的区域、北美洲。【飞行期】6—9月。

[Identification] Male eyes dark blue, face metallic green. Thorax metallic dark green, sides with bluish grey pruinescence. Abdomen metallic dark green, S1-S3 and S8-S10 with white pruinescence. Female coppery green with

足尾丝螅 雄,吉林 | 金洪光 摄
Lestes dryas, male from Jilin | Photo by Hongguang Jin

足尾丝螅 连结产卵,黑龙江 | 莫善濂 摄
Lestes dryas, laying eggs in tandem from Heilongjiang | Photo by Shanlian Mo

足尾丝螅 雄，黑龙江│莫善濂 摄
Lestes dryas, male from Heilongjiang | Photo by Shanlian Mo

足尾丝螅 雌，黑龙江│莫善濂 摄
Lestes dryas, female from Heilongjiang | Photo by Shanlian Mo

yellow stripes. Ovipositor exceeding the end of S10. [Measurements] Total length 38-41 mm, abdomen 29-32 mm, hind wing 22-25 mm. [Habitat] Standing water with abundant emergent plants below 1000 m elevation. [Distribution] Heilongjiang, Jilin, Liaoning, Inner Mongolia, Hebei; Widespread across the northern part of Eurasia, from Ireland to China and Japan, North America. [Flight Season] June to September.

多罗丝螅 *Lestes dorothea* Fraser, 1924

多罗丝螅 雄，云南（德宏）
Lestes dorothea, male from Yunnan (Dehong)

【形态特征】雄性复眼蓝色，面部褐色，上唇蓝色；胸部覆盖蓝灰色粉霜，具黑斑；腹部黑色侧缘具白色条纹，第10节覆盖白色粉霜。雌性黄褐色具黑色和白色斑纹。【长度】体长 41～48 mm，腹长 33～39 mm，后翅 22～26 mm。【栖息环境】海拔 1000 m以下挺水植物茂盛的静水环境。【分布】云南（德宏、西双版纳）；不丹、印度、尼泊尔、泰国、越南、马来半岛。【飞行期】3—10月。

[Identification] Male eyes blue, face brown, labrum blue. Thorax with bluish grey pruinescence and black markings. Abdomen mainly black with lateral white stripes, S10 with white pruinescence. Female yellowish brown with black and white markings. [Measurements] Total length 41-48 mm, abdomen 33-39 mm, hind wing 22-26 mm. [Habitat] Standing water with plenty of emergent plants below 1000 m elevation. [Distribution] Yunnan (Dehong, Xishuangbanna); Bhutan, India, Nepal, Thailand, Vietnam, Peninsular Malaysia. [Flight Season] March to October.

多罗丝螅 雄，云南（德宏）
Lestes dorothea, male from Yunnan (Dehong)

多罗丝螅 连结产卵，云南（西双版纳）｜莫善濂 摄
Lestes dorothea, laying eggs in tandem from Yunnan (Xishuangbanna) | Photo by Shanlian Mo

日本丝螅 *Lestes japonicus* Selys, 1883

　　【形态特征】雄性复眼蓝色，面部金属绿色；胸部墨绿色具金属光泽，侧面具2条黄色宽条纹；腹部墨绿色具金属光泽，第9～10节覆盖蓝灰色粉霜。雌性胸部青铜色，侧面具黄色条纹；腹部黑褐色，侧面具黄色条纹。【长度】体长 38～39 mm，腹长 29～31 mm，后翅 19～20 mm。【栖息环境】海拔 1000 m以下挺水植物茂盛的湿地。【分布】黑龙江、吉林；朝鲜半岛、日本、俄罗斯远东。【飞行期】6—9月。

[Identification] Male eyes blue, face metallic green. Thorax metallic dark green, sides with two broad yellow stripes. Abdomen metallic dark green, S9-S10 with bluish grey pruinescence. Female thorax coppery green, sides with yellow stripes. Abdomen blackish brown, sides with yellow stripes. **[Measurements]** Total length 38-39 mm, abdomen 29-31 mm, hind wing 19-20 mm. **[Habitat]** Standing water with plenty of emergent plants below 1000 m elevation. **[Distribution]** Heilongjiang, Jilin; Korean peninsula, Japan, Russian Far East. **[Flight Season]** June to September.

日本丝螅 雄, 吉林 | 金洪光 摄
Lestes japonicus, male from Jilin | Photo by Hongguang Jin

日本丝螅 雄, 吉林 | 金洪光 摄
Lestes japonicus, male from Jilin | Photo by Hongguang Jin

日本丝螅 雌, 吉林 | 金洪光 摄
Lestes japonicus, female from Jilin | Photo by Hongguang Jin

大痣丝螅 *Lestes macrostigma* (Eversmann, 1836)

【形态特征】雄性复眼蓝色，面部灰褐色；胸部具蓝灰色粉霜；腹部黑褐色，基方和端方具粉霜。雌性与雄性相似。【长度】体长 39～48 mm，腹长 31～38 mm，后翅 24～27 mm。【栖息环境】海拔 500 m 以下挺水植物茂盛的静水环境。【分布】新疆、内蒙古；欧洲西部至蒙古广布。【飞行期】6—8月。

[Identification] Male eyes blue, face greyish brown. Thorax with bluish grey pruinescence. Abdomen dark brown with the base the tip pruinosed. Female similar to male. [Measurements] Total length 39-48 mm, abdomen 31-38 mm, hind wing 24-27 mm. [Habitat] Standing water with plenty of emergent plants below 500 m elevation. [Distribution] Xinjiang, Inner Mongolia; Widespread from West Europe to Mongolia. [Flight Season] June to August.

大痣丝螅 雌，俄罗斯 | Oleg E. Kosterin 摄
Lestes macrostigma, female from Russia | Photo by Oleg E. Kosterin

蕾尾丝螅 *Lestes nodalis* Selys, 1891

蕾尾丝螅 雄,云南(普洱)
Lestes nodalis, male from Yunnan (Pu'er)

蕾尾丝螅 雄,广东
Lestes nodalis, male from Guangdong

蕾尾丝螅 雌,云南(保山)
Lestes nodalis, female from Yunnan (Baoshan)

【形态特征】雄性复眼蓝色,面部褐色;胸部灰色具蓝色条纹;腹部大面积蓝色,第7~8节黑色。雌性通体黄褐色。【长度】体长 38~41 mm,腹长 28~33 mm,后翅 19~20 mm。【栖息环境】海拔 1500 m以下挺水植物茂盛的静水环境。【分布】云南(普洱、保山)、广西、广东、香港;印度、缅甸、泰国、柬埔寨、老挝。【飞行期】2—12月。

[Identification] Male eyes blue, face brown. Thorax grey with blue markings. Abdomen largely blue, S7-S8 black. Female yellowish brown throughout. [Measurements] Total length 38-41 mm, abdomen 28-33 mm, hind wing 19-20 mm. [Habitat] Standing water with plenty of emergent plants below 1500 m elevation. [Distribution] Yunnan (Pu'er, Baoshan), Guangxi, Guangdong, Hong Kong; India, Myanmar, Thailand, Cambodia, Laos. [Flight Season] February to December.

舟尾丝螅 *Lestes praemorsus* Hagen, 1862

舟尾丝螅 雄,云南(德宏)
Lestes praemorsus, male from Yunnan (Dehong)

舟尾丝螅 雌,广东
Lestes praemorsus, female from Guangdong

舟尾丝螅 雄,云南(德宏)
Lestes praemorsus, male from Yunnan (Dehong)

【形态特征】雄性复眼蓝色,面部褐色;胸部覆盖蓝灰色粉霜,具甚小的圆形黑斑;腹部主要黑色,侧缘具白色条纹,第9~10节覆盖白色粉霜。雌性胸部覆盖蓝灰色粉霜,合胸脊两侧具黑褐色斑纹,腹部黑褐色具白色条纹。本种与多罗丝螅相似,但雄性腹部第9~10节两节白色,而多罗丝螅仅第10节白色。【长度】体长 37~42 mm,腹长 29~35 mm,后翅 20~23 mm。【栖息环境】海拔 2500 m以下挺水植物茂盛的湿地。【分布】云南、四川、福建、海南、广东、香港、台湾;南亚、东南亚、新几内亚。【飞行期】2—11月。

[Identification] Male eyes blue, face brown. Thorax with bluish grey pruinescence and small rounded black spots. Abdomen mainly black with lateral white markings, S9-S10 with white pruinescence. Female thorax with bluish grey pruinescence and dark brown spots beside dorsal carina, abdomen blackish brown with white markings. Similar to *L. dorothea* but male S9-S10 white, whereas male of *L. dorothea* possesses white only on S10. [Measurements] Total length 37-42 mm, abdomen 29-35 mm, hind wing 20-23 mm. [Habitat] Standing water with plenty of emergent plants below 2500 m elevation. [Distribution] Yunnan, Sichuan, Fujian, Hainan, Guangdong, Hong Kong, Taiwan; South and Southeast Asia, New Guinea. [Flight Season] February to November.

桨尾丝螅 *Lestes sponsa* (Hansemann, 1823)

【形态特征】雄性复眼深蓝色，面部金属绿色；胸部墨绿色具金属光泽，侧面覆盖蓝灰色粉霜；腹部墨绿色具金属光泽，第1~3节、第9~10节覆盖白色粉霜。雌性青铜色具黄色条纹；产卵管未超出第10节末端。本种与足尾丝螅相似，但雄性的下肛附器较长，末端平直，雌性产卵管未超出第10节末端。【长度】体长36~41 mm，腹长 29~33 mm，后翅 20~23 mm。【栖息环境】海拔 2000 m以下挺水植物茂盛的湿地。【分布】黑龙江、吉林、辽宁、内蒙古、湖北；欧洲西部至日本广布。【飞行期】6—9月。

[Identification] Male eyes dark blue, face metallic green. Thorax metallic dark green, sides with bluish grey pruinescence. Abdomen metallic dark green, S1-S3 and S9-S10 with white pruinescence. Female coppery green with yellow stripes. Ovipositor not exceeding the end of S10. Similar to *L. dryas* but male inferior appendages longer with straight tips and female ovipositor not exceeding the end of S10. [Measurements] Total length 36-41 mm, abdomen 29-33 mm, hind wing 20-23 mm. [Habitat] Standing water with plenty

桨尾丝螅 交尾，黑龙江
Lestes sponsa, mating pair from Heilongjiang

桨尾丝螅 雄，湖北
Lestes sponsa, male from Hubei

桨尾丝螅 雌，湖北
Lestes sponsa, female from Hubei

of emergent plants below 2000 m elevation. [Distribution] Heilongjiang, Jilin, Liaoning, Inner Mongolia, Hubei; Widespread from Western Europe to Japan. [Flight Season] June to September.

蓝绿丝螅 *Lestes temporalis* Selys, 1883

【形态特征】雄性复眼蓝绿色，面部金属绿色；胸部墨绿色具金属光泽，侧面具黄绿色宽条纹；腹部墨绿色具金属光泽，第10节覆盖蓝灰色粉霜。雌性胸部青铜色，侧面具黄色条纹；腹部黑褐色具黄色条纹，产卵管超出第10节末端。本种与日本丝螅相似，但雄性仅在腹部第10节具白色粉霜，两者肛附器构造不同。【长度】体长 43~49 mm，腹长 34~40 mm，后翅 25~26 mm。【栖息环境】海拔 500 m 以下挺水植物茂盛的湿地。【分布】吉林；朝鲜半岛、日本、俄罗斯远东。【飞行期】6—9月。

蓝绿丝螅 交尾，吉林｜金洪光 摄
Lestes temporalis, mating pair from Jilin | Photo by Hongguang Jin

[Identification] Male eyes bluish green, face metallic green. Thorax metallic dark green, sides with broad yellowish green stripes. Abdomen metallic blackish green, S10 with bluish grey pruinescence. Female thorax coppery green, sides with yellow stripes. Abdomen blackish brown with yellow markings, ovipositor exceeding the end of S10. Similar to *L. japonicus*, but males of this species have white pruinescence only on S10, anal appendages also different. [Measurements] Total length 43-49 mm, abdomen 34-40 mm, hind wing 25-26 mm. [Habitat] Standing water with plenty of emergent plants below 500 m elevation. [Distribution] Jilin; Korean peninsula, Japan, Russian Far East. [Flight Season] June to September.

蓝绿丝螅 雄,吉林 |金洪光 摄
Lestes temporalis, male from Jilin | Photo by Hongguang Jin

绿丝螅 *Lestes virens* (Charpentier, 1825)

绿丝螅 雄, 俄罗斯 | Oleg E. Kosterin 摄
Lestes virens, male from Russia | Photo by Oleg E. Kosterin

【形态特征】雄性复眼蓝色, 面部金属绿色; 胸部墨绿色具金属光泽, 侧面具蓝灰色粉霜; 腹部墨绿色具金属光泽, 第9～10节覆盖蓝灰色粉霜。雌性胸部侧面具黄色条纹; 腹部金属绿色, 产卵管伸达第10节末端。本种雄性下肛附器白色, 甚短, 是其显著的区分特征。【长度】体长 30～39 mm, 腹长 25～32 mm, 后翅 19～23 mm。【栖息环境】泥炭藓和芦苇滋生的静水环境。【分布】从欧洲至中国新疆地区广布。【飞行期】4—11月。

[Identification] Male eyes blue, face metallic green. Thorax metallic dark green, sides with bluish grey pruinescence. Abdomen metallic dark green, S9-S10 with bluish grey pruinescence. Female thorax with yellow stripes laterally. Abdomen metallic green, ovipositor reaching the end of S10. Male inferior appendages whitish and short, a key diagnosis for the species. [Measurements] Total length 30-39 mm, abdomen 25-32 mm, hind wing 19-23 mm. [Habitat] Standing water with plenty of peat moss and rushes. [Distribution] Widespread from Europe to Xinjiang of China. [Flight Season] April to November.

丝螅属待定种 *Lestes* sp.

【形态特征】与日本丝螅和蓝绿丝螅相似，但雄性腹部第8～10节具粉霜。【长度】体长 36～42 mm，腹长 29～35 mm，后翅 22～25 mm。【栖息环境】海拔 1500～2000 m挺水植物茂盛的湿地。【分布】湖北。【飞行期】 6—9月。

[Identification] Similar to *L. japonicus* and *L. temporalis*, but male S8-S10 with pruinescence. [Measurements] Total length 36-42 mm, abdomen 29-35 mm, hind wing 22-25 mm. [Habitat] Standing water with plenty of emergent plants at 1500-2000 m elevation. [Distribution] Hubei. [Flight Season] June to September.

丝螅属待定种 雄，湖北
Lestes sp., male from Hubei

丝螅属待定种 雌，湖北
Lestes sp., female from Hubei

丝螅属待定种 连结产卵，湖北
Lestes sp., laying eggs in tandem from Hubei

长痣丝螅属 Genus *Orolestes* McLachlan, 1895

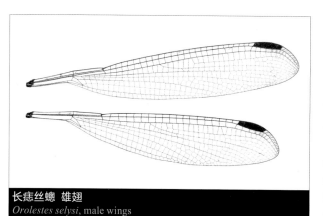

长痣丝螅 雄翅
Orolestes selysi, male wings

本属全球已知4种，仅分布于东洋界。中国已知1种，分布于云南和广西。本属豆娘体型较大，身体主要黄绿色，雄性多型，翅的色彩具有较多的变异。

本属豆娘栖息于森林中的静水潭。雄性通常吊挂在水面上方的树枝上，停歇时翅半张开。交尾在树上完成。雌性将卵产在植物的茎干内，雄性护卫产卵。

This Oriental genus contains four species. Only one species has been recorded from China, found in Yunnan and Guangxi. Species of the genus are fairly large with their body mainly yellowish green, males polymorphic, the wing color is variable.

Orolestes species inhabit pools in forests. Males usually hang from branches above water with wings half spread. When mating they hang from the trees. Females lay eggs into the stalks of plants with the male guarding.

长痣丝螅 雄
Orolestes selysi, male

长痣丝螅 *Orolestes selysi* **McLachlan, 1895**

【形态特征】雄性复眼蓝绿色，面部褐色；胸部黄绿色，透翅型翅透明，斑翅型翅中央具1条黑色宽带，黑带以上有时具1个小白斑；腹部黑色，第3节、第8～10节蓝色。雌性头部和胸部黄绿色，腹部褐色，翅透明。【长度】体长58～71 mm，腹长 47～58 mm，后翅 37～39 mm。【栖息环境】海拔 1000 m以下森林中具有树荫的池塘。【分布】云南（普洱、西双版纳、红河）、广西、海南、台湾；印度、老挝、越南。【飞行期】1—10月。

[Identification] Male eyes bluish green, face brown. Thorax yellowish green, hyaline winged morph wings hyaline, spotted winged morph wings with a median black band, and sometimes with a white spot above the black band. Abdomen black, S3 and S8-S10 blue. Female head and thorax yellowish green, abdomen brown, wings hyaline. [Measurements] Total length 58-71 mm, abdomen 47-58 mm, hind wing 37-39 mm. [Habitat] Shady ponds in forest below 1000 m elevation. [Distribution] Yunnan (Pu'er, Xishuangbanna, Honghe), Guangxi, Hainan, Taiwan; India, Laos, Vietnam. [Flight Season] January to October.

长痣丝螅 雄，透翅型，云南（西双版纳）｜莫善濂 摄
Orolestes selysi, hyaline winged morph male from Yunnan (Xishuangbanna) | Photo by Shanlian Mo

长痣丝螅 雄，斑翅型，云南（普洱）
Orolestes selysi, spotted winged morph male from Yunnan (Pu'er)

长痣丝螅 交尾，云南（普洱）｜莫善濂 摄
Orolestes selysi, mating pair from Yunnan (Pu'er) | Photo by Shanlian Mo

长痣丝螅 雌，云南（普洱）
Orolestes selysi, female from Yunnan (Pu'er)

黄丝螅属 Genus *Sympecma* Burmeister, 1839

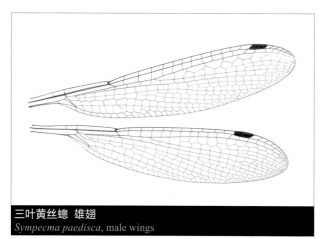

三叶黄丝螅 雄翅
Sympecma paedisca, male wings

本属全球已知3种，分布于古北界。中国已知1种，分布于北方地区，喜欢寒冷的气候。本属豆娘体小型，身体淡褐色具深褐色条纹；前胸后叶分成3叶，与印丝螅属不同；翅透明，后翅的四边室比前翅长，前翅的翅痣比后翅更接近翅端；停歇时翅合并。

本属豆娘栖息于低海拔的静水环境，比较喜欢水草茂盛的湿地和小型水塘。它们以成虫度过严冬，并且在早春开始繁殖。

This genus contains three species, distributed in the Palaearctic region. Only one species is recorded from China where it is common in the north, preferring a cold climate. Damselflies of the genus are usually small-sized with a pale brown body and dark brown markings. The hind lobe of pronotum is tri-lobed, unlike *Indolestes*. Wings hyaline with quadrangle longer in hind wings, the pterostigma in the fore wings is clearly closer to the wing tips than in the hind wings. They perch with the wings closed.

Sympecma species inhabit standing water habitats in lowland, preferring sites with plenty of emergent plants and small pools. They over winter as adults, and reproduce in early spring.

三叶黄丝螅 连结产卵 | 金洪光 摄
Sympecma paedisca, laying eggs in tandem | Photo by Hongguang Jin

三叶黄丝螅 *Sympecma paedisca* (Brauer, 1877)

三叶黄丝螅 雌, 北京 | 陈炜 摄
Sympecma paedisca, female from Beijing | Photo by Wei Chen

三叶黄丝螅 交尾, 北京 | 陈炜 摄
Sympecma paedisca, mating pair from Beijing | Photo by Wei Chen

三叶黄丝螅 雄, 北京 | 陈炜 摄
Sympecma paedisca, male from Beijing | Photo by Wei Chen

【形态特征】两性身体淡褐色具黑褐色斑纹。【长度】体长 31~34 mm, 腹长 25~26 mm, 后翅 19~20 mm。【栖息环境】海拔 1000 m 以下挺水植物茂盛的静水环境。【分布】黑龙江、吉林、内蒙古、新疆、甘肃、北京；从欧洲西部至日本的欧亚大陆北部广布。【飞行期】4—10月。

[Identification] Both sexes pale brown with blackish brown markings. [Measurements] Total length 31-34 mm, abdomen 25-26 mm, hind wing 19-20 mm. [Habitat] Standing water with plenty of emergent plants below 1000 m elevation. [Distribution] Heilongjiang, Jilin, Inner Mongolia, Xinjiang, Gansu, Beijing; Widespread in the northern part of Eurasia, from West Europe to Japan. [Flight Season] April to October.

10 综螅科 Family Synlestidae

　　本科全球已知9属约40种，分布于亚洲、非洲和大洋洲。中国已知2属10余种，主要分布于华中、华南和西南地区。综螅科是体型较大的豆娘，身体通常具有金属光泽；翅较窄且透明，具翅柄，停歇时翅展开；腹部细长，末端常覆盖粉霜。

　　本科栖息于森林中的溪流和水潭。由于飞行能力较差，它们长时间停落在水边的树枝上。雄性会悬挂在水面上方的树枝上等待雌性。有时雌雄连结产卵，有时雌性单独产卵而雄性护卫，卵通常产在树枝上或挺水植物的茎干中。

This family contains about 40 species in nine genera, distributed in Asia, Africa and Oceania. Over ten species in two genera are recorded from China, mainly from the Central, South and Southwest regions. Members of the family are fairly large damselflies with the body usually metallic colored. Wings narrow and hyaline with stalked bases, they perch with wings spread. Abdomen long, the tip usually pruinosed.

Species of the family inhabit streams and pools in forests. Since their flight is weak, they usually perch on trees. Males hang from trees above water and wait for females. Some species oviposit in tandem and in others the female oviposits with the male guarding nearby, eggs are inserted into the stalks of branches or emergent plants.

泰国绿综螅 雄

Megalestes kurahashii, male

绿综蟌属 Genus *Megalestes* Selys, 1862

郝氏绿综蟌 雄
Megalestes haui, male

本属全球已知18种,仅分布于东洋界。中国已知12种,分布于华中、华南和西南地区。

本属豆娘身体墨绿色具金属光泽和黄色条纹,胸部无肩前条纹,翅透明,翅痣为平行四边形。本属种类形态上十分近似,容易混淆,雄性可通过肛附器的形状和阳茎的构造来区分,雌性较难区分。

本属豆娘栖息于具有一定海拔高度的森林中,喜欢清澈的林荫小溪和森林中水草匮乏的小型水潭。雄性终日吊挂在水面上的藤条和树枝上,雄性之间会展开争斗以争夺领地。雌雄连结产卵。

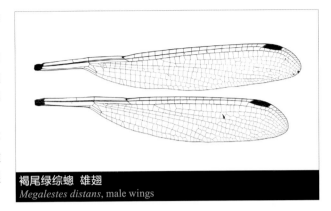

褐尾绿综螆 雄翅
Megalestes distans, male wings

This genus contains 18 species confined to Oriental region. 12 species are recorded from China, distributed in the Central, South and Southwest regions.

Species of the genus have a metallic dark green body with yellow stripes, the antehumeral stripes absent, wings hyaline, pterostigma parallelogram-shaped. These damselflies are similar in appearance and easily confused, males can be distinguished by the shape of anal appendages and penis, females are more difficult.

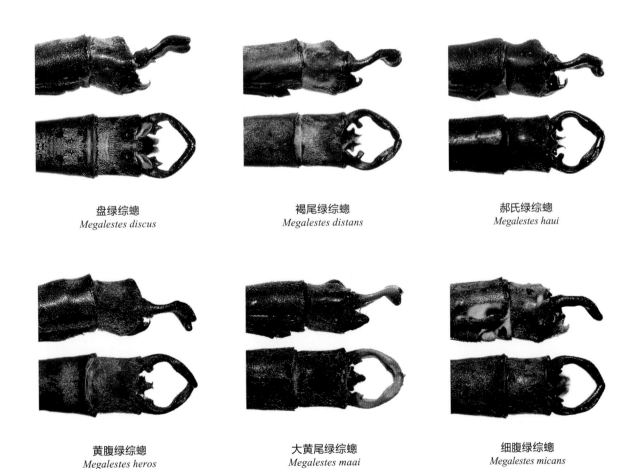

盘绿综螆
Megalestes discus

褐尾绿综螆
Megalestes distans

郝氏绿综螆
Megalestes haui

黄腹绿综螆
Megalestes heros

大黄尾绿综螆
Megalestes maai

细腹绿综螆
Megalestes micans

绿综螆属 雄性肛附器
Genus *Megalestes*, male anal appendages

泰国绿综蟌
Megalestes kurahashii

峨眉绿综蟌
Megalestes omeiensis

绿综蟌属 雄性肛附器
Genus *Megalestes,* male anal appendages

Megalestes species inhabit forests at moderate altitudes, favouring the clean shady streams and small pools with few aquatic plants. Males hang on canes and branches above the water, sometimes fighting for territory. They oviposit in tandem.

盘绿综蟌 *Megalestes discus* Wilson, 2004

盘绿综蟌 雄，湖南
Megalestes discus, male from Hunan

【形态特征】雄性复眼蓝色；面部和胸部墨绿色具金属光泽，合胸侧面具黄色条纹，随年纪增长逐渐覆盖粉霜；腹部黑褐色，第9～10节覆盖白色粉霜。【长度】雄性体长 56～60 mm，腹长 46～49 mm，后翅 30～32 mm。【栖息环境】海拔 1000～2000 m森林中的水潭和溪流。【分布】中国特有，分布于广东、湖南。【飞行期】6—10月。

[Identification] Male eyes blue. Face and thorax metallic dark green, sides of synthorax with yellow stripes and gradually pruinosed with age. Abdomen blackish brown, S9-S10 whitish pruinosed. [Measurements] Male total length 56-60 mm, abdomen 46-49 mm, hind wing 30-32 mm. [Habitat] Ponds and streams in forest at 1000-2000 m elevation. [Distribution] Endemic to China, recorded from Guangdong, Hunan. [Flight Season] June to October.

褐尾绿综蟌 *Megalestes distans* Needham, 1930

褐尾绿综蟌 雄，湖北
Megalestes distans, male from Hubei

褐尾绿综蟌 雄，湖北
Megalestes distans, male from Hubei

褐尾绿综蟌 交尾，重庆
Megalestes distans, mating pair from Chongqing

褐尾绿综螆 连结产卵，重庆
Megalestes distans, laying eggs in tandem from Chongqing

【形态特征】雄性复眼蓝色；面部和胸部墨绿色具金属光泽，合胸侧面具黄色条纹，随年纪增长逐渐覆盖粉霜；腹部黑褐色，第9～10节覆盖白色粉霜。雌性与雄性相似，色彩稍淡，腹部末端具黄斑。【长度】体长 60～64 mm，腹长 50～52 mm，后翅 32～40 mm。【栖息环境】海拔 500～2000 m森林中的水潭和林荫小溪。【分布】甘肃、湖北、四川、重庆、贵州、江西、广西、广东；越南。【飞行期】6—10月。

[Identification] Male eyes blue. Face and thorax metallic dark green, sides of synthorax with yellow stripes, and gradually pruinosed with age. Abdomen blackish brown, S9-S10 whitish pruinosed. Female similar to male but paler, abdominal tip with yellow spots. [Measurements] Total length 60-64 mm, abdomen 50-52 mm, hind wing 32-40 mm. [Habitat] Ponds and shady streams in forest at 500-2000 m elevation. [Distribution] Gansu, Hubei, Sichuan, Chongqing, Guizhou, Jiangxi, Guangxi, Guangdong; Vietnam. [Flight Season] June to October.

郝氏绿综螆 *Megalestes haui* Wilson & Reels, 2003

【形态特征】雄性复眼蓝绿色；面部和胸部墨绿色具金属光泽，合胸侧面具黄色条纹，随年纪增长逐渐覆盖蓝白色粉霜；腹部第1～2节墨绿色，第3～6节橙褐色，第7～10节黑褐色，第9～10节稍微覆盖粉霜。雌性铜褐色具黄色条纹。【长度】体长 60～68 mm，腹长 47～58 mm，后翅 35～41 mm。【栖息环境】海拔 1000～2000 m的林荫小溪。【分布】云南（文山、红河、普洱）、广西；越南。【飞行期】6—10月。

[Identification] Male eyes bluish green. Face and thorax metallic dark green, sides of synthorax with yellow stripes, and gradually bluish white pruinosed with age. Abdomen S1-S2 dark green, S3-S6 orange brown, S7-S10 blackish brown, S9-S10 slightly pruinosed. Female coppery brown with yellow stripes. **[Measurements]** Total length 60-68 mm, abdomen 47-58 mm, hind wing 35-41 mm. **[Habitat]** Shady streams in forest at 1000-2000 m elevation. **[Distribution]** Yunnan (Wenshan, Honghe, Pu'er), Guangxi; Vietnam. **[Flight Season]** June to October.

郝氏绿综螅 雄, 云南 (红河)
Megalestes haui, male from Yunnan (Honghe)

郝氏绿综螅 雌, 云南 (红河)
Megalestes haui, female from Yunnan (Honghe)

郝氏绿综螅 雄, 云南 (文山)
Megalestes haui, male from Yunnan (Wenshan)

黄腹绿综蟌 *Megalestes heros* Needham, 1930

【形态特征】雄性复眼深蓝色；面部和胸部墨绿色具金属光泽，合胸侧面具蓝灰色粉霜；腹部第1~2节背面金属绿色，侧面黄色，第3~7节橙褐色，第8~10节墨绿色，第9~10节稍微覆盖粉霜。雌性与雄性相似。【长度】体长71~80 mm，腹长58~66 mm，后翅43~44 mm。【栖息环境】海拔500~2000 m森林中的林荫小溪、沟渠和水潭。【分布】中国特有，分布于四川、浙江、福建、广东。【飞行期】5—10月。

[Identification] Male eyes dark blue. Face and thorax metallic dark green, sides of synthorax with bluish grey pruinescence. S1-S2 metallic green dorsally and yellow laterally, S3-S7 orange brown, S8-S10 metallic dark green, S9-S10 slightly pruinosed. Female similar to male. [Measurements] Total length 71-80 mm, abdomen 58-66 mm, hind wing 43-44 mm. [Habitat] Shady streams, ditches and ponds in forest at 500-2000 m elevation. [Distribution] Endemic to China, recorded from Sichuan, Zhejiang, Fujian, Guangdong. [Flight Season] May to October.

黄腹绿综蟌 雄，浙江
Megalestes heros, male from Zhejiang

黄腹绿综蟌 雌，广东
Megalestes heros, female from Guangdong

大黄尾绿综蟌 *Megalestes maai* Chen, 1947

大黄尾绿综蟌 雄,台湾 | 嘎嘎 摄
Megalestes maai, male from Taiwan | Photo by Gaga

【形态特征】雄性复眼绿色；面部和胸部墨绿色具金属光泽，合胸侧面具2条宽阔的黄色条纹；腹部褐色，上肛附器白色。【长度】体长 65~75 mm。【栖息环境】海拔 500~3000 m森林中的林荫小溪和水潭。【分布】中国台湾特有。【飞行期】3—10月。

[Identification] Male eyes green. Face and thorax metallic dark green, sides of synthorax with two broad yellow stripes. Abdomen brown, superior appendages white. [Measurements] Total length 65-75 mm. [Habitat] Shady streams and ponds in forest at 500-3000 m elevation. [Distribution] Endemic to Taiwan of China. [Flight Season] March to October.

细腹绿综蟌 *Megalestes micans* Needham, 1930

【形态特征】雄性复眼蓝绿色；面部和胸部墨绿色具金属光泽，前胸黄色，胸部侧面具黄色条纹；腹部第1~2节金属绿色，第3~10节黄褐色或深褐色，第9~10节覆盖粉霜。雌性与雄性相似，合胸脊具黄色条纹。【长度】体长 57~64 mm，腹长 46~52 mm，后翅 33~36 mm。【栖息环境】海拔 1500~3000 m森林中的溪流、沟渠和水潭。【分布】云南（德宏、大理、保山）、四川、广西；印度、老挝、越南。【飞行期】5—10月。

[Identification] Male eyes bluish green. Face and thorax metallic dark green, prothorax yellow, sides of synthorax with yellow stripes. S1-S2 metallic green, S3-S10 yellowish brown or dark brown, S9-S10 pruinosed. Female similar to male, thorax with yellow stripes along dorsal carina. [Measurements] Total length 57-64 mm, abdomen 46-52 mm, hind wing 33-36 mm. [Habitat] Streams, ditches and ponds in forest at 1500-3000 m elevation. [Distribution] Yunnan (Dehong, Dali, Baoshan), Sichuan, Guangxi; India, Laos, Vietnam. [Flight Season] May to October.

细腹绿综蟌 雄，云南（大理）
Megalestes micans, male from Yunnan (Dali)

细腹绿综蟌 雄，云南（大理）
Megalestes micans, male from Yunnan (Dali)

细腹绿综蟌 雌，云南（大理）
Megalestes micans, female from Yunnan (Dali)

泰国绿综蟌 *Megalestes kurahashii* Asahina, 1985

【形态特征】雄性复眼蓝绿色；面部和胸部墨绿色具金属光泽，合胸侧面具黄色条纹，随年纪增长逐渐覆盖粉霜；腹部第1~2节金属绿色，第3~10节褐色。雌性青铜色具黄色条纹。【长度】体长 60~72 mm，腹长 48~60 mm，后翅 35~38 mm。【栖息环境】海拔 1000~2500 m森林中的溪流、沟渠和水潭。【分布】云南（德宏、大理、保山、普洱、西双版纳）；印度、泰国。【飞行期】5—12月。

泰国绿综蟌 雌，云南（大理）
Megalestes kurahashii, female from Yunnan (Dali)

泰国绿综蟌 雌，未熟，云南（大理）
Megalestes kurahashii, immature female from Yunnan (Dali)

泰国绿综蟌 雄，云南（大理）
Megalestes kurahashii, male from Yunnan (Dali)

泰国绿综螅 连结产卵,云南(大理)
Megalestes kurahashii, laying eggs in tandem from Yunnan (Dali)

[Identification] Male eyes bluish green. Face and thorax metallic dark green, sides of synthorax with yellow stripes, gradually pruinosed with age. S1-S2 metallic green, S3-S10 brown. Female coppery green with yellow stripes. [Measurements] Total length 60-72 mm, abdomen 48-60 mm, hind wing 35-38 mm. [Habitat] Streams, ditches and ponds in forest at 1000-2500 m elevation. [Distribution] Yunnan (Dehong, Dali, Baoshan, Pu'er, Xishuangbanna); India, Thailand. [Flight Season] May to December.

峨眉绿综螅 *Megalestes omeiensis* Chao, 1965

【形态特征】雄性复眼蓝色;面部和胸部墨绿色具金属光泽,胸部侧面具黄色条纹,随年纪增长逐渐覆盖粉霜;腹部金属墨绿色具淡黄色条纹,第9~10节覆盖白色粉霜。【长度】雄性体长 73~75 mm,腹长 59~61 mm,后翅 39~42 mm。【栖息环境】海拔500~2000 m森林中的溪流、沟渠和水潭。【分布】中国特有,分布于湖北、四川。【飞行期】6—10月。

[Identification] Male eyes blue. Face and thorax metallic dark green, sides of synthorax with yellow stripes, gradually pruinosed with age. Abdomen metallic dark green with pale yellow stripes, S9-S10 whitish

峨眉绿综螅 雄,湖北
Megalestes omeiensis, male from Hubei

pruinosed. [Measurements] Male total length 73-75 mm, abdomen 59-61 mm, hind wing 39-42 mm. [Habitat] Streams, ditches and ponds in forest at 500-2000 m elevation. [Distribution] Endemic to China, recorded from Hubei, Sichuan. [Flight Season] June to October.

峨眉绿综蟌 雄，湖北
Megalestes omeiensis, male from Hubei

白尾绿综蟌 *Megalestes riccii* Navás, 1935

白尾绿综蟌 雄，广东 | 祁麟峰 摄
Megalestes riccii, male from Guangdong | Photo by Linfeng Qi

【形态特征】雄性复眼绿色；面部和胸部墨绿色具金属光泽，胸部侧面具2条宽阔的黄色条纹；腹部黑褐色，上肛附器白色。本种与大黄尾绿综蟌相似，后者可能是异名，但本种体型小于后者。【长度】腹长 44～55 mm，后翅 31～32 mm。【栖息环境】海拔 500～2000 m森林中的溪流和水潭。【分布】中国特有，分布于江西、浙江、广东。【飞行期】6—10月。

[Identification] Male eyes green. Face and thorax metallic dark green, sides of synthorax with two broad yellow stripes. Abdomen blackish brown, superior appendages white. Similar to *M. maai*, it is possible that *M. maai* is a synonym, but this species is smaller. [Measurements] Abdomen 44-55 mm, hind wing 31-32 mm. [Habitat] Streams and ponds in forest at 500-2000 m elevation. [Distribution] Endemic to China, recorded from Jiangxi, Zhejiang, Guangdong. [Flight Season] June to October.

华综蟌属 Genus *Sinolestes* Needham, 1930

本属豆娘分布于中国和越南。此处认为本属仅1种，并将 *Sinolestes editus* Needham, 1930 作为本属唯一的有效名，广泛分布于华中、华南和西南地区。本属与绿综蟌属的主要区别在于具有黄色的肩前条纹，翅痣色彩较浅且非平行四边形。本属雄性多型，翅的色彩变异较大，有时透明，有时中央具黑褐色宽带。

本属豆娘栖息于森林中的小型静水潭和沟渠。早春发生，在贵州中部的高山地区早于四月下旬就进入繁殖期，是当地一年最早出现的蜻蜓。雌性产卵于挺水植物的茎干中，有时雄性护卫。

This genus is distributed in China and Vietnam. *Sinolestes editus* Needham, 1930 is here considered the only species in this genus, widely distributed in the Central, South and Southwest regions. The genus can be distinguished from *Megalestes* by the presence of yellow antehumeral stripes, pterostigma pale but not parallelogram-shaped. Male polymorphic, wing color variable, hyaline or with broad median blackish brown cross band.

Sinolestes species frequents small pools and ditches in forest. It is an early season species, in some high mountains in central Guizhou fully mature individuals appear in the end of April, the first odonate of the year in the region. Females oviposit into stalks of emergent plants, sometimes with the male guarding.

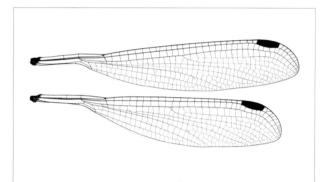

黄肩华综蟌
Sinolestes editus

黄肩华综蟌 雄翅
Sinolestes editus, male wings

华综蟌属 雄性肛附器
Genus *Sinolestes*, male anal appendages

黄肩华综蟌 雄
Sinolestes editus, male

黄肩华综螅 *Sinolestes editus* Needham, 1930

黄肩华综螅 雄，广东 | 宋睿斌 摄
Sinolestes editus, male from Guangdong | Photo by Ruibin Song

黄肩华综蟌 雄，湖北
Sinolestes editus, male from Hubei

黄肩华综蟌 雌，广东 | 宋睿斌 摄
Sinolestes editus, female from Guangdong | Photo by Ruibin Song

【形态特征】雄性复眼蓝色；面部和胸部墨绿色具金属光泽，胸部具肩前条纹，合胸侧面具2条宽阔的黄色条纹，翅透明或具黑褐色带；腹部黑褐色稍带金属光泽，第1~8节侧面具淡黄色斑，第9~10节覆盖白色粉霜。雌性与雄性相似，翅透明。【长度】体长 60~71 mm，腹长 49~58 mm，后翅 37~40 mm。【栖息环境】海拔 500~2000 m 森林中的溪流、沟渠和水潭。【分布】四川、贵州、湖北、安徽、浙江、福建、广西、广东、海南、台湾；越南。【飞行期】4—6月。

黄肩华综蟌 雄，广东 | 宋睿斌 摄
Sinolestes editus, male from Guangdong | Photo by Ruibin Song

[Identification] Male eyes blue. Face and thorax metallic dark green, antehumeral stripes present, sides of synthorax with two broad yellow stripes, wings hyaline or with broad blackish brown bands. Abdomen slightly matallic blackish brown, S1-S8 with pale yellow markings laterally, S9-S10 whitish pruinosed. Female similar to male, wings hyaline. [Measurements] Total length 60-71 mm, abdomen 49-58 mm, hind wing 37-40 mm. [Habitat] Streams, ditches and ponds in forest at 500-2000 m elevation. [Distribution] Sichuan, Guizhou, Hubei, Anhui, Zhejiang, Fujian, Guangxi, Guangdong, Hainan, Taiwan; Vietnam. [Flight Season] April to June.

黄肩华综螅 护卫产卵，广东 | 宋睿斌 摄
Sinolestes editus, female laying eggs with male guarding from Guangdong | Photo by Ruibin Song

黄肩华综螅 交尾，广东 | 宋睿斌 摄
Sinolestes editus, mating pair from Guangdong | Photo by Ruibin Song

11 扇螅科 Family Platycnemididae

　　本科全球已知40属超过400种，世界性分布。中国已知7属40余种，全国广布。本科是一类体小型至中型、腹部细长且体色艳丽的豆娘；翅通常透明并具较长的翅柄，翅痣很短，呈平行四边形，四边室长矩形；足上具长刺。

　　本科豆娘栖息于湿地、溪流和具渗流的石壁等多种生境。很多种类栖息于较阴暗的环境。雄性经常停立在水边的叶片上或者悬挂在树枝上占据领地。多数种类雌雄连结产卵。

This family contains over 400 species in 40 genera, widely distributed all over the world. Over 40 species in seven genera are recorded from China where they are widespread throughout the country. Members of the family are small to medium sized species with a colorful body and a slender, long abdomen. Wings usually hyaline with a long stalk at bases, pterostigma short, parallelogram shaped, quadrangle long and rectangular. Legs have long spines.

Species of the family inhabit wetlands, streams and precipices with trickles, occupying many kinds of habitats. Many species prefer to perch in shade. Males usually stand on the leaves or hang on the branches near water for territory. Most species oviposit in tandem.

黑袜丽扇螅 雄
Calicnemia soccifera, male

金脊长腹扇螅 雄
Coeliccia chromothorax, male

丽扇螅属 Genus *Calicnemia* Strand, 1928

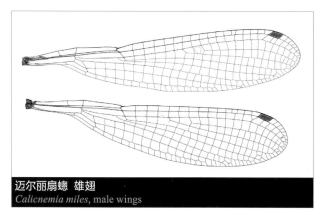

迈尔丽扇螅 雄翅
Calicnemia miles, male wings

本属全世界已知20余种,分布于亚洲的热带和亚热带地区。中国已知12种,分布于华中、华南和西南地区。本属是一类小型豆娘,雄性身体通常具有非常鲜艳的红色;前翅四边室的前边明显短于后边。多数种类的雄性仅通过身体色彩即可区分。

本属豆娘主要生活在森林中的狭窄小溪、渗流地和具有滴流的石壁。雄性停落水边的树枝或叶片上。雌雄连结产卵。

This genus contains over 20 species, distributed in tropical and subtropical Asia. 12 species are recorded from China, found in the Central, South and Southwest regions. They are small-sized species, with the males usually bright red. The upper side of the quadrangle in the fore wings is clearly shorter than the lower side. Most males can be easily distinguished by the body color pattern.

Calicnemia species inhabit narrow streams, seepages and precipices with trickles in forest. Males perch on leaves or low branches beside the water. They oviposit in tandem.

赭腹丽扇螅 交尾
Calicnemia erythromelas, mating pair

赵氏丽扇螅 *Calicnemia chaoi* Wilson, 2004

赵氏丽扇螅 雄，广东 ｜宋黎明 摄
Calicnemia chaoi, male from Guangdong | Photo by Liming Song

【形态特征】雄性面部黑色具红色条纹；胸部黑色，具红色的肩前条纹，合胸侧面具2条白色条纹，翅琥珀色，足红褐色；腹部红色。雌性胸部的条纹为淡黄色，足和腹部黄褐色。【长度】体长 34~38 mm，腹长 27~31 mm，后翅 22~23 mm。【栖息环境】海拔 1000~2000 m森林中具有滴流的陡坡和石壁。【分布】中国特有，分布于湖南、福建、广东。【飞行期】6—8月。

[Identification] Male face black with red markings. Thorax black with red antehumeral stripes, sides of synthorax with two white stripes, wings amber, legs reddish brown. Abdomen red. Female thorax with pale yellow stripes, legs and abdomen yellowish brown. [Measurements] Total length 34-38 mm, abdomen 27-31 mm, hind wing 22-23 mm. [Habitat] Slopes and precipices with trickles in forest at 1000-2000 m elevation. [Distribution] Endemic to China, recorded from Hunan, Fujian, Guangdong. [Flight Season] June to August.

赵氏丽扇螅 连结产卵，广东 ｜莫善濂 摄
Calicnemia chaoi, laying eggs in tandem from Guangdong | Photo by Shanlian Mo

赭腹丽扇螅 *Calicnemia erythromelas* (Selys, 1891)

赭腹丽扇螅 雄，云南（德宏）
Calicnemia erythromelas, male from Yunnan (Dehong)

【形态特征】雄性面部黑色具红色和黄色斑条；胸部黑色具橙红色肩前条纹，合胸侧面具2条橙黄色条纹；腹部第1~6节红色，第7~10节黑色。雌性胸部的条纹为淡黄色；腹部第1~6节橙红色。【长度】体长 34~39 mm，腹长 29~32 mm，后翅 22~24 mm。【栖息环境】海拔 1000~1500 m森林中的渗流地、狭窄小溪和小型水潭。【分布】云南（德宏、普洱、西双版纳）；缅甸、泰国、老挝、越南。【飞行期】4—11月。

[Identification] Male face black with red and yellow markings. Thorax black with orange red antehumeral stripes, sides of synthorax with two orange yellow stripes. S1-S6 red, S7-S10 black. Female thorax with pale yellow stripes. S1-S6 orange red. [Measurements] Total length 34-39 mm, abdomen 29-32 mm, hind wing 22-24 mm. [Habitat]

赭腹丽扇螅 雌，云南（西双版纳）
Calicnemia erythromelas, female from Yunnan (Xishuangbanna)

赭腹丽扇螅 雌雄连结，云南（德宏）
Calicnemia erythromelas, pair in tandem from Yunnan (Dehong)

Seepages, narrow streams and small pools in forest at 1000-1500 m elevation. **[Distribution]** Yunnan (Dehong, Pu'er, Xishuangbanna); Myanmar, Thailand, Laos, Vietnam. **[Flight Season]** April to November.

朱腹丽扇螅 *Calicnemia eximia* (Selys, 1863)

朱腹丽扇螅 雌，云南（大理）
Calicnemia eximia, female from Yunnan (Dali)

朱腹丽扇螅 连结产卵，云南（大理）
Calicnemia eximia, laying eggs from Yunnan (Dali)

【形态特征】雄性面部黑色具红色条纹；胸部黑色，具橙红色的肩前条纹，合胸侧面具2条黄色条纹；腹部红色。雌性胸部黑色具黄色条纹；腹部黄褐色。【长度】体长 34～41 mm，腹长 27～34 mm，后翅 21～25 mm。【栖息环境】海拔 2500 m以下森林中的渗流地和具有滴流的石壁。【分布】西藏、四川、贵州、云南、广西、台湾；孟加拉国、不丹、印度、尼泊尔、缅甸、泰国、老挝、越南。【飞行期】4—11月。

[Identification] Male face black with red markings. Thorax black with orange red antehumeral stripes, sides of synthorax with two yellow stripes. Abdomen red. Female thorax black with yellow stripes. Abdomen yellowish brown. [Measurements] Total length 34-41 mm, abdomen 27-34 mm, hind wing 21-25 mm. [Habitat] Seepages and precipices with trickles in forest below 2500 m elevation. [Distribution] Tibet, Sichuan, Guizhou, Yunnan, Guangxi, Taiwan; Bangladesh, Bhutan, India, Nepal, Myanmar, Thailand, Laos, Vietnam. [Flight Season] April to November.

朱腹丽扇蟌 雄, 云南 (红河)
Calicnemia eximia, male from Yunnan (Honghe)

古蔺丽扇蟌 *Calicnemia gulinensis* Yu & Bu, 2008

【形态特征】雄性面部黑色具红色和黄色斑纹；胸部黑色，具红色的肩前条纹，合胸侧面具2条黄白色条纹；腹部粉红色。雌性胸部黑色具淡黄色条纹；腹部黄褐色。【长度】雄性体长 38 mm，腹长 31 mm，后翅 23 mm。【栖息环境】海拔 1500 m以下森林中的渗流地和具有滴流的石壁。【分布】中国特有，分布于四川、重庆、湖北。【飞行期】6—9月。

古蔺丽扇蟌 雄，重庆
Calicnemia gulinensis, male from Chongqing

古蔺丽扇蟌 雌，湖北 | 莫善濂 摄
Calicnemia gulinensis, female from Hubei | Photo by Shanlian Mo

[Identification] Male face black with red and yellow markings. Thorax black with red antehumeral stripes, sides of synthorax with two yellowish white stripes. Abdomen pink red. Female thorax black with pale yellow stripes. Abdomen yellowish brown. [Measurements] Male total length 38 mm, abdomen 31 mm, hind wing 23 mm. [Habitat] Seepages and precipices with trickles in forest below 1500 m elevation. [Distribution] Endemic to China, recorded from Sichuan, Chongqing, Hubei. [Flight Season] June to September.

古蔺丽扇蟌 雄，重庆
Calicnemia gulinensis, male from Chongqing

黑丽扇螅 *Calicnemia haksik* Wilson & Reels, 2003

【形态特征】雄性面部黑色；胸部黑色，具灰色的肩前条纹，合胸侧面具2条黄色条纹；腹部第1～3节红色，第4～10节黑色。雌性胸部黑色具淡黄色条纹；腹部黑色。【长度】雄性体长 39～41 mm，腹长 32～33 mm，后翅 24～25 mm。【栖息环境】海拔 500～1500 m森林中的渗流地和具有滴流的石壁。【分布】贵州、湖南、广西；越南。【飞行期】5—8月。

[Identification] Male face black. Thorax black with grey antehumeral stripes, sides of synthorax with two yellow stripes. S1-S3 red, S4-S10 black. Female thorax black with pale yellow stripes. Abdomen black. [Measurements] Male total length 39-41 mm, abdomen 32-33 mm, hind wing 24-25 mm. [Habitat] Seepages and precipices with trickles in forest at 500-1500 m elevation. [Distribution] Guizhou, Hunan, Guangxi; Vietnam. [Flight Season] May to August.

黑丽扇螅 雄，广西
Calicnemia haksik, male from Guangxi

黑丽扇螅 雌雄连结，广西
Calicnemia haksik, pair in tandem from Guangxi

灰丽扇螅 *Calicnemia imitans* Lieftinck, 1948

灰丽扇螅 雄，云南（德宏）
Calicnemia imitans, male from Yunnan (Dehong)

【形态特征】雄性面部黑色；胸部黑色，具灰色的肩前条纹，合胸侧面具2条灰色条纹；腹部黑色，稍微覆盖灰色粉霜。雌性胸部黑色具黄色条纹；腹部黑色。【长度】体长 30~37 mm，腹长 24~30 mm，后翅 19~23 mm。【栖息环境】海拔 1000 m以下森林中具有滴流的石壁。【分布】云南（德宏）；印度、缅甸、泰国、老挝、越南。【飞行期】4—10月。

[Identification] Male face black. Thorax black with grey antehumeral stripes, sides of synthorax with two grey stripes. Abdomen black and slightly greyish pruinosed. Female thorax black with yellow stripes. Abdomen black. [Measurements] Total length 30-37 mm, abdomen 24-30 mm, hind wing 19-23 mm. [Habitat] Precipices with trickles in forest below 1000 m elevation. [Distribution] Yunnan (Dehong); India, Myanmar, Thailand, Laos, Vietnam. [Flight Season] April to October.

灰丽扇螅 雌，产卵，云南（德宏）
Calicnemia imitans, female laying eggs from Yunnan (Dehong)

迈尔丽扇螅 *Calicnemia miles* (Laidlaw, 1917)

【形态特征】雄性面部黑色具红色和黄色条纹；胸部黑色具红色的肩前条纹，侧面具2条橙黄色或黄白色条纹；腹部红色，第9~10节具黑斑。雌性胸部黑色具黄色条纹；腹部红褐色，第8~10节具黑斑。【长度】体长37~39 mm，腹长 29~31 mm，后翅 23~26 mm。【栖息环境】海拔 2500 m以下森林中的渗流地和具有滴流的石壁。【分布】西藏、云南、广西；印度、缅甸、泰国、老挝、越南。【飞行期】4—11月。

迈尔丽扇螅 雌, 云南(西双版纳)
Calicnemia miles, female from Yunnan (Xishuangbanna)

迈尔丽扇螅 雌雄连结, 云南(普洱)
Calicnemia miles, pair in tandem from Yunnan (Pu'er)

[Identification] Male face black with red and yellow markings. Thorax black with red antehumeral stripes, sides of synthorax with two orange yellow or yellowish white stripes. Abdomen red, S9-S10 with black spots. Female thorax black with yellow stripes. Abdomen reddish brown, S8-S10 with black spots. [Measurements] Total length 37-39 mm, abdomen 29-31 mm, hind wing 23-26 mm. [Habitat] Seepages and precipices with trickles in forest below 2500 m elevation. [Distribution] Tibet, Yunnan, Guangxi; India, Myanmar, Thailand, Laos, Vietnam. [Flight Season] April to November.

迈尔丽扇螅 雄, 云南(红河)
Calicnemia miles, male from Yunnan (Honghe)

华丽扇螅 *Calicnemia sinensis* Lieftinck, 1984

【形态特征】雄性面部黑色具褐色条纹；胸部黑色，具灰色的肩前条纹，合胸侧面具2条蓝灰色或黄白色条纹；腹部粉红色，有时具黑色斑。雌性胸部黑色具淡黄色条纹；腹部褐色具黑色条纹。【长度】体长 35～42 mm，腹长 27～34 mm，后翅 20～24 mm。【栖息环境】海拔 1500 m以下森林中的渗流地、沟渠和具有滴流的石壁。【分布】中国特有，分布于湖南、浙江、福建、广东、香港。【飞行期】5—9月。

[Identification] Male face black with brown markings. Thorax black with grey antehumeral stripes, sides of synthorax with two bluish grey or yellowish white stripes. Abdomen pink red, sometimes with black spots. Female thorax black with pale yellow stripes. Abdomen brown with black spots. [Measurements] Total length 35-42 mm, abdomen 27-34 mm, hind wing 20-24 mm. [Habitat] Seepages, ditches and precipices with trickles in forest below 1500 m elevation. [Distribution] Endemic to China, recorded from Hunan, Zhejiang, Fujian, Guangdong, Hong Kong. [Flight Season] May to September.

华丽扇螅 雄，广东｜宋黎明 摄
Calicnemia sinensis, male from Guangdong｜Photo by Liming Song

华丽扇螅 雄，广东
Calicnemia sinensis, male from Guangdong

华丽扇螅 连结产卵，广东
Calicnemia sinensis, laying eggs in tandem from Guangdong

黑袜丽扇螅 *Calicnemia soccifera* Yu & Chen, 2013

黑袜丽扇螅 雄,云南(红河)
Calicnemia soccifera, male from Yunnan (Honghe)

【形态特征】雄性面部黑色具红褐色条纹；胸部黑色，具甚细的灰色肩前条纹，合胸侧面具2条蓝灰色条纹；腹部第1~6节红色，第7~10节黑色。雌性胸部黑色具淡黄色条纹；腹部红褐色。【长度】体长 34~36 mm，腹长 27~29 mm，后翅 21~22 mm。【栖息环境】海拔 1500 m 以下森林中的渗流地、沟渠和具有滴流的石壁。【分布】云南(红河)；越南。【飞行期】4—6月。

黑袜丽扇螅 连结产卵,云南(红河)
Calicnemia soccifera, laying eggs from Yunnan (Honghe)

[Identification] Male face black with reddish brown markings. Thorax black with narrow grey antehumeral stripes, sides of synthorax with two bluish grey stripes. S1-S6 red, S7-S10 black. Female thorax black with pale yellow stripes. Abdomen reddish brown. [Measurements] Total length 34-36 mm, abdomen 27-29 mm, hind wing 21-22 mm. [Habitat] Seepages, ditches and precipices with trickles in forest below 1500 m elevation. [Distribution] Yunnan (Honghe); Vietnam. [Flight Season] April to June.

丽扇螅属待定种 *Calicnemia* sp.

【形态特征】本种与黑丽扇螅相似，但体型稍小，腹部第3节的红色区域较长。【长度】雄性体长 36 mm，腹长 29 mm，后翅 21 mm。【栖息环境】海拔 1000 m以下森林中的渗流地、沟渠和具有滴流的石壁。【分布】广西。【飞行期】5—8月。

[Identification] Similar to *C. haksik* but size smaller, S3 with longer red stripe. [Measurements] Male total length 36 mm, abdomen 29 mm, hind wing 21 mm. [Habitat] Seepages, ditches and precipices with trickles in forest below 1000 m elevation. [Distribution] Guangxi. [Flight Season] May to August.

丽扇螅属待定种 雄，广西
Calicnemia sp., male from Guangxi

长腹扇蟌属 Genus *Coeliccia* Kirby, 1890

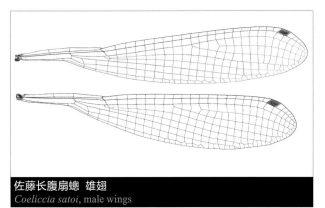

佐藤长腹扇蟌 雄翅
Coeliccia satoi, male wings

本属全球已知60余种，分布于亚洲的热带和亚热带地区。中国已知10余种，分布于华中、华南和西南地区。本属是一类体中型、腹部极为细长的豆娘，身体通常具有鲜艳的色彩；翅透明，四边室的前边明显短于后边，前翅更显著。

本属主要栖息于森林中的狭窄小溪、渗流地和具有滴流的石壁。雄性通常停落在小溪边缘较低处的树枝和叶片上。雌雄连结产卵。

The genus contains over 60 known species, found in tropical and subtropical Asia. Over ten species are recorded from China in the Central, South and Southwest regions. This is a group of medium-sized species with a very long abdomen, their body usually with bright coloration. Wings hyaline with the upper side of quadrangle clearly shorter than the lower side, more readily seen in the fore wings.

Coeliccia species inhabit narrow streams, seepages and precipices with trickles in forest. Males usually perch on the marginal leaves or branches very low. They oviposit in tandem.

海南长腹扇蟌 雄
Coeliccia hainanense, male

金脊长腹扇螅 *Coeliccia chromothorax* (Selys, 1891)

金脊长腹扇螅 雄,云南(西双版纳)
Coeliccia chromothorax, male from Yunnan (Xishuangbanna)

【形态特征】雄性面部黑色;胸部黑色具甚阔的橙色肩前条纹,合胸侧面具2条黄色条纹;腹部黑色,肛附器黄色。雌性胸部条纹为黄色,腹部黑色,第8~10节具白斑。【长度】体长 47~55 mm,腹长 39~47 mm,后翅 24~29 mm。【栖息环境】海拔 2000 m以下森林中的林荫小溪、渗流地和小型水潭。【分布】云南;缅甸、泰国、老挝、越南。【飞行期】3—12月。

[Identification] Male face black. Thorax black with broad orange antehumeral stripes, sides of synthorax with two yellow stripes. Abdomen black, anal appendages yellow. Female thorax with yellow markings, abdomen black with white spots on S8-S10. [Measurements] Total length 47-55 mm, abdomen 39-47 mm, hind wing 24-29 mm. [Habitat] Shady streams, seepages and small pools in forest below 2000 m elevation. [Distribution] Yunnan; Myanmar, Thailand, Laos, Vietnam. [Flight Season] March to December.

金脊长腹扇螅 雄,云南(西双版纳)
Coeliccia chromothorax, male from Yunnan (Xishuangbanna)

金脊长腹扇螅 雌,云南(西双版纳)
Coeliccia chromothorax, female from Yunnan (Xishuangbanna)

金脊长腹扇螅 连结产卵,云南(西双版纳) |莫善濂 摄
Coeliccia chromothorax, laying eggs in tandem from Yunnan (Xishuangbanna) | Photo by Shanlian Mo

黄纹长腹扇螅 *Coeliccia cyanomelas* Ris, 1912

【形态特征】雄性面部黑色具蓝色斑纹;胸部黑色,背面具4个淡蓝色斑,侧面具2条淡蓝色条纹;腹部黑色,第1~7节侧面具蓝白色斑,第8~10节和肛附器淡蓝色。雌性胸部具黄色条纹;腹部黑色,第8~9节具白斑。【长度】体长 46~51 mm,腹长 39~44 mm,后翅 24~27 mm。【栖息环境】海拔 2000 m以下森林中的林荫小溪、渗流地和小型水潭。【分布】中国西北、华中、华南、西南地区广布;老挝、越南。【飞行期】4—10月。

[Identification] Male face black with blue markings. Thorax black with four pale blue spots dorsally, sides with two pale blue stripes. Abdomen black, S1-S7 with bluish white spots, S8-S10 and anal appendages pale blue. Female thorax with yellow stripes. Abdomen black, S8-S9 with white spots. [Measurements] Total length 46-51 mm, abdomen 39-44 mm, hind wing 24-27 mm. [Habitat] Shady streams, seepages and small pools in forest below 2000 m elevation. [Distribution] Widespread in the Northwest, Central, South and Southwest regions; Laos, Vietnam. [Flight Season] April to October.

黄纹长腹扇螅 雌雄连结，贵州
Coeliccia cyanomelas, pair in tandem from Guizhou

黄纹长腹扇螅 雌雄连结，浙江
Coeliccia cyanomelas, pair in tandem from Zhejiang

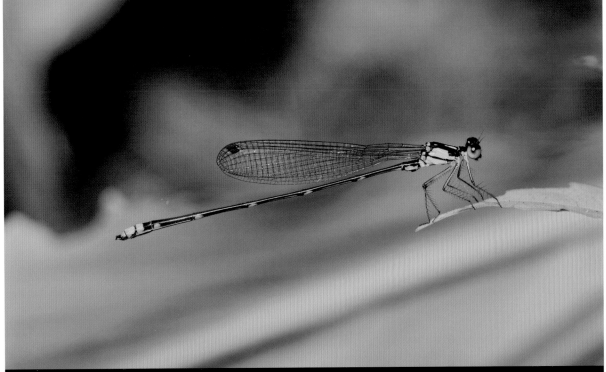

黄纹长腹扇螅 雄，海南
Coeliccia cyanomelas, male from Hainan

四斑长腹扇蟌 *Coeliccia didyma* (Selys, 1863)

【形态特征】雄性与黄纹长腹扇蟌相似，但腹部第8节无蓝斑。雌性多型，分为蓝色型和黄色型。【长度】体长 44～53 mm，腹长 38～45 mm，后翅 23～28 mm。【栖息环境】海拔 1500 m以下森林中的林荫小溪、渗流地和小型水潭。【分布】云南（德宏）；印度、缅甸、泰国、老挝、马来半岛。【飞行期】全年可见。

[Identification] Male similar to *C. cyanomelas* but S8 without blue spots. Female polymorphic with blue morph and yellow morph. [Measurements] Total length 44-53 mm, abdomen 38-45 mm, hind wing 23-28 mm. [Habitat] Shady streams, seepages and small pools in forest below 1500 m elevation. [Distribution] Yunnan (Dehong); India, Myanmar, Thailand, Laos, Peninsular Malaysia. [Flight Season] Throughout the year.

四斑长腹扇蟌 雄，云南（德宏）
Coeliccia didyma, male from Yunnan (Dehong)

四斑长腹扇蟌 雌雄连结，云南（德宏）
Coeliccia didyma, pair in tandem from Yunnan (Dehong)

四斑长腹扇蟌 雌雄连结，云南（德宏）
Coeliccia didyma, pair in tandem from Yunnan (Dehong)

黄尾长腹扇螆 *Coeliccia flavicauda* Ris, 1912

黄尾长腹扇螆 雄,台湾 | 嘎嘎 摄
Coeliccia flavicauda, male from Taiwan | Photo by Gaga

【形态特征】雄性面部黑色具蓝色斑纹；胸部黑色，胸部背面具2个淡蓝色斑，侧面具3条淡蓝色条纹；腹部黑色，肛附器黄色。雌性与雄性相似，腹部第8~10节具甚大的黄斑。【长度】体长 48~53 mm。【栖息环境】海拔 1000 m以下森林中的林荫小溪、渗流地和小型水潭。【分布】中国台湾；日本。【飞行期】2—11月。

[Identification] Male face black with blue markings. Thorax black with two pale blue spots dorsally, sides with three pale blue stripes. Abdomen black, appendages yellow. Female similar to male, abdomen with large yellow spots on S8-S10. [Measurements] Total length 48-53 mm. [Habitat] Shady streams, seepages and small pools in forest below 1000 m elevation. [Distribution] Taiwan of China; Japan. [Flight Season] February to November.

黄尾长腹扇螆 雌雄连结,台湾 | 嘎嘎 摄
Coeliccia flavicauda, pair in tandem from Taiwan | Photo by Gaga

蓝黑长腹扇螅 *Coeliccia furcata* Hämäläinen, 1986

蓝黑长腹扇螅 雄, 云南 (德宏)
Coeliccia furcata, male from Yunnan (Dehong)

蓝黑长腹扇螅 雌, 云南 (德宏)
Coeliccia furcata, female from Yunnan (Dehong)

【形态特征】雄性面部黑色；胸部黑色，具蓝黄色的肩前条纹，侧面具2条蓝黄色条纹；腹部黑色，第1节侧面具1个蓝黄色斑。雌性与雄性相似。【长度】体长 44~48 mm，腹长 36~40 mm，后翅 26~27 mm。【栖息环境】海拔1000~1500 m森林中的渗流地。【分布】云南（德宏）；缅甸。【飞行期】9—11月。

[Identification] Male face black. Thorax black with bluish yellow antehumeral stripes, sides with two bluish yellow stripes. Abdomen black, S1 with a lateral bluish yellow spot. Female similar to male. [Measurements] Total length 44-48 mm, abdomen 36-40 mm, hind wing 26-27 mm. [Habitat] Seepages in forest at 1000-1500 m elevation. [Distribution] Yunnan (Dehong); Myanmar. [Flight Season] September to November.

海南长腹扇蟌 *Coeliccia hainanense* Laidlaw, 1932

【形态特征】雄性面部黑色；胸部黑色，背面具1对黄色斑，侧面具2条黄色条纹；腹部黑色，第1~7节具白斑，第8节后缘、第9~10节及肛附器黄色。雌性胸部黑色具黄绿色条纹；腹部黑色具白斑。【长度】体长 56~58 mm，腹长 47~50 mm，后翅 29~33 mm。【栖息环境】海拔 1500 m以下森林中的林荫小溪、渗流地和小型水潭。【分布】中国海南特有。【飞行期】4—10月。

[Identification] Male face black. Thorax black with a pair of yellow spots dorsally, sides with two yellow stripes. Abdomen black, S1-S7 with white spots, the end of S8, S9-S10 and appendages yellow. Female thorax black with yellowish green stripes. Abdomen black with white spots. [Measurements] Total length 56-58 mm, abdomen 47-

海南长腹扇蟌 雌雄连结，海南
Coeliccia hainanense, pair in tandem from Hainan

50 mm, hind wing 29-33 mm. [Habitat] Shady streams, seepages and small pools in forest below 1500 m elevation. [Distribution] Endemic to Hainan of China. [Flight Season] April to October.

海南长腹扇螅 雄，海南
Coeliccia hainanense, male from Hainan

海南长腹扇螅 雌，海南
Coeliccia hainanense, female from Hainan

蓝斑长腹扇螅 *Coeliccia loogali* Fraser, 1932

【形态特征】雄性面部黑色具蓝色条纹；胸部黑色具淡蓝色的肩前条纹，侧面具2条淡蓝色条纹；腹部黑色。雌性胸部黑色具黄色条纹；腹部黑色，第8～9节具较大的白斑。【长度】体长 49～51 mm，腹长 41～42 mm，后翅 27～28 mm。【栖息环境】海拔 1500 m以下森林中的溪流和小型水潭。【分布】云南（德宏、临沧、普洱、西双版纳）；印度、尼泊尔、缅甸、泰国、老挝。【飞行期】4—11月。

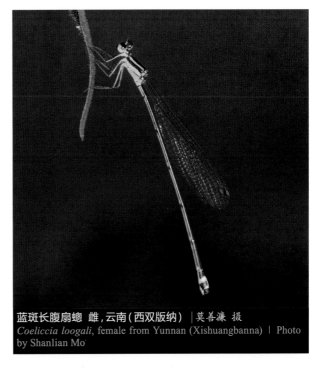

[Identification] Male face black with blue markings. Thorax black with pale blue antehumeral stripes, sides with two pale blue stripes. Abdomen black. Female thorax black with yellow markings. Abdomen black with large white spots on S8-S9. [Measurements] Total length 49-51 mm, abdomen 41-42 mm, hind wing 27-28 mm. [Habitat] Streams and small pools in forest below 1500 m elevation. [Distribution] Yunnan (Dehong, Lincang, Pu'er, Xishuangbanna); India, Nepal, Myanmar, Thailand, Laos. [Flight Season] April to November.

蓝斑长腹扇螅 雌，云南（西双版纳）｜莫善濂 摄
Coeliccia loogali, female from Yunnan (Xishuangbanna)｜Photo by Shanlian Mo

蓝斑长腹扇螅 雄，云南（西双版纳）｜莫善濂 摄
Coeliccia loogali, male from Yunnan (Xishuangbanna)｜Photo by Shanlian Mo

蓝脊长腹扇螅 *Coeliccia poungyi* Fraser, 1924

【形态特征】雄性面部黑色；胸部黑色具甚阔的天蓝色肩前条纹，侧面具2条天蓝色条纹；腹部黑色，第9~10节和肛附器黄色。雌性胸部黑色具黄色条纹；腹部第8~10节大面积黄色。【长度】体长 46~50 mm，腹长 38~42 mm，后翅 24~29 mm。【栖息环境】海拔 1500 m以下森林中的林荫小溪、渗流地和小型水潭。【分布】云南（临沧、普洱、西双版纳）、广西；缅甸、泰国、老挝。【飞行期】4—12月。

蓝脊长腹扇螅 雄，云南（临沧）
Coeliccia poungyi, male from Yunnan (Lincang)

[Identification] Male face black. Thorax black with broad sky-blue antehumeral stripes, sides with two sky-blue stripes. Abdomen black, S9-S10 and anal appendages yellow. Female thorax black with yellow stripes. S8-S10 largely yellow. [Measurements] Total length 46-50 mm, abdomen 38-42 mm, hind wing 24-29 mm. [Habitat] Shady streams, seepages and small pools in forest below 1500 m elevation. [Distribution] Yunnan (Lincang, Pu'er, Xishuangbanna), Guangxi; Myanmar, Thailand, Laos. [Flight Season] April to December.

蓝脊长腹扇螅 雌，云南（临沧）
Coeliccia poungyi, female from Yunnan (Lincang)

黄蓝长腹扇螅 *Coeliccia pyriformis* Laidlaw, 1932

【形态特征】雄性面部黑色具蓝色斑纹；胸部黑色具蓝色的肩前条纹，侧面具2条蓝色条纹；腹部黑色，第1~7节具甚小的白斑，第9~10节和肛附器黄色。雌性胸部黑色具黄色条纹；腹部第8~9节具黄斑。【长度】体长40~47 mm，腹长 34~40 mm，后翅 20~25 mm。【栖息环境】海拔 1000 m以下森林中的林荫溪流和小型水潭。【分布】云南（红河）、广西；越南。【飞行期】4—6月。

[Identification] Male face black with blue markings. Thorax black with blue antehumeral stripes, sides with two blue stripes. Abdomen black with small white spots on S1-S7, S9-S10 and anal appendages yellow. Female thorax black

黄蓝长腹扇螅 雄，云南（红河）
Coeliccia pyriformis, male from Yunnan (Honghe)

with yellow stripes, S8-S9 with yellow spots. [Measurements] Total length 40-47 mm, abdomen 34-40 mm, hind wing 20-25 mm. [Habitat] Shady streams and small pools in forest below 1000 m elevation. [Distribution] Yunnan (Honghe), Guangxi; Vietnam. [Flight Season] April to June.

黄蓝长腹扇螅 雄,云南(红河)
Coeliccia pyriformis, male from Yunnan (Honghe)

黄蓝长腹扇螅 雌雄连结,云南(红河)
Coeliccia pyriformis, pair in tandem from Yunnan (Honghe)

佐藤长腹扇螅 *Coeliccia satoi* Asahina, 1997

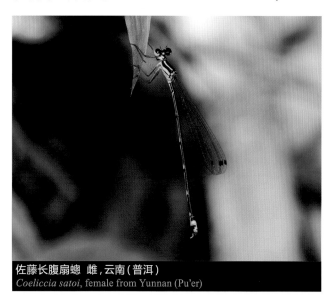

佐藤长腹扇螅 雌,云南(普洱)
Coeliccia satoi, female from Yunnan (Pu'er)

【形态特征】雄性面部黑色具蓝色斑纹;胸部黑色,背面具3对淡蓝色斑,侧面具2条淡蓝色条纹;腹部黑色,第9~10节和肛附器黄色。雌性胸部黑色具黄色条纹;腹部第8~10节具黄斑。【长度】体长43~48 mm,腹长 36~39 mm,后翅 23~28 mm。【栖息环境】海拔 1500 m以下森林中的林荫小溪、渗流地和小型水潭。【分布】云南(西双版纳、普洱);越南。【飞行期】4—10月。

[Identification] Male face black with blue markings. Thorax black with three pairs of pale blue spots dorsally, sides with two pale blue stripes. Abdomen black, S9-S10 and anal appendages yellow. Female thorax black with yellow stripes, S8-S10 with yellow spots. [Measurements] Total length 43-48 mm,

佐藤长腹扇螅 交尾，云南（普洱）
Coeliccia satoi, mating pair from Yunnan (Pu'er)

佐藤长腹扇螅 雄，云南（普洱）
Coeliccia satoi, male from Yunnan (Pu'er)

abdomen 36-39 mm, hind wing 23-28 mm. [Habitat] Shady streams, seepages and small pools in forest below 1000 m elevation. [Distribution] Yunnan (Xishuangbanna, Pu'er); Vietnam. [Flight Season] April to October.

截斑长腹扇螅 *Coeliccia scutellum* Laidlaw, 1932

截斑长腹扇螅 雌，云南（红河）｜莫善濂 摄
Coeliccia scutellum, female from Yunnan (Honghe) | Photo by Shanlian Mo

【形态特征】本种与海南长腹扇螅相似，但雄性腹部第8节无黄斑，雌性肩前条纹中央间断。【长度】体长 48~56 mm，腹长 41~47 mm，后翅 27~30 mm。【栖息环境】海拔 500 m以下森林中的林荫小溪、渗流地和小型水潭。【分布】云南（红河）；老挝、越南。【飞行期】3—12月。

[Identification] Similar to *C. hainanense*, but male S8 without yellow spots, female antehumeral stripes interrupted at mid point. [Measurements] Total length 48-56 mm, abdomen 41-47 mm, hind wing 27-30 mm. [Habitat] Shady streams, seepages and small pools in forest below 500 m elevation. [Distribution] Yunnan (Honghe); Laos, Vietnam. [Flight Season] March to December.

截斑长腹扇螅 雄,云南(红河)
Coeliccia scutellum, male from Yunnan (Honghe)

截斑长腹扇螅 连结产卵,云南(红河) | 莫善濂 摄
Coeliccia scutellum, laying eggs in tandem from Yunnan (Honghe) | Photo by Shanlian Mo

双色长腹扇螅 *Coeliccia svihleri* **Asahina, 1970**

【形态特征】本种与截斑长腹扇螅相似，但雄性胸部侧面的条纹色彩不同，分别为黄色和淡蓝色，雌性肩前条纹中央未间断。【长度】体长 41~43 mm，腹长 35~38 mm，后翅 21~24 mm。【栖息环境】海拔 500 m 以下森林中的林荫溪流、渗流地和小型水潭。【分布】云南（德宏）；印度、缅甸。【飞行期】9—12月。

[**Identification**] Similar to *C. scutellum*, but male thorax with stripes of different color, yellow and pale blue respectively, female antehumeral stripes not interrupted at mid point. [**Measurements**] Total length 41-43 mm, abdomen 35-38 mm, hind wing 21-24 mm. [**Habitat**] Shady streams, seepages and small pools in forest below 500 m elevation. [**Distribution**] Yunnan (Dehong); India, Myanmar. [**Flight Season**] September to December.

双色长腹扇螅 雄，云南（红河）
Coeliccia svihleri, male from Yunnan (Dehong)

双色长腹扇螅 雌，云南（红河）
Coeliccia svihleri, female from Yunnan (Dehong)

长腹扇螺属待定种1 *Coeliccia* sp. 1

长腹扇螺属待定种1 雄,云南(德宏)
Coeliccia sp. 1, male from Yunnan (Dehong)

　　【形态特征】雄性面部黑色具淡蓝色条纹；胸部黑色,背面具1对绿色斑点,侧面具2条宽阔的天蓝色条纹；腹部黑色,肛附器黄色。【长度】雄性体长 40~42 mm,腹长 33~34 mm,后翅 21~22 mm。【栖息环境】海拔500~1500 m森林中的渗流地。【分布】云南(德宏)。【飞行期】5—7月。

　　[Identification] Male face black with pale blue markings. Thorax black with a pair of green spots dorsally, sides with two broad sky-blue stripes. Abdomen black, anal appendages yellow. [Measurements] Total length 40-42 mm, abdomen 33-34 mm, hind wing 21-22 mm. [Habitat] Seepages in forest at 500-1500 m elevation. [Distribution] Yunnan (Dehong). [Flight Season] May to July.

长腹扇螆属待定种2 *Coeliccia* sp. 2

长腹扇螆属待定种2 雄, 云南（德宏）
Coeliccia sp. 2, male from Yunnan (Dehong)

【形态特征】本种与黄纹长腹扇螆和四斑长腹扇螆相似, 但雄性胸部背面和腹部第9~10节的蓝色斑纹不同。
【长度】雄性体长 46 mm, 腹长 38 mm, 后翅 25 mm。【栖息环境】海拔 800 m左右森林中的林荫溪流。【分布】
云南（德宏）。【飞行期】5—6月。

[Identification] Similar to *C. cyanomelas* and *C. didyma*, but male with different dorsal spots of thorax and blue
spots on S9-S10. [Measurements] Male total length 46 mm, abdomen 38 mm, hind wing 25 mm. [Habitat] Shady
streams in forest at about 800 m elevation. [Distribution] Yunnan (Dehong). [Flight Season] May to June.

长腹扇螅属待定种3 *Coeliccia* sp. 3

【形态特征】本种被认为是印扇螅，但与后者差异较大。【长度】雌性体长 62 mm，腹长 50 mm，后翅 34 mm。【栖息环境】海拔 1000 m以下森林中的林荫溪流。【分布】海南。【飞行期】4—7月。

[Identification] The species was regarded as *Indocnemis orang* but clearly different from that species. [Measurements] Female total length 62 mm, abdomen 50 mm, hind wing 34 mm. [Habitat] Shady streams in forest below 1000 m elevation. [Distribution] Hainan. [Flight Season] April to July.

长腹扇螅属待定种3 雌，海南
Coeliccia sp. 3, female from Hainan

长腹扇螅属待定种4 *Coeliccia* sp. 4

【形态特征】雄性面部黑色具甚小的黄斑；胸部黑色，背面具1对圆形黄斑，侧面大面积黄色；腹部黑色，第10节和肛附器黄色。【长度】雄性体长 41~46 mm，腹长 36~40 mm，后翅 24~26 mm。【栖息环境】海拔 1000~1500 m 森林中的渗流地。【分布】云南（德宏）。【飞行期】5—6月。

[Identification] Male face black with small yellow spots. Thorax black with a pair of yellow rounded spots, sides largely yellow. Abdomen black, S10 and anal appendages yellow. [Measurements] Male total length 41-46 mm, abdomen 36-40 mm, hind wing 24-26 mm. [Habitat] Seepages in forest at 1000-1500 m elevation. [Distribution] Yunnan (Dehong). [Flight Season] May to June.

长腹扇螅属待定种4 雄，云南（德宏）
Coeliccia sp. 4, male from Yunnan (Dehong)

长腹扇螅属待定种5 *Coeliccia* sp. 5

【形态特征】雌性面部黑色具绿色斑纹；胸部黑色具较长的肩前条纹，侧面具2条绿黄色条纹；腹部黑色，第1~8节侧面具黄白色条纹，第9~10节金黄色，产卵管甚长。【长度】雌性体长 45 mm，腹长 38 mm，后翅 28 mm。【栖息环境】海拔 1000 m左右森林中的林荫溪流。【分布】云南（普洱）。【飞行期】5—6月。

[Identification] Female face black with green markings. Thorax black with long antehumeral stripes, sides with two greenish yellow stripes. Abdomen black, S1-S8 with lateral yellowish white stripes, S9-S10 orange yellow, ovipositor long. [Measurements] Female total length 45 mm, abdomen 38 mm, hind wing 28 mm. [Habitat] Shady streams in forest at about 1000 m elevation. [Distribution] Yunnan (Pu'er). [Flight Season] May to June.

长腹扇螅属待定种5 雌，云南（普洱）
Coeliccia sp. 5, female from Yunnan (Pu'er)

长腹扇螅属待定种6 *Coeliccia* sp. 6

【形态特征】本种与蓝斑长腹扇螅相似,但雄性肛附器和雌性的前胸构造不同。【长度】体长 48~51 mm,腹长 41~44 mm,后翅 27~28 mm。【栖息环境】海拔 1500~2000 m 森林中的溪流。【分布】云南(丽江)。【飞行期】5—7月。

[Identification] The species is similar to *C. loogali* but different in the male anal appendages and female prothorax. [Measurements] Total length 48-51 mm, abdomen 41-44 mm, hind wing 27-28 mm. [Habitat] Streams in forest at 1500-2000 m elevation. [Distribution] Yunnan (Lijiang). [Flight Season] May to July.

长腹扇螅属待定种6 雄,云南(丽江)
Coeliccia sp. 6, male from Yunnan (Lijiang)

长腹扇螅属待定种6 雌雄连结,云南(丽江)
Coeliccia sp. 6, pair in tandem from Yunnan (Lijiang)

长腹扇螅属待定种6 雌,云南(丽江)
Coeliccia sp. 6, female from Yunnan (Lijiang)

狭扇螺属 Genus *Copera* Kirby, 1890

此处将伪狭扇螺属的4种全部归入本属处理。本属全球已知10余种,分布于亚洲和非洲的热带和亚热带地区。中国已知5种,主要分布于华北、华中、华南和西南地区。本属豆娘小型至中型,腹部细长;翅透明,四边室的前边与后边近等长或稍短;雄性足胫节稍微膨大;下肛附器明显长于上肛附器。

本属豆娘主要栖息于溪流、河流、渗流地和池塘。雄性停落在距离水面较低处的树枝和叶片上。雌雄连结产卵。雌性将卵产于水生植物的茎干中。

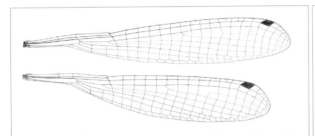

褐狭扇螺 雄翅
Copera vittata, male wings

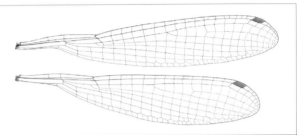

毛狭扇螺 雄翅
Copera ciliata, male wings

Four species from genus *Pseudocopera* are still placed in this genus. The genus contains over ten species distributed in tropical and subtropical Asia and Africa. Five species are recorded from China, mainly found in the North, Central, South and Southwest regions. Species of the genus are small to medium sized damselflies with a slender abdomen. Wings hyaline, the upper side of the quadrangle is as long as the lower side or slightly shorter. Male legs with the tibiae often slighted expanded. The inferior appendages much longer than the superiors.

Copera species inhabit streams, rivers, seepages and ponds. Males perch on the leaves or branches low above water. They oviposit in tandem. Females lay eggs into the stalks of aquatic plants.

黄狭扇螺 连结产卵
Copera marginipes, laying eggs in tandem

白狭扇蟌 *Copera annulata* (Selys, 1863)

白狭扇蟌 雄, 贵州
Copera annulata, male from Guizhou

白狭扇蟌 雌, 贵州
Copera annulata, female from Guizhou

【形态特征】雄性面部黑色具淡蓝色条纹；胸部黑色具蓝白色的肩前条纹，侧面具2条蓝白色条纹，足主要白色，腿节端方1/2黑色，胫节稍微膨大；腹部黑色，第9～10节和肛附器大面积蓝白色。雌性黑褐色具黄色或蓝白色条纹。【长度】体长 43～45 mm，腹长 37～38 mm，后翅 22～24 mm。【栖息环境】海拔 1500 m以下水草茂盛的湿地。【分布】北京、陕西、四川、重庆、云南、贵州、湖北、浙江、福建、广西；朝鲜半岛、日本。【飞行期】5—9月。

[Identification] Male face black with pale blue markings. Thorax black with bluish white antehumeral stripes, sides with two bluish white stripes, legs mainly white, femora with apical half black, tibiae slightly expanded. Abdomen black, S9-S10 and anal appendages largely bluish white. Female blackish brown with yellow or bluish white markings. [Measurements] Total length 43-45 mm, abdomen 37-38 mm, hind wing 22-24 mm. [Habitat] Wetland with abundant emergent plants below 1500 m elevation. [Distribution] Beijing, Shaanxi, Sichuan, Chongqing, Yunnan, Guizhou, Hubei, Zhejiang, Fujian, Guangxi; Korean peninsula, Japan. [Flight Season] May to September.

白狭扇螅　交尾，贵州
Copera annulata, mating pair from Guizhou

毛狭扇螅 *Copera ciliata* (Selys, 1863)

【形态特征】本种与白狭扇螅近似但可通过足的色彩区分。本种雄性腿节的黑色区域明显少于后者，腿节仅在端方1/4~1/3黑色。【长度】体长 42~47 mm，腹长 34~39 mm，后翅 20~24 mm。【栖息环境】海拔 1500 m以下水草茂盛的湿地和溪流。【分布】云南、贵州、广西、广东、香港、台湾；南亚、东南亚。【飞行期】全年可见。

[Identification] The species is similar to *C. annulata* but can be distinguished from the color pattern of legs. The black area of male femora in this species is much smaller than in *C. annulata*, femora with only apical one fourth to one third black. [Measurements] Total length 42-47 mm, abdomen 34-39 mm, hind wing 20-24 mm. [Habitat] Wetland with abundant emergent plants and streams below 1500 m elevation. [Distribution] Yunnan, Guizhou, Guangxi, Guangdong, Hong Kong, Taiwan; South and Southeast Asia. [Flight Season] Throughout the year.

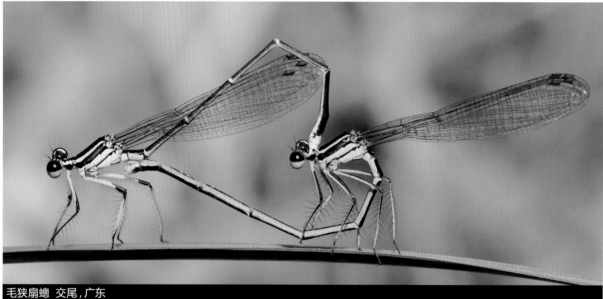

毛狭扇螅 交尾，广东
Copera ciliata, mating pair from Guangdong

毛狭扇螅 雄，广东
Copera ciliata, male from Guangdong

黄狭扇螅 *Copera marginipes* (Rambur, 1842)

【形态特征】雄性面部黑色具黄色条纹；胸部黑色具黄色条纹，足黄色，胫节稍微膨大；腹部黑色，第8节末端至第10节及肛附器白色，上肛附器短。雌性未熟时为白色，成熟以后有较多色型；胫节未膨大。【长度】体长 34～39 mm，腹长 28～31 mm，后翅 16～20 mm。【栖息环境】海拔 1500 m以下的湿地、河流和溪流。【分布】云南、贵州、浙江、福建、广东、广西、海南、香港、台湾；南亚、东南亚。【飞行期】全年可见。

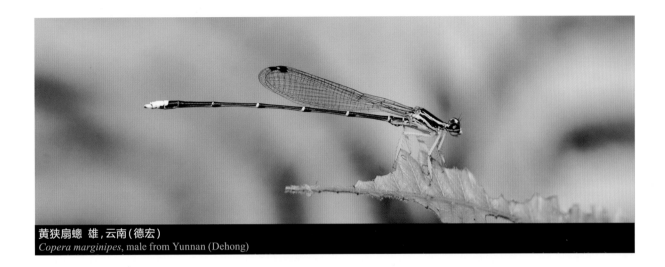

黄狭扇螅 雄，云南（德宏）
Copera marginipes, male from Yunnan (Dehong)

[Identification] Male face black with yellow markings. Thorax black with yellow stripes, legs yellow, tibiae slightly expanded. Abdomen black, the end of S8 to S10 and anal appendages white, superiors short. Female white when immature, and polymorphic when mature. Tibiae not expanded. [Measurements] Total length 34-39 mm, abdomen 28-31 mm, hind wing 16-20 mm. [Habitat] Wetlands, rivers and streams below 1500 m elevation. [Distribution] Yunnan, Guizhou, Zhejiang, Fujian, Guangdong, Guangxi, Hainan, Hong Kong, Taiwan; South and Southeast Asia. [Flight Season] Throughout the year.

黄狭扇螅 连结产卵，云南（德宏）
Copera marginipes, laying eggs in tandem from Yunnan (Dehong)

黑狭扇螅 *Copera tokyoensis* Asahina, 1948

黑狭扇螅 雄，湖北
Copera tokyoensis, male from Hubei

黑狭扇螅 雌，湖北
Copera tokyoensis, female from Hubei

【形态特征】雄性面部黑色具白色条纹；胸部黑色，侧面具2条白色条纹，足黑色和白色，腿节端方2/3黑色，胫节稍微膨大；腹部黑色，第10节及肛附器蓝白色。雌性黑褐色具黄色条纹；胸部具淡黄色的肩前条纹。【长度】体长33~36 mm，腹长 22~28 mm，后翅 16~17 mm。【栖息环境】海拔 1000 m以下水草茂盛的湿地。【分布】北京、天津、安徽、江苏、湖北；朝鲜半岛、日本、俄罗斯远东。【飞行期】4—9月。

[Identification] Male face black with white markings. Thorax black, sides with two white stripes, legs black and white, femora with apical two thirds black, tibiae slightly expanded. Abdomen black, S10 and anal appendages bluish white. Female dark brown with yellow markings. Thorax with pale yellow antehumeral stripes. [Measurements] Total length 33-36 mm, abdomen 22-28 mm, hind wing 16-17 mm. [Habitat] Wetland with abundant emergent plants below 1000 m elevation. [Distribution] Beijing, Tianjin, Anhui, Jiangsu, Hubei; Korean peninsula, Japan, Russian Far East. [Flight Season] April to September.

褐狭扇螅 *Copera vittata* (Selys, 1863)

【形态特征】本种与黄狭扇螅相似，但上肛附器的长度超过下肛附器长度的1/2，而后者不足1/3。【长度】体长37~40 mm，腹长 31~33 mm，后翅 17~18 mm。【栖息环境】海拔 1500 m以下森林中的小型水潭和流速缓慢的溪流。【分布】云南（德宏、西双版纳）、海南；南亚、东南亚。【飞行期】全年可见。

[Identification] The species is similar to *C. marginipes*, but the superior appendages slightly longer than half length of the inferiors but less than one third in *C. marginipes*. [Measurements] Total length 37-40 mm, abdomen

褐狭扇螅 雄，云南（西双版纳）
Copera vittata, male from Yunnan (Xishuangbanna)

31-33 mm, hind wing 17-18 mm. [Habitat] Small pools and slow flowing streams in forest below 1500 m elevation. [Distribution] Yunnan (Dehong, Xishuangbanna), Hainan; South and Southeast Asia. [Flight Season] Throughout the year.

褐狭扇蟌 雄, 云南(西双版纳)
Copera vittata, male from Yunnan (Xishuangbanna)

褐狭扇蟌 雌, 云南(西双版纳)
Copera vittata, female from Yunnan (Xishuangbanna)

印扇螅属 Genus *Indocnemis* Laidlaw, 1917

本属全球已知2种，分布于亚洲的热带和亚热带地区，并都分布于中国华中、华南和西南地区。本属是体中型的豆娘，腹部细长；翅透明，四边室的前边明显短于后边。

本属豆娘栖息于森林中的小型静水潭和沟渠。雄性停落在水边较低处的树叶上或者趴在岩石表面。

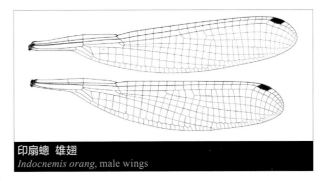

印扇螅 雄翅
Indocnemis orang, male wings

This genus contains two species, distributed in tropical and subtropical Asia, both are recorded from Central, South and Southwest China. Species of the genus are medium-sized damselflies with long and slender abdomen. Wings hyaline, the upper side of quadrangle much shorter than the lower side.

Indocnemis species inhabit small pools and ditches in forests. Males perch low on leaves or the surface of rocks.

印扇螅，雄
Indocnemis orang, male

黑背印扇螆 *Indocnemis ambigua* (Asahina, 1997)

【形态特征】雄性面部黑色；胸部黑色稍微覆盖粉霜，具灰色的肩前条纹，侧面具2条蓝色条纹；腹部黑色，第2～8节侧面具白斑，第9～10节和肛附器蓝色。雌性黑色具黄色和淡蓝色斑纹。【长度】体长 53～58 mm，腹长 45～50 mm，后翅 31～32 mm。【栖息环境】海拔 1500 m以下森林中的小型水潭和溪流。【分布】广西；越南。【飞行期】5—7月。

[Identification] Male face black. Thorax black and slightly pruinosed, with grey antehumeral stripes, sides with two blue stripes. Abdomen black with white spots laterally on S2-S8, S9-S10 and anal appendages blue. Female black with yellow and pale blue markings. [Measurements] Total length 53-58 mm, abdomen 45-50 mm, hind wing 31-32 mm. [Habitat] Small pools and streams in forest below 1500 m elevation. [Distribution] Guangxi; Vietnam. [Flight Season] May to July.

黑背印扇螆 雄，广西
Indocnemis ambigua, male from Guangxi

黑背印扇螅 雌雄连结，广西
Indocnemis ambigua, pair in tandem from Guangxi

印扇螅 *Indocnemis orang* (Förster, 1907)

【形态特征】雄性面部黑色；胸部黑色具蓝紫色肩前条纹，侧面具2条淡蓝色条纹；腹部黑色，第9节末端和第10节具蓝色斑。雌性肩前条纹为黄色，胸部和腹部具淡蓝色条纹。【长度】雄性体长 60～62 mm，腹长 49～52 mm，后翅 33～35 mm。【栖息环境】海拔 1500 m以下森林中的小型水潭。【分布】云南、贵州、湖北、湖南、浙江、福建、广东、广西；孟加拉国、印度、泰国、老挝、越南、马来半岛。【飞行期】4—10月。

印扇螅 雄，云南（德宏）
Indocnemis orang, male from Yunnan (Dehong)

[Identification] Male face black. Thorax black with bluish violet antehumeral stripes, sides with two pale blue stripes. Abdomen black, the end of S9 and S10 with blue spots. Female with yellow antehumeral stripes, thorax and abdomen with pale blue markings. [Measurements] Male total length 60-62 mm, abdomen 49-52 mm, hind wing 33-35 mm. [Habitat] Small pools in forest below 1500 m elevation. [Distribution] Yunnan, Guizhou, Hubei, Hunan, Zhejiang, Fujian, Guangdong, Guangxi; Bangladesh, India, Thailand, Laos, Vietnam, Peninsular Malaysia. [Flight Season] April to October.

印扇螅 雌雄连结，广东 ｜莫善濂 摄
Indocnemis orang, pair in tandem from Guangdong ｜ Photo by Shanlian Mo

印扇螅 雌，云南（红河） ｜莫善濂 摄
Indocnemis orang, female from Yunnan (Honghe) ｜ Photo by Shanlian Mo

同痣蟌属 Genus *Onychargia* Selys, 1865

本属全球已知2种，分布于亚洲的热带和亚热带地区。中国已知1种，主要分布于华南和西南地区。本属是体小型的豆娘，腹部较短；翅透明，前翅四边室的前边明显短于后边，仅为后边的1/3，后翅约为1/2，具较长的臀脉；腹部第10节后缘具1对短突起，肛附器甚短。

本属豆娘栖息于森林中水草茂盛的池塘。雄性经常隐藏在杂草丛中。雌雄连结产卵。

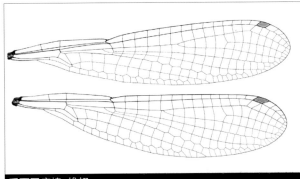

毛面同痣蟌 雄翅
Onychargia atrocyana, male wings

The genus contains two species, distributed in tropical and subtropical Asia. Only one species is recorded from China, mainly found in the South and Southwest regions. Species of the genus are small-sized damselflies with an unusually short abdomen. Wings hyaline, the upper side of quadrangle much shorter than the lower side, about one third as long as the lower side in the fore wings and half length in hind wings, anal vein long. S10 with a pair of short prominences on the hind margin, anal appendages very short.

Onychargia species inhabit well-vegetated ponds in forest. Males usually hide in the weed. They oviposit in tandem.

毛面同痣蟌 雌雄连结
Onychargia atrocyana, pair in tandem

毛面同痣螅 *Onychargia atrocyana* Selys, 1865

【形态特征】广东的雄性面部黑色具蓝色条纹；胸部的肩前条纹和侧面的条纹覆盖白色粉霜；腹部黑色具白色斑纹，第8~10节覆盖白色粉霜。雌性与雄性相似。云南、海南的雄性通体黑色，胸部具蓝紫色光泽；雌性黑褐色具黄色条纹。【长度】体长 28~35 mm，腹长 23~28 mm，后翅 16~20 mm。【栖息环境】海拔 2000 m 以下挺水植物茂盛的静水环境。【分布】云南、海南、广东、香港、台湾；南亚、东南亚。【飞行期】3—11月。

毛面同痣螅 雄，云南（德宏）
Onychargia atrocyana, male from Yunnan (Dehong)

毛面同痣螅 雌，云南（德宏）
Onychargia atrocyana, female from Yunnan (Dehong)

毛面同痣螅 雌，广东｜莫善濂 摄
Onychargia atrocyana, female from Guangdong ｜ Photo by Shanlian Mo

毛面同痣蟌 雄，广东 | 莫善濂 摄
Onychargia atrocyana, male from Guangdong | Photo by Shanlian Mo

[Identification] In Guangdong male face black with blue markings. Thoracic antehumeral stripes and lateral stripes whitish pruinosed. Abdomen black with white spots, S8-S10 whitish pruinosed. Female similar to male. In Yunnan and Hainan male black throughout, thorax shining bluish violet. Female blackish brown with yellow markings. [Measurements] Total length 28-35 mm, abdomen 23-28 mm, hind wing 16-20 mm. [Habitat] Standing water with plenty of emergent plants below 2000 m elevation. [Distribution] Yunnan, Hainan, Guangdong, Hong Kong, Taiwan; South and Southeast Asia. [Flight Season] March to November.

扇蟌属 Genus *Platycnemis* Burmeister, 1839

本属全球已知约12种，主要分布于亚洲古北区和欧洲。中国已知2种，分布于东北、华北、华中和西南地区。本属豆娘体小型，腹部细长；翅透明，四边室的前边与后边近等长或稍短，具臀脉；中后足的胫节极度膨大成叶片状。

本属豆娘栖息于溪流、河流和水草茂盛的池塘。雄性停落在水面上的叶片或树枝上占据领地。雌雄连结产卵。雌性将卵产于水生植物的茎干中。

The genus contains about 12 species, most of them distributed in Palaearctic Asia and Europe. Two species

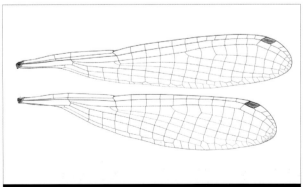

叶足扇蟌 雄翅
Platycnemis phyllopoda, male wings

白扇蟌 雄
Platycnemis foliacea, male

are recorded from China, distributed in the Northeast, North, Central and Southwest regions. This is a group of small species with a long, slender abdomen. Wings hyaline, the upper side of quadrangle as long as the lower side or slightly shorter, anal vein present. The tibiae in the middle and hind legs strongly expanded to form a leaf-shape.

Platycnemis species inhabit streams, rivers and well-vegetated ponds. Males perch on the leaves or branches above water for territory. They oviposit in tandem. Female lay eggs into the stalks of aquatic plants.

白扇蟌 *Platycnemis foliacea* Selys, 1886

【形态特征】雄性面部和胸部覆盖蓝白色粉霜，中足和后足的胫节叶片状；腹部黑色具白色斑纹，肛附器白色。雌性黑褐色具黄白色条纹，足的胫节未膨大。【长度】体长 33~35 mm，腹长 26~28 mm，后翅 18~19 mm。【栖息环境】海拔 1000 m以下流速缓慢的溪流。【分布】北京、陕西、河北、山东、天津、上海；日本。【飞行期】6—9月。

[Identification] Male face and thorax with bluish white pruinosity, legs with median and hind tibiae leaf-shaped. Abdomen black with white markings, anal appendages white. Female blackish brown with yellowish white markings, tibiae not expanded.

白扇蟌 雌，北京
Platycnemis foliacea, female from Beijing

[Measurements] Total length 33-35 mm, abdomen 26-28 mm, hind wing 18-19 mm. [Habitat] Slow flowing streams below 1000 m elevation. [Distribution] Beijing, Shaanxi, Hebei, Shandong, Tianjin, Shanghai; Japan. [Flight Season] June to September.

白扇蟌 雄，北京
Platycnemis foliacea, male from Beijing

叶足扇螅 *Platycnemis phyllopoda* Djakonov, 1926

【形态特征】雄性面部黑色，上唇和唇基淡蓝色；胸部黑色具淡黄色的肩条纹和肩前条纹，侧面具2条黄色条纹，中足和后足的胫节叶片状；腹部黑色具白色斑纹，肛附器白色。雌性黑色具黄色条纹，足的胫节未膨大。【长度】体长 33~34 mm，腹长 26~27 mm，后翅 16~17 mm。【栖息环境】海拔 2000 m 以下流速缓慢的溪流和湿地。【分布】黑龙江、辽宁、北京、云南、山东、天津、重庆、湖北、江苏、江西、浙江；朝鲜半岛、俄罗斯远东。【飞行期】4—10月。

叶足扇螅 雄，湖北
Platycnemis phyllopoda, male from Hubei

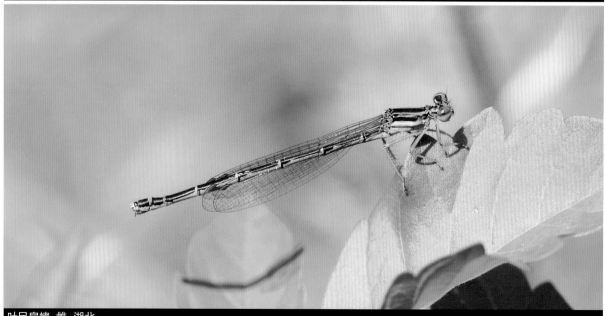

叶足扇螅 雌，湖北
Platycnemis phyllopoda, female from Hubei

叶足扇螅 交尾，湖北
Platycnemis phyllopoda, mating pair from Hubei

[Identification] Male face black, labrum and clypeus pale blue. Thorax black with pale yellow humeral and antehumeral stripes, sides with two yellow stripes, legs with median and hind tibiae leaf-shaped. Abdomen black with white markings, anal appendages white. Female black with yellow markings, tibiae not expanded. [Measurements] Total length 33-34 mm, abdomen 26-27 mm, hind wing 16-17 mm. [Habitat] Slow flowing streams and wetlands below 2000 m elevation. [Distribution] Heilongjiang, Liaoning, Beijing, Yunnan, Shandong, Tianjin, Chongqing, Hubei, Jiangsu, Jiangxi, Zhejiang; Korean peninsula, Russian Far East. [Flight Season] April to October.

微桥原螅属 Genus *Prodasineura* Cowley, 1934

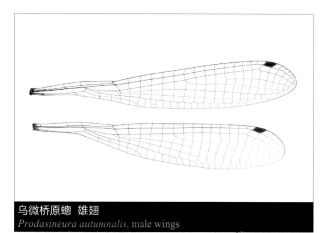

乌微桥原螅 雄翅
Prodasineura autumnalis, male wings

本属全球已知约40种，广泛分布于亚洲的热带和亚热带地区。中国已知9种，分布于华中、华南和西南地区。本属是一类小型豆娘，腹部十分细长；翅透明，四边室的前边与后边近等长，臀脉退化。

本属豆娘栖息于森林中的溪流和池塘。雄性停落在水面附近较低处的树枝和叶片上。雄性可以长时间悬停飞行。雌雄连结产卵。雌性将卵产在水生植物的茎干中。

The genus contains about 40 species, distributed in tropical and subtropical Asia. Nine species are recorded from China, found in the Central, South and Southwest regions. This is a group of small species with a long, slender abdomen. Wings hyaline, the upper side of quadrangle as long as the lower side, anal vein vestigial.

Prodasineura species inhabit streams and ponds in forest. Males perch on branches or leaves low above water. Males can perform lengthy hovering flights. They oviposit in tandem. Females lay eggs into the stalks of aquatic plants.

朱背微桥原螅 雄 | 宋睿斌 摄
Prodasineura croconota, male | Photo by Ruibin Song

乌微桥原螅 *Prodasineura autumnalis* (Fraser, 1922)

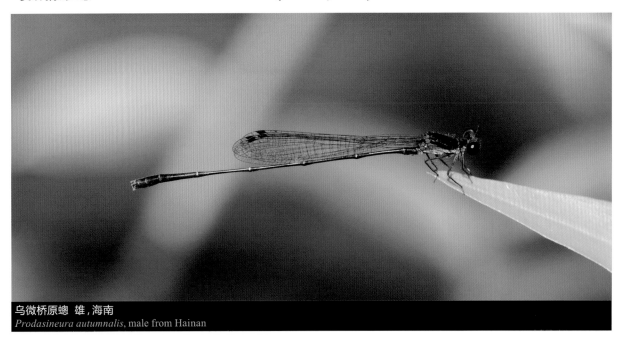

乌微桥原螅 雄, 海南
Prodasineura autumnalis, male from Hainan

【形态特征】雄性通体黑褐色, 胸部侧面稍微覆盖灰色粉霜。雌性黑色具黄色条纹。【长度】体长 38～40 mm, 腹长 31～34 mm, 后翅 19～22 mm。【栖息环境】海拔 1000 m以下的溪流和池塘。【分布】云南、广西、广东、海南、香港、台湾; 东南亚。【飞行期】全年可见。

[Identification] Male blackish brown throughout, sides of thorax slightly greyish pruinosed. Female black with yellow markings. [Measurements] Total length 38-40 mm, abdomen 31-34 mm, hind wing 19-22 mm. [Habitat]

乌微桥原螅 雌, 云南 (西双版纳)
Prodasineura autumnalis, female from Yunnan (Xishuangbanna)

乌微桥原螅 连结飞行, 海南
Prodasineura autumnalis, pair in tandem in flight from Hainan

Streams and ponds below 1000 m elevation. [Distribution] Yunnan, Guangxi, Guangdong, Hainan, Hong Kong, Taiwan; Southeast Asia. [Flight Season] Throughout the year.

朱背微桥原螅 *Prodasineura croconota* (Ris, 1916)

朱背微桥原螅 雌雄连结, 海南
Prodasineura croconota, pair in tandem from Hainan

【形态特征】雄性面部黑色；胸部黑色具宽阔的橙色肩前条纹，侧面具2条橙色条纹；腹部黑色具甚小的白斑，上肛附器淡蓝色。雌性黑色具黄色条纹。【长度】体长 38~40 mm，腹长 32~34 mm，后翅 20~21 mm。【栖息环境】海拔 1000 m以下流速缓慢的溪流。【分布】广西、广东、海南、台湾、香港；老挝、越南。【飞行期】4—10月。

[Identification] Male face black. Thorax black with broad orange antehumeral stripes, sides with two orange stripes. Abdomen black with tiny white spots, superior appendages pale blue. Female black with yellow markings. [Measurements] Total length 38-40 mm, abdomen 32-34 mm, hind wing 20-21 mm. **[Habitat]** Slow flowing streams below 1000 m elevation. **[Distribution]** Guangxi, Guangdong, Hainan, Taiwan, Hong Kong; Laos, Vietnam. **[Flight Season]** April to October.

朱背微桥原螅 连结产卵,广东 | 宋睿斌 摄
Prodasineura croconota, laying eggs in tandem from Guangdong | Photo by Ruibin Song

朱背微桥原螅 雄,海南
Prodasineura croconota, male from Hainan

福建微桥原螅 *Prodasineura fujianensis* Xu, 2006

【形态特征】雄性面部大面积黑色，头顶具淡蓝色条纹；胸部黑色，具蓝色的肩前条纹，侧面具2条蓝色条纹；腹部黑色，第9~10节具蓝色斑，上肛附器蓝色。【长度】雄性体长 36~37 mm，腹长 30~31 mm，后翅 18~19 mm。【栖息环境】海拔 1000 m以下流速缓慢的溪流。【分布】中国特有，分布于重庆、江西、福建。【飞行期】6—9月。

福建微桥原螅 雄，重庆
Prodasineura fujianensis, male from Chongqing

[Identification] Male face largely black, vertex with pale blue stripes. Thorax black with blue antehumeral stripes, sides with two blue stripes. Abdomen black, S9-S10 with blue spots, superior appendages blue. [Measurements] Male total length 36-37 mm, abdomen 30-31 mm, hind wing 18-19 mm. [Habitat] Slow flowing streams below 1000 m elevation. [Distribution] Endemic to China, recorded from Chongqing, Jiangxi, Fujian. [Flight Season] June to September.

福建微桥原螅 雌雄连结，江西 | 钟彦蔡 摄
Prodasineura fujianensis, pair in tandem from Jiangxi | Photo by Yankui Zhong

"赤微桥原螅"的分类疑难 Complex of "*Prodasineura verticalis* (Selys, 1860)"

　　真正的赤微桥原螅产自婆罗洲,但目前在中国已经发现几种与赤微桥原螅近似的物种,它们可能是新种。关于亚洲地区微桥原螅的真正身份需要开展更多细致的分类学研究。

The true *Prodasineura verticalis* is a species confined to Borneo, but some species with similar appearance have been discovered in China, some of them might be new species. More taxonomic study of this group, widespread in southeastern Asia, is needed.

微桥原螅属待定种1 *Prodasineura* sp. 1

　　【形态特征】雄性面部黑色,头顶具橙红色条纹;胸部黑色,具橙红色的肩前条纹,侧面具2条橙红色条纹;腹部黑色具白斑,上肛附器白色。雌性黑色具黄色条纹。【长度】雄性体长 41~42 mm,腹长 32~33 mm,后翅 21~22 mm。【栖息环境】海拔 1000 m以下流速缓慢的溪流和河流。【分布】云南、贵州、广西、广东。【飞行期】4—10月。

[Identification] Male face black, vertex with orange red stripes. Thorax black with orange red antehumeral stripes, sides with two orange red stripes. Abdomen black with white spots, superior appendages white. Female black with yellow markings. [Measurements] Male total length 41-42 mm, abdomen 32-33 mm, hind wing 21-22 mm. [Habitat]

微桥原螅属待定种1　雄,云南(西双版纳)
Prodasineura sp.1, male from Yunnan (Xishuangbanna)

Slow flowing streams and rivers below 1000 m elevation. **[Distribution]** Yunnan, Guizhou, Guangxi, Guangdong. **[Flight Season]** April to October.

微桥原螅属待定种1 雄，广东
Prodasineura sp.1, male from Guangdong

微桥原螅属待定种1 雌雄连接，云南（西双版纳）
Prodasineura sp.1, pair in tandem from Yunnan (Xishuangbanna)

微桥原螅属待定种2 *Prodasineura* sp. 2

【形态特征】雄性面部黑色；胸部黑色，具橙色的肩前条纹，侧面具2条橙色条纹；腹部黑色，上肛附器白色。【长度】雄性体长 35 mm，腹长 30 mm，后翅 18 mm。【栖息环境】海拔 500 m以下的池塘。【分布】云南（德宏）。【飞行期】4—7月。

[Identification] Male face black. Thorax black with orange antehumeral stripes, sides with two orange stripes. Abdomen black, superior appendages white. [Measurements] Male total length 35 mm, abdomen 30 mm, hind wing 18 mm. [Habitat] Ponds below 500 m elevation. [Distribution] Yunnan (Dehong). [Flight Season] April to July.

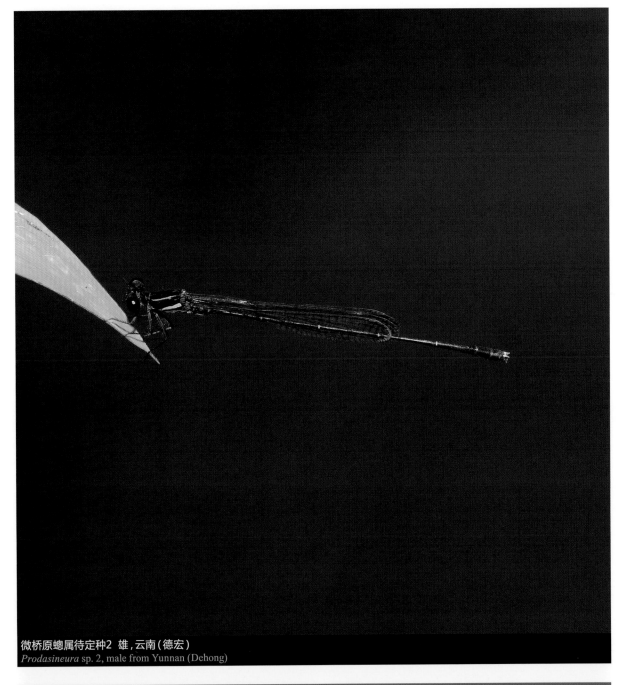

微桥原螅属待定种2 雄，云南（德宏）
Prodasineura sp. 2, male from Yunnan (Dehong)

微桥原螅属待定种3 *Prodasineura* sp. 3

【形态特征】与待定种1相似,但肛附器黑色。【长度】雄性体长 35~38 mm,腹长 29~32 mm,后翅 19~20 mm。【栖息环境】海拔 1000~1500 m的山区溪流。【分布】云南(临沧)。【飞行期】5—8月。

[Identification] The species is similar to *Prodasineura* sp.1, but anal appendages black. [Measurements] Male total length 35-38 mm, abdomen 29-32 mm, hind wing 19-20 mm. [Habitat] Montane streams at 1000-1500 m elevation. [Distribution] Yunnan (Lincang). [Flight Season] May to August.

微桥原螅属待定种3 雄,云南(临沧)
Prodasineura sp. 3, male from Yunnan (Lincang)

微桥原螅属待定种4 *Prodasineura* sp. 4

微桥原螅属待定种4 雄,云南(德宏)
Prodasineura sp. 4, male from Yunnan (Dehong)

【形态特征】雄性面部黑色;胸部黑色,具极细的肩前条纹,侧面具2条橙色条纹;腹部黑色具白斑,上肛附器大面积白色,末端黑色。【长度】雄性体长 37 mm,腹长 31 mm,后翅 19 mm。【栖息环境】海拔 500 m以下的溪流。【分布】云南(德宏)。【飞行期】10—12月。

[Identification] Male face black. Thorax black with extremely narrow antehumeral stripes, sides with two orange stripes. Abdomen black with white spots, superior appendages largely white with black tips. [Measurements] Male total length 37 mm, abdomen 31 mm, hind wing 19 mm. [Habitat] Streams below 500 m elevation. [Distribution] Yunnan (Dehong). [Flight Season] October to December.

12 蟌科 Family Coenagrionidae

　　本科是蜻蜓目中最庞大的一类，已知114个属超过1250种，世界性分布，包括了全世界最小和最长的蜻蜓。中国已知13属70余种，多数是体型较小的种类。本属豆娘体色艳丽；翅透明，基方具较长的翅柄，四边室的前边通常短于后边。

　　本科豆娘主要栖息于水草茂盛的静水环境和流速缓慢具有丰富水生植物的溪流。雄性经常停落在水面的水草上占据领地。许多种类雌雄连结产卵。

This is the largest family of the order, containing about 114 genera with over 1250 species globally, distributed all over the world. The family includes both the smallest and largest of all odonates. Over 70 species in 13 genera are recorded from China, most of them are small and slender species. Species of the family are colorful. The wings are hyaline with a long stalk at the base, the upper side of quadrangle usually shorter than the lower.

Species of the family inhabit standing water with plenty of emergent plants and slow flowing streams with plenty of aquatic plants. Males usually perch on the plants low above water and defend territories. Many species oviposit in tandem.

捷尾螅 雄
Paracercion v-nigrum, male

褐斑螅 雄
Pseudagrion spencei, male

狭翅螅属 Genus *Aciagrion* Selys, 1891

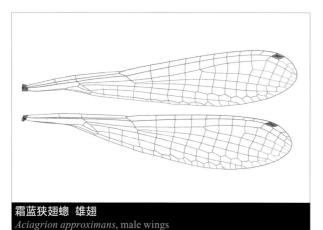

霜蓝狭翅螅 雄翅
Aciagrion approximans, male wings

本属全球已知约30种，分布于亚洲、非洲和澳新界。中国已知5种，分布于华中、华南和西南地区。本属是一类体型较小的豆娘，腹部细长；翅狭长而透明，弓脉位于第2条结前横脉以外，结后横脉10条或更多，四边室的前边大约为后边长度的1/2或更短；雄性上肛附器较长，下肛附器退化。

本属豆娘主要栖息于水草茂盛的静水环境包括水稻田。雄性经常停落在挺水植物上。雌雄连结产卵。

The genus contains about 30 species, distributed in Asia, Africa and Australasia. Five species are recorded from China, found in the Central, South and Southwest regions. Species of the genus are small-sized with a slender and long abdomen. Wings narrow and hyaline, arculus beginning after the second antenodal, postnodals 10 or even more, the upper side of the quadrangle about half the length of the lower or even shorter. Male superior appendages moderately long but the inferiors reduced.

Aciagrion species inhabit standing water with plenty of emergent plants including paddy fields. Males usually perch on emergent plants. They oviposit in tandem.

蓝尾狭翅螅 雌雄连结
Aciagrion olympicum, pair in tandem

霜蓝狭翅螅 *Aciagrion approximans* (Selys, 1876)

【形态特征】雄性面部黑褐色具蓝紫色条纹；胸部具蓝紫色的肩前条纹，侧面具蓝白色粉霜；腹部主要黑色，第8~9节蓝色。雌性身体黑色具黄色条纹。【长度】体长 28~31 mm，腹长 22~25 mm，后翅 13~16 mm。【栖息环境】海拔 2000 m 以下森林中的小型水潭、水稻田和水草茂盛的静水环境。【分布】云南（红河、德宏）、广西、广东、海南、香港；印度、泰国、柬埔寨、越南、印度尼西亚。【飞行期】3—11月。

霜蓝狭翅螅 雌，广东｜宋黎明 摄
Aciagrion approximans, female from Guangdong ｜ Photo by Liming Song

霜蓝狭翅螅 交尾，广东｜宋黎明 摄
Aciagrion approximans, mating pair from Guangdong ｜ Photo by Liming Song

[Identification] Male face dark brown with bluish violet stripes. Thorax with bluish violet antehumeral stripes, sides with bluish white pruinescence. Abdomen mainly black, S8-S9 blue. Female black with yellow markings. [Measurements] Total length 28-31 mm, abdomen 22-25 mm, hind wing 13-16 mm. [Habitat] Small pools, paddy fields and standing water with plenty of emergent plants below 2000 m elevation. [Distribution] Yunnan (Honghe, Dehong), Guangxi, Guangdong, Hainan, Hong Kong; India, Thailand, Cambodia, Vietnam, Indonesia. [Flight Season] March to November.

霜蓝狭翅螅 雄，广东
Aciagrion approximans, male from Guangdong

针尾狭翅螅 *Aciagrion migratum* (Selys, 1876)

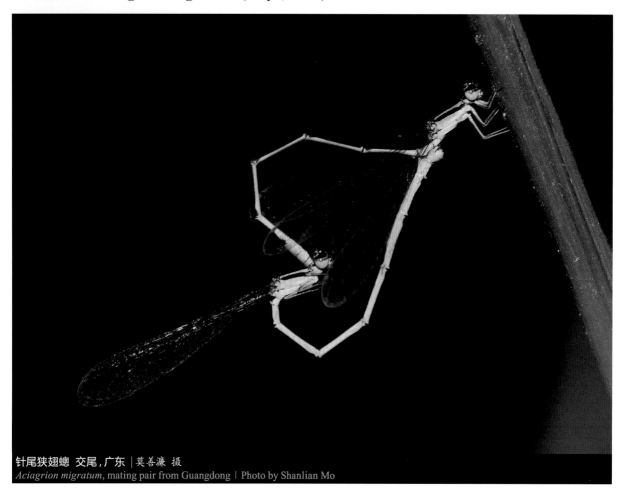

针尾狭翅螅 交尾，广东 | 莫善濂 摄
Aciagrion migratum, mating pair from Guangdong | Photo by Shanlian Mo

针尾狭翅螅 雌，广西
Aciagrion migratum, female from Guangxi

【形态特征】雄性面部褐色具淡蓝色条纹；胸部具黄绿色的肩前条纹，侧面蓝绿色；腹部黑色具淡蓝色条纹，第8~10节淡蓝色。雌性黑色具黄色条纹。【长度】体长 28~34 mm，腹长 24~29 mm，后翅 15~21 mm。【栖息环境】海拔 2000 m以下水草茂盛的静水环境包括水稻田。【分布】云南、贵州、四川、江西、浙江、福建、广东、广西、海南、台湾；朝鲜半岛、日本。【飞行期】全年可见。

[Identification] Male face brown with pale blue stripes. Thorax with yellowish green antehumeral stripes, sides bluish green. Abdomen black with pale blue markings, S8-S10 pale blue. Female black with yellow markings. [Measurements] Total length 28-34 mm, abdomen 24-29 mm, hind wing 15-21 mm. [Habitat] Standing water with plenty of emergent plants including paddy fields below 2000 m elevation.

针尾狭翅蟌 雄, 广东 | 莫善濂 摄
Aciagrion migratum, male from Guangdong | Photo by Shanlian Mo

[Distribution] Yunnan, Guizhou, Sichuan, Jiangxi, Zhejiang, Fujian, Guangdong, Guangxi, Hainan, Taiwan; Korean peninsula, Japan. [Flight Season] Throughout the year.

蓝尾狭翅蟌 *Aciagrion olympicum* Laidlaw, 1919

　　【形态特征】雄性面部黑褐色具淡蓝色条纹；胸部具宽阔的淡蓝色肩前条纹，侧面淡蓝色；腹部黑色具淡蓝色条纹，第8~10节淡蓝色。雌性与雄性相似。本种与针尾狭翅蟌相似，但肩前条纹更宽阔，雌性的体色为蓝色而针尾狭翅蟌为黄色。【长度】体长 34~36 mm，腹长 29~30 mm，后翅 20~22 mm。【栖息环境】海拔 2500 m以下水草

蓝尾狭翅蟌 雌, 云南 (保山)
Aciagrion olympicum, female from Yunnan (Baoshan)

蓝尾狭翅蟌 雌雄连结飞行, 云南 (保山)
Aciagrion olympicum, pair in tandem in flight from Yunnan (Baoshan)

茂盛的静水环境包括水稻田。【分布】西藏、云南（德宏、保山、大理、普洱）、广东；不丹、印度、尼泊尔。【飞行期】全年可见。

[Identification] Male face dark brown with pale blue stripes. Thorax with broad pale blue antehumeral stripes, sides pale blue. Abdomen black with pale blue markings, S8-S10 pale blue. Female similar to male. Similar to *A. migratum* but antehumeral stripes broader, female body color blue but yellow in *A. migratum*. [Measurements] Total length 34-36 mm, abdomen 29-30 mm, hind wing 20-22 mm. [Habitat] Standing water with plenty of emergent plants including paddy fields below 2500 m elevation. [Distribution] Tibet, Yunnan (Dehong, Baoshan, Dali, Pu'er), Guangdong; Bhutan, India, Nepal. [Flight Season] Throughout the year.

蓝尾狭翅螅 雄，云南（保山）
Aciagrion olympicum, male from Yunnan (Baoshan)

森狭翅螅 *Aciagrion pallidum* Selys, 1891

　　【形态特征】雄性身体大面积黄褐色具蓝色和黑色斑纹。雌性头部和胸部与雄性相似，腹部红色。【长度】体长 32~34 mm，腹长 27~28 mm，后翅 16~17 mm。【栖息环境】海拔 1500 m以下森林中的静水环境。【分布】云南（德宏）、广东；南亚、东南亚。【飞行期】全年可见。

森狭翅蟌 雄,云南(德宏)
Aciagrion pallidum, male from Yunnan (Dehong)

[Identification] Male body largely yellowish brown with blue and black stripes. Female head and thorax similar to male, abdomen red. [Measurements] Total length 32-34 mm, abdomen 27-28 mm, hind wing 16-17 mm. [Habitat] Standing water in forest below 1500 m elevation. [Distribution] Yunnan (Dehong), Guangdong; South and Southeast Asia. [Flight Season] Throughout the year.

森狭翅蟌 雌,云南(德宏)
Aciagrion pallidum, female from Yunnan (Dehong)

小螅属 Genus *Agriocnemis* Selys, 1877

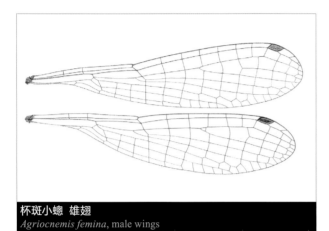

杯斑小螅 雄翅
Agriocnemis femina, male wings

本属全球已知超过40种，分布于亚洲、非洲和大洋洲。中国已知9种，分布于华中、华南和西南地区。本属是一类侏儒型豆娘；翅的弓脉位于第2条结前横脉以外，结后横脉6～8条，四边室的前边为后边长度的1/2～2/3。

本属豆娘栖息于水草茂盛的湿地，通常它们的种群数量庞大。雄性和雌性都停落在挺水植物上或者水边的茂盛草丛中。

The genus contains over 40 species, distributed in Asia, Africa and Oceania. Nine species are recorded from China, found in the Central, South and Southwest regions. This is a group of tiny species. Wings with the arculus beginning after the second antenodal and 6-8 postnodals and the upper side of quadrangle about half to two thirds length of the lower.

Agriocnemis species inhabit wetlands with plenty of emergent plants, usually at high population densities. Both sexes perch on emergent plants or thick grass nearby.

杯斑小螅 交尾
Agriocnemis femina, mating pair

蓝斑小蟌 *Agriocnemis clauseni* Fraser, 1922

蓝斑小蟌 雌,云南(德宏)
Agriocnemis clauseni, female from Yunnan (Dehong)

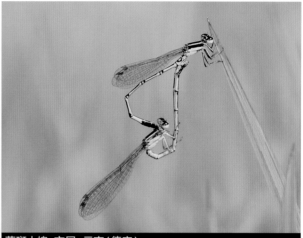

蓝斑小蟌 交尾,云南(德宏)
Agriocnemis clauseni, mating pair from Yunnan (Dehong)

蓝斑小蟌 雄,云南(德宏)
Agriocnemis clauseni, male from Yunnan (Dehong)

【形态特征】雄性身体黑色具蓝白色条纹。雌性身体黑色具黄绿色条纹。【长度】体长 22~23 mm,腹长 18~19 mm,后翅 10~13 mm。【栖息环境】海拔 1000~2500 m水草茂盛的静水环境。【分布】云南(德宏);孟加拉国、不丹、印度、尼泊尔、缅甸、泰国。【飞行期】4—8月。

[Identification] Male body black with bluish white markings. Female body black with yellowish green markings. [Measurements] Total length 22-23 mm, abdomen 18-19 mm, hind wing 10-13 mm. [Habitat] Standing water with

蓝斑小蟌 未熟雌，云南（德宏）
Agriocnemis clauseni, immature female from Yunnan (Dehong)

emergent plants at 1000-2500 m elevation. [**Distribution**] Yunnan (Dehong); Bangladesh, Bhutan, India, Nepal, Myanmar, Thailand. [**Flight Season**] April to August.

杯斑小蟌 *Agriocnemis femina* (Brauer, 1868)

　　【形态特征】雄性胸部覆盖浓密的白色粉霜；腹部未熟时末端橙黄色，成熟后黑色。雌性腹部未熟时为红色，成熟后绿色或黄色具黑色条纹。【长度】体长 21~25 mm，腹长 16~18 mm，后翅 10~11 mm。【栖息环境】海拔 2000 m 以下水草茂盛的湿地和水稻田。【分布】华中、华南、西南地区广布；日本、南亚、东南亚、大洋洲。【飞行期】全年可见。

杯斑小蟌 雌，云南（德宏）
Agriocnemis femina, female from Yunnan (Dehong)

杯斑小蟌 雌，未熟，云南（德宏）
Agriocnemis femina, immature female from Yunnan (Dehong)

[Identification] Male thorax with white pruinescence. Abdomen tip orange yellw when immature, becoming black when mature. Female red when immature, green or yellow with black markings when fully mature. [Measurements] Total length 21-25 mm, abdomen 16-18 mm, hind wing 10-11 mm. [Habitat] Wetlands with emergent plants and paddy fields below 2000 m elevation. [Distribution] Widespread in the Central, South and Southwest regions; Japan, South and Southeast Asia, Oceania. [Flight Season] Throughout the year.

杯斑小蟌 雄，云南（德宏）
Agriocnemis femina, male from Yunnan (Dehong)

白腹小蟌 *Agriocnemis lacteola* Selys, 1877

【形态特征】雄性身体大面积白色，头部、胸部背面和腹部第1~3节具黑斑。雌性身体黑色具黄绿色斑纹。【长度】腹长 16~18 mm，后翅 10~11 mm。【栖息环境】海拔 500 m以下水草茂盛的静水环境，包括水稻田。【分布】云南、广西、广东、海南、香港；孟加拉国、印度、尼泊尔、泰国、柬埔寨、越南。【飞行期】3—11月。

[Identification] Male body largely black, head, dorsal part of thorax and S1-S3 with black markings. Female body black with yellowish green markings. [Measurements] Abdomen 16-18 mm, hind wing 10-11 mm. [Habitat] Standing water with emergent plants including paddy fields below 500 m elevation. [Distribution] Yunnan, Guangxi, Guangdong, Hainan, Hong Kong; Bangladesh, India, Nepal, Thailand, Cambodia, Vietnam. [Flight Season] March to November.

白腹小蟌 雄，广东
Agriocnemis lacteola, male from Guangdong

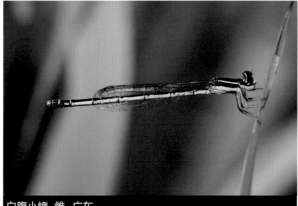

白腹小蟌 雌，广东
Agriocnemis lacteola, female from Guangdong

白腹小蟌 雄，广东 | 宋黎明 摄
Agriocnemis lacteola, male from Guangdong | Photo by Liming Song

樽斑小蟌 *Agriocnemis nana* Laidlaw, 1914

　　【形态特征】雄性身体黑色具白色条纹。雌性身体黑色具黄绿色条纹。本种与蓝斑小蟌相似，但本种雄性腹部第8节具白色斑，雌性后头具蓝色斑，而后者雌性具绿色条纹。【长度】体长 19~22 mm，腹长 16~17 mm，后翅 9~11 mm。【栖息环境】海拔 1500 m以下水草茂盛的静水环境。【分布】云南（保山、德宏）；东南亚。【飞行期】全年可见。

樽斑小蟌 雌，云南（德宏）
Agriocnemis nana, female from Yunnan (Dehong)

樽斑小蟌 雄，云南（德宏）
Agriocnemis nana, male from Yunnan (Dehong)

　　[Identification] Male body black with white markings. Female body black with yellowish green markings. Similar to *A. clauseni* but male of this species with white spots on S8, female occiput with blue spots but green stripes in *A. clauseni*. [Measurements] Total length 19-22 mm, abdomen 16-17 mm, hind wing 9-11 mm. [Habitat] Standing water with emergent plants below 1500 m elevation. [Distribution] Yunnan (Baoshan, Dehong); Southeast Asia. [Flight Season] Throughout the year.

樽斑小蟌 雄，云南（德宏）
Agriocnemis nana, male from Yunnan (Dehong)

黄尾小蟌 *Agriocnemis pygmaea* (Rambur, 1842)

【形态特征】雄性黑色具黄绿色条纹，腹部末端橙黄色。雌性黑色具黄绿色或绿色条纹。本种与杯斑小蟌相似，但雄性下肛附器短于上肛附器，而杯斑小蟌下肛附器明显长于上肛附器。【长度】体长 21~25 mm，腹长 16~18 mm，后翅 9~12 mm。【栖息环境】海拔 1000 m 以下水草茂盛的静水环境，包括水稻田。【分布】华中、华南、西南地区；日本、南亚、东南亚、澳新界。【飞行期】全年可见。

黄尾小蟌 雌，云南（德宏）
Agriocnemis pygmaea, female from Yunnan (Dehong)

黄尾小蟌 交尾，云南（德宏）
Agriocnemis pygmaea, mating pair from Yunnan (Dehong)

[Identification] Male black with yellowish green stripes, abdominal tip orange yellow. Female black with yellowish green or green markings. Similar to *A. femina* but male inferior appendages shorter than the superiors, in *A. femina* the inferior appendages clearly longer than the superiors. [Measurements] Total length 21-25 mm, abdomen 16-18 mm, hind wing 9-12 mm. [Habitat] Standing water with emergent plants including paddy fields below 1000 m elevation. [Distribution] Widespread in the Central, South and Southwest regions; Japan, South and Southeast Asia, Australasia. [Flight Season] Throughout the year.

黄尾小蟌 雄，云南（德宏）
Agriocnemis pygmaea, male from Yunnan (Dehong)

印度小螅 *Agriocnemis splendidissima* Laidlaw, 1919

【形态特征】雄性黑色具蓝色条纹，胸部侧面覆盖粉霜。雌性与雄性相似。本种与真正的印度小螅稍有差异，需要进一步确定其身份。【长度】体长 23~24 mm，腹长 19~20 mm，后翅 11~12 mm。【栖息环境】海拔 500 m 以下水草茂盛的静水环境和流速缓慢的溪流。【分布】云南（德宏、临沧）；印度、巴基斯坦。【飞行期】全年可见。

[Identification] Male black with blue stripes, sides of throax pruinosed. Female similar to male. Slightly different from the true *A. splendissima*, more study is needed for its true identity. [Measurements] Total length 23-24 mm,

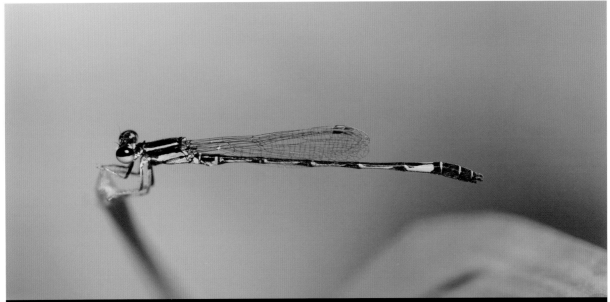

印度小螅 雄，云南（德宏）
Agriocnemis splendissima, male from Yunnan (Dehong)

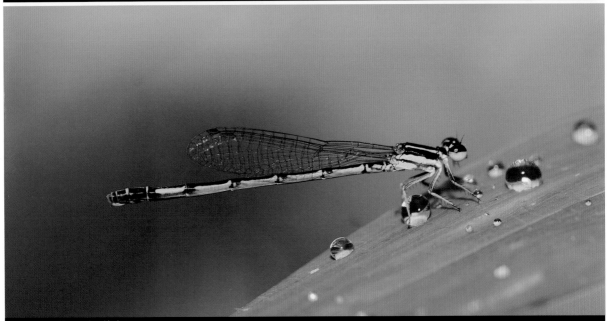

印度小螅 雌，云南（德宏）
Agriocnemis splendissima, female from Yunnan (Dehong)

印度小蟌 交尾,云南(德宏)
Agriocnemis splendissima, mating pair from Yunnan (Dehong)

abdomen 19-20 mm, hind wing 11-12 mm. [Habitat] Standing water with emergent plants and slow flowing streams below 500 m elevation. [Distribution] Yunnan (Dehong, Lincang); India, Pakistan. [Flight Season] Throughout the year.

印度小蟌 雌雄连结,云南(德宏)
Agriocnemis splendissima, pair in tandem from Yunnan (Dehong)

安螅属 Genus *Amphiallagma* Kennedy, 1920

本属全球仅1种，分布于亚洲的热带和亚热带地区。它是一种体型较小的蓝色豆娘；翅的弓脉位于第2条结前横脉下方，结后横脉6～8条，前翅四边室的前边大约为后边长度的1/5，在后翅为1/2；雄性上肛附器与下肛附器近等长。

本属豆娘栖息于低海拔处水草茂盛的静水环境。雄性和雌性都停落在挺水植物上或者水边的草丛中。雌性单独在水草上产卵。

This genus contains a single species distributed in tropical and subtropical Asia. It is a small and largely blue species. Wings with the arculus just below the second antenodal and with 6-8 postnodals and the upper side of quadrangle about one fifth length of the lower in fore wings and half in hind wings. The male superior appendages same length as the inferiors.

Amphiallagma species inhabits standing water with plenty of emergent plants in lowland. Both sexes perch on emergent plants or in grass nearby. Females lay eggs alone on aquatic plants.

天蓝安螅 雄
Amphiallagma parvum, male

天蓝安螅 *Amphiallagma parvum* (Selys, 1876)

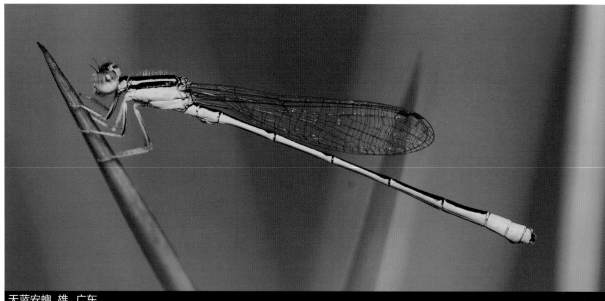

天蓝安螅 雄，广东
Amphiallagma parvum, male from Guangdong

天蓝安螅 雌，广东
Amphiallagma parvum, female from Guangdong

【形态特征】雄性大面积天蓝色，头部、胸部背面和腹部背面具黑色斑纹。雌性大面积蓝白色具黑色条纹。【长度】体长 21~22 mm，腹长 17~18 mm，后翅 10~11 mm。【栖息环境】海拔 500 m以下水草茂盛的静水环境。【分布】广东；印度、斯里兰卡、尼泊尔、缅甸、泰国、柬埔寨。【飞行期】4—9月。

[Identification] Male largely sky-blue, head, thorax and abdomen with black markings dorsally. Female largely bluish white with black markings. [Measurements] Total length 21-22 mm, abdomen 17-18 mm, hind wing 10-11 mm. [Habitat] Standing water with emergent plants below 500 m elevation. [Distribution] Guangdong; India, Sri Lanka, Nepal, Myanmar, Thailand, Cambodia. [Flight Season] April to September.

黑蟌属 Genus *Argiocnemis* Selys, 1877

本属全球已知3种,分布于南亚、东南亚、澳新界和毛里求斯。中国已知1种,仅分布于云南。本属豆娘不同年纪的个体体色存在明显的差异。翅的弓脉位于第2条结前横脉以外,结后横脉10~12条,四边室的前边大约为后边长度的1/3,雄性上肛附器为下肛附器长度的2倍。

本属豆娘主要栖息于水草茂盛的静水环境。雄性具领域行为,经常停落在水面的植物上。雌性单独产卵于水草的茎干中。

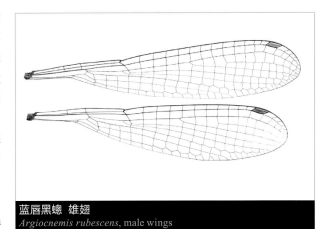

蓝唇黑蟌 雄翅
Argiocnemis rubescens, male wings

The genus contains three species, distributed in South and Southeast Asia, Australasia and Mauritius. Only one species is recorded from China but confined to Yunnan. The body color of the species changes dramatically with age. Wings with arculus beyond the second antenodal, 10-12 postnodals, upper side of quadrangle about one third length of the lower side, male superior appendages twice as long as the inferiors.

Argiocnemis species inhabit standing water with plenty of emergent plants. Males exhibit territorial behavior when they perch on the plants above water. Females lay eggs unaccompanied on aquatic plants.

蓝唇黑蟌 雄
Argiocnemis rubescens, male

蓝唇黑螅 *Argiocnemis rubescens* Selys, 1877

蓝唇黑螅 雄,云南(德宏)
Argiocnemis rubescens, male from Yunnan (Dehong)

蓝唇黑螅 雄,云南(德宏)
Argiocnemis rubescens, male from Yunnan (Dehong)

【形态特征】雄性未熟时腹部红色；成熟后身体黑色具蓝色条纹；老熟后身体黑色具黄褐色条纹，胸部和腹部稍微覆盖粉霜。雌性与雄性相似，随年纪的不同色彩有明显差异。【长度】体长 33～36 mm，腹长 26～30 mm，后翅 16～20 mm。【栖息环境】海拔 1000 m以下水草茂盛的静水环境。【分布】云南（德宏、临沧、普洱、西双版纳）；南亚、东南亚、澳新界。【飞行期】全年可见。

[Identification] Male abdomen red when immature. Body black with blue markings when mature. Body black with yellowish brown markings when old, thorax and abdomen slightly pruinosed. Female similar to male, color changing with age. [Measurements] Total length 33-36 mm, abdomen 26-30 mm, hind wing 16-20 mm. [Habitat] Standing water with emergent plants below 1000 m elevation. [Distribution] Yunnan (Dehong, Lincang, Pu'er, Xishuangbanna); South and Southeast Asia, Australasia. [Flight Season] Throughout the year.

蓝唇黑螅 雌，云南（德宏）
Argiocnemis rubescens, female from Yunnan (Dehong)

蓝唇黑螅 未熟雌，云南（德宏）
Argiocnemis rubescens, immature female from Yunnan (Dehong)

黄蟌属 Genus *Ceriagrion* Selys, 1876

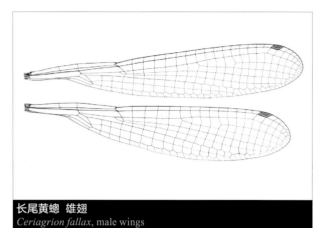

长尾黄蟌 雄翅
Ceriagrion fallax, male wings

本属全球已知50余种，分布于亚洲、欧洲、非洲和大洋洲。中国已知约10种，分布于华北、华中、华南和西南地区。本属豆娘通常色彩均一而缺乏斑纹；翅的弓脉位于第2条结前横脉下方，结后横脉12条以上，四边室的前边大约为后边长度的1/2。

本属豆娘主要生活在水草茂盛的静水环境。雄性和雌性都会停落在挺水植物上或者水边的草丛中。雄性具领域行为。雌雄在水草上连结产卵。

The genus contains over 50 species, distributed in Asia, Europe, Africa and Oceania. About ten species are recorded from China, found in the North, Central, South and Southwest regions. Species of the genus are usually uniform in body color without markings. Wings with the arculus just below the second antenodal, postnodals over 12, the upper side of quadrangle about half length of the lower.

Ceriagrion species inhabit standing water with plenty of emergent plants. Both sexes often perch on emergent plants or grass nearby. Males exhibit territorial behavior. They oviposit in tandem on aquatic plants.

长尾黄蟌 雄，交尾
Ceriagrion fallax, mating pair

翠胸黄螅 *Ceriagrion auranticum ryukyuanum* Asahina, 1967

翠胸黄螅 雄, 海南
Ceriagrion auranticum ryukyuanum, male from Hainan

翠胸黄螅 雄, 贵州
Ceriagrion auranticum ryukyuanum, male from Guizhou

【形态特征】雄性复眼绿色，面部红褐色；胸部绿色；腹部橙红色。雌性头部和胸部与雄性相似，腹部浅褐色或橙红色。【长度】体长 33~41 mm，腹长 28~35 mm，后翅 17~23 mm。【栖息环境】海拔 1500 m 以下水草茂盛的静水环境。【分布】云南（红河）、湖北、湖南、浙江、福建、广西、广东、海南、香港、台湾；朝鲜半岛、日本。【飞行期】全年可见。

翠胸黄蟌 雌雄连结，湖北
Ceriagrion auranticum ryukyuanum, pair in tandem from Hubei

翠胸黄蟌 交尾，广东｜宋睿斌 摄
Ceriagrion auranticum ryukyuanum, mating pair from Guangdong｜Photo by Ruibin Song

[Identification] Male eyes green, face reddish brown. Thorax green. Abdomen orange red. Female head and thorax similar to male, abdomen pale brown or orange red. [Measurements] Total length 33-41 mm, abdomen 28-35 mm, hind wing 17-23 mm. [Habitat] Standing water with emergent plants below 1500 m elevation. [Distribution] Yunnan (Honghe), Hubei, Hunan, Zhejiang, Fujian, Guangxi, Guangdong, Hainan, Hong Kong, Taiwan; Korean peninsula, Japan. [Flight Season] Throughout the year.

翠胸黄蟌 雌雄连结，贵州
Ceriagrion auranticum ryukyuanum, pair in tandem from Guizhou

天蓝黄蟌 *Ceriagrion azureum* (Selys, 1891)

【形态特征】雄性通体蓝色；腹部第8~10节黑色。雌性通体绿褐色。广东雄性色彩较淡，翅稍染褐色。两者是否同种有待确定。【长度】体长 43~45 mm，腹长 35~38 mm，后翅 23~24 mm。【栖息环境】海拔1500 m 以下水草茂盛的静水环境包括水稻田。【分布】云南、广西、广东；印度、尼泊尔、孟加拉国、缅甸、泰国、柬埔寨、老挝、越南。【飞行期】全年可见。

天蓝黄蟌 雌，云南（德宏）
Ceriagrion azureum, female from Yunnan (Dehong)

[Identification] Male blue throughout. S8-S10 black. Female greenish brown throughout. Males from Guangdong paler, wings slightly tinted with amber brown. Not absolutely sure if they are the same species. [Measurements] Total length 43-45 mm, abdomen 35-38 mm, hind wing 23-24 mm. [Habitat] Standing water with emergent plants including paddy fields below 1500 m elevation. [Distribution] Yunnan, Guangxi, Guangdong; India, Nepal, Bangladesh, Myanmar, Thailand, Cambodia, Laos, Vietnam. [Flight Season] Throughout the year.

天蓝黄蟌 雄，云南（普洱）
Ceriagrion azureum, male from Yunnan (Pu'er)

天蓝黄蟌 雄，广东 | 宋睿斌 摄
Ceriagrion azureum, male from Guangdong | Photo by Ruibin Song

橙黄蟌 *Ceriagrion chaoi* Schmidt, 1964

橙黄蟌 雄,云南(德宏)
Ceriagrion chaoi, male from Yunnan (Dehong)

橙黄蟌 连结产卵,云南(西双版纳) | 莫善濂 摄
Ceriagrion chaoi, laying eggs in tandem from Yunnan (Xishuangbanna) | Photo by Shanlian Mo

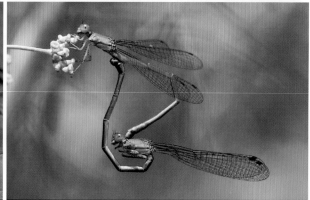

橙黄蟌 交尾,云南(德宏) | 莫善濂 摄
Ceriagrion chaoi, mating pair from Yunnan (Dehong) | Photo by Shanlian Mo

【形态特征】雄性面部和胸部橙黄色;腹部橙红色,末端具黑褐色斑。雌性胸部黄绿色,腹部褐色。【长度】体长 32~40 mm,腹长 21~31 mm,后翅 17~33 mm。【栖息环境】海拔 1000 m 以下水草茂盛的静水环境。【分布】云南;缅甸、泰国、柬埔寨、老挝、越南、马来半岛、新加坡。【飞行期】全年可见。

[Identification] Male face and thorax orange yellow. Abdomen orange red, the tip with dark brown spots. Female thorax yellowish green, abdomen brown. [Measurements] Total length 32-40 mm, abdomen 21-31 mm, hind wing 17-33 mm. [Habitat] Standing water with emergent plants below 1000 m elevation. [Distribution] Yunnan; Myanmar, Thailand, Cambodia, Laos, Vietnam, Peninsular Malaysia, Singapore. [Flight Season] Throughout the year.

长尾黄螅 *Ceriagrion fallax* Ris, 1914

【形态特征】雄性面部和胸部黄绿色；腹部黄色，第7～10节具黑斑。雌性头部和胸部黄绿色；腹部褐色，末端黑色。【长度】体长 37～47 mm，腹长 30～38 mm，后翅 20～24 mm。【栖息环境】海拔 2500 m以下水草茂盛的静水环境。【分布】华中、华南、西南地区广布；南亚、东南亚。【飞行期】3—12月。

[Identification] Male face and thorax yellowish green. Abdomen yellow, S7-S10 with black spots. Female head and thorax yellowish green. Abdomen brown with black tip. [Measurements] Total length 37-47 mm, abdomen 30-38 mm, hind wing 20-24 mm. [Habitat] Standing water with emergent plants below 2500 m elevation. [Distribution] Widespread in the Central, South and Southwest regions; South and Southeast Asia. [Flight Season] March to December.

长尾黄螅 雄，云南（德宏）
Ceriagrion fallax, male from Yunnan (Dehong)

长尾黄螅 雌雄连结，云南（德宏）
Ceriagrion fallax, pair in tandem from Yunnan (Dehong)

长尾黄螅 雄，贵州
Ceriagrion fallax, male from Guizhou

柠檬黄螅 *Ceriagrion indochinense* Asahina, 1967

柠檬黄螅 雌，云南（德宏）
Ceriagrion indochinense, female from Yunnan (Dehong)

【形态特征】雄性复眼绿色，面部黄色；胸部黄绿色；腹部黄色。雌性头部和胸部黄绿色；腹部黄褐色。【长度】体长 35~38 mm，腹长 28~30 mm，后翅 18~20 mm。【栖息环境】海拔 1000 m以下水草茂盛的静水环境。【分布】云南（德宏）、海南；泰国、柬埔寨、老挝、越南。【飞行期】4—10月。

[Identification] Male eyes green, face yellow. Thorax yellowish green. Abdomen yellow. Female head and thorax yellowish green. Abdomen yellowish brown. [Measurements] Total length 35-38 mm, abdomen 28-30 mm, hind wing 18-20 mm. [Habitat] Standing water with emergent plants below 1000 m elevation. [Distribution] Yunnan (Dehong), Hainan; Thailand, Cambodia, Laos, Vietnam. [Flight Season] April to October.

柠檬黄螅 雄，云南（德宏）
Ceriagrion indochinense, male from Yunnan (Dehong)

短尾黄螅 *Ceriagrion melanurum* Selys, 1876

短尾黄螅 雄，湖北
Ceriagrion melanurum, male from Hubei

　　【形态特征】雄性复眼绿色，面部黄色；胸部黄绿色；腹部黄色，第7～10节背面具黑斑。雌性主要黄绿色。本种与长尾黄螅相似，但雄性上肛附器甚短，约为第10节长度的1/3，而长尾黄螅的上肛附器约为第10节长度的2/3。【长度】体长40～44 mm，腹长33～35 mm，后翅23～24 mm。【栖息环境】海拔2500 m以下水草茂盛的静水环境。【分布】华中、华南、西南地区广布；朝鲜半岛、日本。【飞行期】5—10月。

[Identification] Male eyes green, face yellow. Thorax yellowish green. Abdomen yellow, S7-S10 with dorsal black spots. Female mainly yellowish green. Similar to *C. fallax*, but the superior appendages shorter, about one third length of S10, superior appendages of *C. fallax* about two thirds length of S10. [Measurements] Total length 40-44 mm, abdomen 33-35 mm, hind wing 23-24 mm. [Habitat] Standing water with emergent plants below 2500 m elevation. [Distribution] Widespread in the Central, South and Southwest regions; Korean peninsula, Japan. [Flight Season] May to October.

短尾黄螅 雌，贵州，宋睿斌 摄
Ceriagrion melanurum, female from Guizhou ｜ Photo by Ruibin Song

赤黄蟌 *Ceriagrion nipponicum* Asahina, 1967

【形态特征】雄性头部和胸部红褐色；腹部红色。雌性头部和胸部绿色；腹部褐色。【长度】体长 36～41 mm，腹长 29～33 mm，后翅 20～21 mm。【栖息环境】海拔 1500 m以下水草茂盛的静水环境。【分布】北京、四川、贵州、湖北、江苏、浙江、福建、广东、台湾；朝鲜半岛、日本。【飞行期】4—10月。

[Identification] Male head and thorax reddish brown. Abdomen red. Female head and thorax green. Abdomen brown. [Measurements] Total length 36-41 mm, abdomen 29-33 mm, hind wing 20-21 mm. [Habitat] Standing water with emergent plants below 1500 m elevation. [Distribution] Beijing, Sichuan, Guizhou, Hubei, Jiangsu, Zhejiang, Fujian, Guangdong, Taiwan; Korean peninsula, Japan. [Flight Season] April to October.

赤黄蟌 交尾, 贵州 | 宋黎明 摄
Ceriagrion nipponicum, mating pair from Guizhou | Photo by Liming Song

赤黄蟌 雌雄连结, 贵州 | 宋黎明 摄
Ceriagrion nipponicum, pair in tandem from Guizhou | Photo by Liming Song

赤黄蟌 雄, 贵州
Ceriagrion nipponicum, male from Guizhou

钩尾黄螅 *Ceriagrion olivaceum* Laidlaw, 1914

【形态特征】雄性和雌性通体黄褐色。【长度】体长49～50 mm，腹长40～41 mm，后翅25～26 mm。【栖息环境】海拔2000 m以下水草茂盛的静水环境。【分布】云南（西双版纳、德宏）；南亚、东南亚。【飞行期】全年可见。

钩尾黄螅 雄，云南（西双版纳）
Ceriagrion olivaceum, male from Yunnan (Xishuangbanna)

钩尾黄螅 雌，云南（德宏）
Ceriagrion olivaceum, female from Yunnan (Dehong)

[Identification] Both sexes yellowish brown throughout. [Measurements] Total length 49-50 mm, abdomen 40-41 mm, hind wing 25-26 mm. [Habitat] Standing water with emergent plants below 2000 m elevation. [Distribution] Yunnan (Xishuangbanna, Dehong); South and Southeast Asia. [Flight Season] Throughout the year.

钩尾黄螅 雄，云南（德宏）
Ceriagrion olivaceum, male from Yunnan (Dehong)

中华黄蟌 *Ceriagrion sinense* Asahina, 1967

中华黄蟌 雄,贵州
Ceriagrion sinense, male from Guizhou

【形态特征】雄性复眼红色和绿色,面部黑褐色;胸部背面黑褐色,侧面黄绿色;腹部红色,第5～6节背面具黑色条纹。雌性头部和胸部与雄性相似;腹部红色或黄褐色。【长度】雄性体长 42 mm,腹长 35 mm,后翅 24 mm。【栖息环境】海拔 1500 m以下水草茂盛的静水环境。【分布】中国特有,分布于浙江、贵州。【飞行期】4—8月。

中华黄蟌 雄,贵州 | 宋黎明 摄
Ceriagrion sinense, male from Guizhou | Photo by Liming Song

中华黄蟌 雌雄连结,贵州 | 莫善濂 摄
Ceriagrion sinense, pair in tandem from Guizhou | Photo by Shanlian Mo

[Identification] Male eyes red and green, face dark brown. Thorax blackish brown dorsally, yellowish green laterally. Abdomen red, S5-S6 with black stripes dorsally. Female head and thorax similar to male. Abdomen red or yellowish brown. [Measurements] Male total length 42 mm, abdomen 35 mm, hind wing 24 mm. [Habitat] Standing water with emergent plants below 1500 m elevation. [Distribution] Endemic to China, recorded from Zhejiang, Guizhou. [Flight Season] April to August.

黄蟌属待定种1 *Ceriagrion* sp. 1

【形态特征】本种与短尾黄蟌相似，但体型较小，雄性肛附器构造和雌性色彩不同。【长度】体长 34~37 mm，腹长 28~30 mm，后翅 18~20 mm。【栖息环境】海拔 1500 m左右水草茂盛的小型水塘。【分布】云南（德宏）。【飞行期】5—11月。

[Identification] Similar to *C. melanurum* but size smaller, male anal appendages and female body color different. [Measurements] Total length 34-37 mm, abdomen 28-30 mm, hind wing 18-20 mm. [Habitat] Small pools with emergent plants at about 1500 m elevation. [Distribution] Yunnan (Dehong). [Flight Season] May to November.

黄蟌属待定种1 雌，云南（德宏）
Ceriagrion sp. 1, female from Yunnan (Dehong)

黄蟌属待定种1 交尾，云南（德宏）
Ceriagrion sp. 1, mating pair from Yunnan (Dehong)

黄蟌属待定种1 雄，云南（德宏）
Ceriagrion sp. 1, male from Yunnan (Dehong)

黄蟌属待定种2 *Ceriagrion* sp. 2

【形态特征】本种与翠胸黄蟌相似，但雄性胸部色彩为黄褐色。【长度】雄性体长 37～41 mm，腹长 30～33 mm，后翅 19～20 mm。【栖息环境】海拔 500 m以下水草茂盛的静水环境。【分布】云南（德宏）。【飞行期】4—10月。

[Identification] Similar to *C. auranticum ryukyuanum* but male thoracic color yellowish brown. [Measurements] Male total length 37-41 mm, abdomen 30-33 mm, hind wing 19-20 mm. [Habitat] Standing water with emergent plants below 500 m elevation. [Distribution] Yunnan (Dehong). [Flight Season] April to October.

黄蟌属待定种2 雄，云南（德宏）
Ceriagrion sp. 2, male from Yunnan (Dehong)

黄螅属待定种3 *Ceriagrion* sp. 3

黄螅属待定种3 雄，云南（德宏）
Ceriagrion sp. 3, male from Yunnan (Dehong)

【形态特征】本种与翠胸黄螅相似但体型稍小，雄性胸部色彩为黄褐色，腹部第6~7节背面色彩加深。【长度】体长 35~38 mm，腹长 29~30 mm，后翅 17~18 mm。【栖息环境】海拔 500 m以下水草茂盛的静水环境。【分布】云南（德宏）。【飞行期】全年可见。

[Identification] Similar to *C. auranticum ryukyuanum* but smaller, male thorax yellowish brown, S6-S7 color darkened dorsally. [Measurements] Total length 35-38 mm, abdomen 29-30 mm, hind wing 17-18 mm. [Habitat] Standing water with emergent plants below 500 m elevation. [Distribution] Yunnan (Dehong). [Flight Season] Throughout the year.

黄螅属待定种3 雄，云南（德宏）
Ceriagrion sp. 3, male from Yunnan (Dehong)

黄蟌属待定种4 *Ceriagrion* sp. 4

【形态特征】雄性通体蓝白色。雌性绿褐色。【长度】体长 45~47 mm，腹长 38~40 mm，后翅 24~26 mm。【栖息环境】海拔 1000~1500 m森林中的水潭。【分布】云南（普洱）。【飞行期】5—7月。

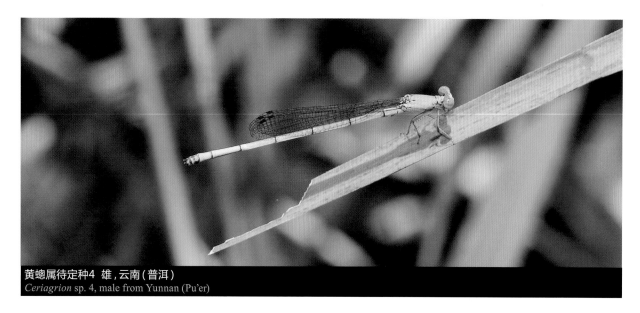

黄蟌属待定种4　雄，云南（普洱）
Ceriagrion sp. 4, male from Yunnan (Pu'er)

[Identification] Male bluish white throughout. Female greenish brown throughout. [Measurements] Total length 45-47 mm, abdomen 38-40 mm, hind wing 24-26 mm. [Habitat] Pools in forest at 1000-1500 m elevation. [Distribution] Yunnan (Pu'er). [Flight Season] May to July.

黄蟌属待定种4　雌雄连结，云南（普洱）
Ceriagrion sp. 4, pair in tandem from Yunnan (Pu'er)

螅属 Genus *Coenagrion* Kirby, 1890

本属全球已知约30种，主要分布于亚洲、欧洲和北美洲的温带地区。中国已知约10种，这类豆娘喜欢较寒冷的气候，在北方较常见。本属豆娘体型较小，通常体色黑色具发达的蓝色斑纹；翅的弓脉位于第2条结前横脉下方，结后横脉12条以上，四边室的前边大约为后边长度的1/2或更长；雄性的下肛附器长于上肛附器或近等长。

本属豆娘主要栖息于水草茂盛的湿地。雄性经常停落在挺水植物上或者水边的草丛中。雌雄连结产卵。

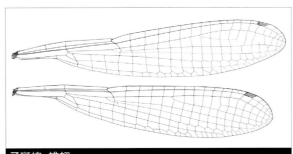

矛斑螅 雄翅
Coenagrion lanceolatum, male wings

The genus contains about 30 species, mainly distributed in the temperate parts of Asia, Europe and North America. About ten species are recorded from China, these species prefer cold weather and are common in the north of China. Species of the genus are small-sized damselflies, body black with extensive blue markings. Wings with arculus just below the second antenodal and over 12 postnodals, the upper side of quadrangle about half length of the lower or even longer. Male inferior appendages longer or same length as the superiors.

腾冲螅 雌雄连结
Coenagrion exclamationis, pair in tandem

Coenagrion species inhabit wetlands with plenty of emergent plants. Males often perch on emergent plants or grass nearby. They oviposit in tandem.

盃纹螅
Coenagrion ecornutum

黑格螅
Coenagrion hylas

纤腹螅
Coenagrion johanssoni

矛斑螅
Coenagrion lanceolatum

月斑螅
Coenagrion lunulatum

腾冲螅
Coenagrion exclamationis

螅属 雄性肛附器
Genus *Coenagrion*, male anal appendages

多棘螅 *Coenagrion aculeatum* Yu & Bu, 2007

多棘螅 雌,安徽 | 陈尽 摄
Coenagrion aculeatum, female from Anhui | Photo by Jin Chen

【形态特征】雄性面部黑色具蓝色斑纹；胸部具天蓝色的肩前条纹，侧面蓝色；腹部黑色具蓝斑。雌性黑色具淡蓝色和蓝色斑纹。【长度】雄性腹长 27～28 mm，后翅 18～19 mm。【栖息环境】森林中的池塘。【分布】中国特有，分布于重庆、贵州、安徽、浙江。【飞行期】5—7月。

[Identification] Male face black with blue markings. Thorax with sky-blue antehumeral stripes, sides blue. Abdomen black with blue markings. Female black with pale blue and blue markings. [Measurements] Male abdomen 27-28 mm, hind wing 18-19 mm. [Habitat] Ponds in forest. [Distribution] Endemic to China, recorded from Chongqing, Guizhou, Anhui, Zhejiang. [Flight Season] May to July.

多棘螅 雄,安徽 | 陈尽 摄
Coenagrion aculeatum, male from Anhui | Photo by Jin Chen

盃纹螅 *Coenagrion ecornutum* (Selys, 1872)

盃纹螅 雄,吉林 | 金洪光 摄
Coenagrion ecornutum, male from Jilin | Photo by Hongguang Jin

盃纹螅 雌雄连结，黑龙江 ｜莫善濂 摄
Coenagrion ecornutum, pair in tandem from Heilongjiang ｜ Photo by Shanlian Mo

【形态特征】雄性面部黑色具蓝色斑纹；胸部具淡蓝色肩前条纹，侧面淡蓝色；腹部黑色具天蓝色条纹。雌性多型，身体淡蓝色或黄色具黑色条纹。【长度】体长28～30 mm，腹长23～24 mm，后翅15～17 mm。【栖息环境】海拔 500 m以下水草茂盛的静水环境。【分布】黑龙江、吉林；朝鲜半岛、日本、蒙古、西伯利亚东南部、乌拉尔山脉西南部。【飞行期】5—7月。

[Identification] Male face black with blue markings. Thorax with pale blue antehumeral stripes, sides pale blue. Abdomen black with sky-blue markings. Female polymorphic, body color pale blue or yellow with black markings. [Measurements] Total length 28-30 mm, abdomen 23-24 mm, hind wing 15-17 mm. [Habitat] Standing water with emergent plants below 500 m elevation. [Distribution] Heilongjiang, Jilin; Korean peninsula, Japan, Mongolia, southeastern Siberia, southwestern part of Ural Mountains. [Flight Season] May to July.

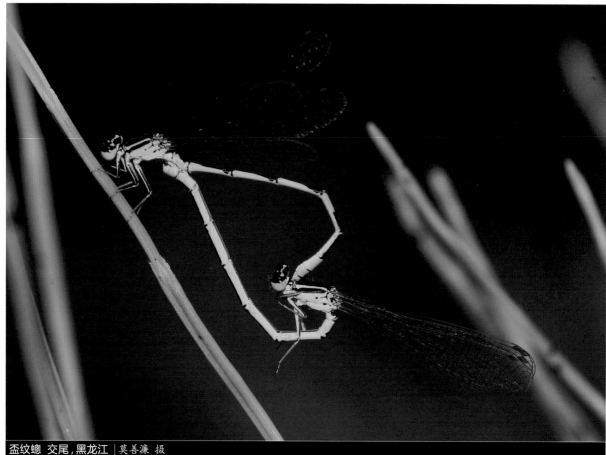

盃纹螅 交尾，黑龙江 ｜莫善濂 摄
Coenagrion ecornutum, mating pair from Heilongjiang ｜ Photo by Shanlian Mo

三纹蟌 *Coenagrion hastulatum* (Charpentier, 1825)

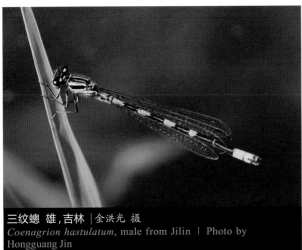

三纹蟌 雄,吉林 | 金洪光 摄
Coenagrion hastulatum, male from Jilin | Photo by Hongguang Jin

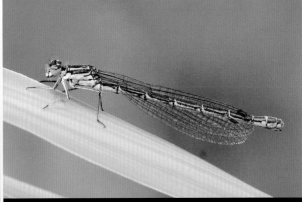

三纹蟌 雌,吉林 | 金洪光 摄
Coenagrion hastulatum, female from Jilin | Photo by Hongguang Jin

三纹蟌 雌雄连结,吉林 | 金洪光 摄
Coenagrion hastulatum, pair in tandem from Jilin | Photo by Hongguang Jin

　　【形态特征】雄性面部黑色具蓝色斑纹；胸部具蓝色肩前条纹，侧面蓝色；腹部黑色具蓝色条纹。雌性黑色具黄褐色条纹。【长度】体长 31～33 mm，腹长 22～26 mm，后翅 16～22 mm。【栖息环境】海拔 500 m 以下水草茂盛的静水环境。【分布】黑龙江、吉林；从欧洲东北部经西伯利亚至朝鲜半岛广布。【飞行期】5—8月。

　　[Identification] Male face black with blue markings. Thorax with blue antehumeral stripes, sides blue. Abdomen black with blue markings. Female black with yellowish brown markings. [Measurements] Total length 31-33 mm, abdomen 22-26 mm, hind wing 16-22 mm. [Habitat] Standing water with emergent plants below 500 m elevation. [Distribution] Heilongjiang, Jilin; Widespread from northeastern Europe throughout Siberia to Korean peninsula in the southeast. [Flight Season] May to August.

黑格螅 *Coenagrion hylas* (Trybom, 1889)

【形态特征】雄性面部黑色具蓝色斑纹；胸部具蓝色肩前条纹，侧面具2条蓝色条纹；腹部黑色具蓝色条纹。雌性黑色具淡蓝色条纹。【长度】体长 33~38 mm，腹长 25~32 mm，后翅 19~28 mm。【栖息环境】海拔 500 m以下水草茂盛的静水环境。【分布】黑龙江、吉林；朝鲜半岛、日本、西伯利亚，欧洲的孤立种群分布于俄罗斯北部和欧洲中部的阿尔卑斯山脉。【飞行期】6—8月。

黑格螅 雌,吉林 | 金洪光 摄
Coenagrion hylas, female from Jilin | Photo by Hongguang Jin

三纹螅 雌雄连结,吉林 | 金洪光 摄
Coenagrion hastulatum, pair in tandem from Jilin | Photo by Hongguang Jin

[Identification] Male face black with blue markings. Thorax with blue antehumeral stripes, sides with two blue stripes. Abdomen black with blue markings. Female black with pale blue markings. [Measurements] Total length 33-38 mm, abdomen 25-32 mm, hind wing 19-28 mm. [Habitat] Standing water with emergent plants below 500 m elevation. [Distribution] Heilongjiang, Jilin; Korean peninsula, Japan, Siberia, isolated populations in European part of northern Russia and in the Alps of central Europe. [Flight Season] June to August.

黑格螅 雄,吉林 | 金洪光 摄
Coenagrion hylas, male from Jilin | Photo by Hongguang Jin

纤腹螅 *Coenagrion johanssoni* (Wallengren, 1894)

纤腹螅 雄, 吉林 | 金洪光 摄
Coenagrion johanssoni, male from Jilin | Photo by Hongguang Jin

【形态特征】雄性面部黑色具蓝色斑纹；胸部具蓝色肩前条纹，侧面蓝色；腹部黑色具蓝色条纹。雌性黑色具淡蓝色条纹。【长度】体长 27~30 mm，腹长 20~24 mm，后翅 15~19 mm。【栖息环境】海拔 500 m以下水草茂盛的静水环境。【分布】黑龙江、吉林、内蒙古、新疆、河北；从欧洲东北部经西伯利亚至朝鲜半岛广布。【飞行期】5—7月。

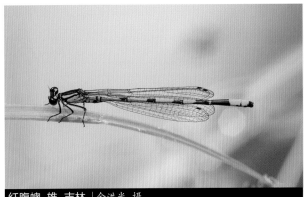

纤腹螅 雄, 吉林 | 金洪光 摄
Coenagrion johanssoni, male from Jilin | Photo by Hongguang Jin

纤腹螅 雌雄连结, 吉林 | 金洪光 摄
Coenagrion johanssoni, pair in tandem from Jilin | Photo by Hongguang Jin

[Identification] Male face black with blue markings. Thorax with blue antehumeral stripes, sides blue. Abdomen black with blue markings. Female black with pale blue markings. [Measurements] Total length 27-30 mm, abdomen 20-24 mm, hind wing 15-19 mm. [Habitat] Standing water with emergent plants below 500 m elevation. [Distribution] Heilongjiang, Jilin, Inner Mongolia, Xinjiang, Hebei; Widespread from northeastern Europe throughout Siberia to Korean peninsula in the southeast. [Flight Season] May to July.

矛斑蟌 *Coenagrion lanceolatum* (Selys, 1872)

【形态特征】雄性面部黑色具蓝色斑纹；胸部具天蓝色肩前条纹，侧面天蓝色；腹部蓝色具黑色条纹。雌性多型，身体淡蓝色、绿色或黄色具黑色条纹。【长度】体长 35~36 mm，腹长 28~29 mm，后翅 21~22 mm。【栖息环境】海拔 500 m 以下水草茂盛的静水环境。【分布】黑龙江、吉林；朝鲜半岛、日本、西伯利亚。【飞行期】5—7月。

矛斑蟌 雄,吉林 | 金洪光 摄
Coenagrion lanceolatum, male from Jilin | Photo by Hongguang Jin

[Identification] Male face black with blue markings. Thorax with sky-blue antehumeral stripes, sides sky-blue. Abdomen blue with black markings. Female polymorphic, body color pale blue, green or yellow with black markings. [Measurements] Total length 35-36 mm, abdomen 28-29 mm, hind wing 21-22 mm. [Habitat] Standing water with emergent plants below 500 m elevation. [Distribution] Heilongjiang, Jilin; Korean peninsula, Japan, Siberia. [Flight Season] May to July.

矛斑螅 雌,黑龙江 | 莫善濂 摄
Coenagrion lanceolatum, female from Heilongjiang | Photo by Shanlian Mo

矛斑螅 雌雄连结,吉林 | 金洪光 摄
Coenagrion lanceolatum, pair in tandem from Jilin | Photo by Hongguang Jin

月斑螅 *Coenagrion lunulatum* (Charpentier, 1840)

【形态特征】雄性面部黑色具蓝色斑纹；胸部具蓝色肩前条纹,侧面蓝色；腹部黑色具蓝色条纹。雌性黑色具蓝色条纹。本种与三纹螅相似,但雄性腹部第3~6节的黑色区域更长。【长度】体长 30~33 mm,腹长 22~26 mm,后翅 16~22 mm。【栖息环境】海拔 500 m以下水草茂盛的静水环境。【分布】黑龙江、吉林；从欧洲东北部经西伯利亚至俄罗斯远东地区广布。【飞行期】5—7月。

[Identification] Male face black with blue markings. Thorax with blue antehumeral stripes, sides blue. Abdomen black with blue markings. Female black with blue markings. Similar to *C. hastulatum* but male with larger black area on

月斑螅 雄,黑龙江 | 莫善濂 摄
Coenagrion lunulatum, male from Heilongjiang | Photo by Shanlian Mo

月斑螅 雌,黑龙江 | 莫善濂 摄
Coenagrion lunulatum, female from Heilongjiang | Photo by Shanlian Mo

S3-S6. [Measurements] Total length 30-33 mm, abdomen 22-26 mm, hind wing 16-22 mm. [Habitat] Standing water with emergent plants below 500 m elevation. [Distribution] Heilongjiang, Jilin; Widespread from northeastern Europe throughout Siberia to Russian Far East. [Flight Season] May to July.

月斑蟌 雄,黑龙江 | 莫善濂 摄
Coenagrion lunulatum, male from Heilongjiang | Photo by Shanlian Mo

腾冲蟌 *Coenagrion exclamationis* (Fraser, 1919)

【形态特征】本种与多棘蟌相似,但雄性腹部第7节的蓝斑更大。雌性多型,身体黑色具黄色或淡蓝色斑。【长度】体长 33~34 mm,腹长 27~28 mm,后翅 18~22 mm。【栖息环境】海拔 1500~2500 m森林中的池塘。【分布】云南(保山、德宏);不丹、印度、尼泊尔。【飞行期】4—11月。

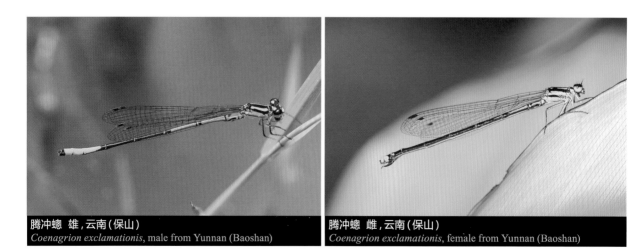

腾冲蟌 雄,云南(保山)
Coenagrion exclamationis, male from Yunnan (Baoshan)

腾冲蟌 雌,云南(保山)
Coenagrion exclamationis, female from Yunnan (Baoshan)

腾冲螅 雄，云南（德宏）
Coenagrion tengchongensis, male from Yunnan (Dehong)

[Identification] The species is similar to *C. aculeatum*, but male with larger blue spot on S7. Female polymorphic, body black with yellow or pale blue markings. [Measurements] Total length 33-34 mm, abdomen 27-28 mm, hind wing 18-22 mm. [Habitat] Pools in forest at 1500-2500 m elevation. [Distribution] Yunnan (Baoshan, Dehong); Bhutan, India, Nepal. [Flight Season] April to November.

绿螅属 Genus *Enallagma* Charpentier, 1840

本属全球已知近50种，主要分布在北美洲，少数分布于欧亚大陆。中国已知1种，主要分布于西北、华北和东北地区。本属豆娘体型较小，身体通常大面积蓝色具黑色斑纹；翅的弓脉位于第2条结前横脉下方，结后横脉10～12条，四边室的前边大约为后边长度的2/3；雌性第8节腹面具1个刺状突起。

本属豆娘栖息于水草茂盛的静水环境。雄性经常停落在水面的漂浮物上或水边的植物丛中。它们在水草上连结产卵。

The genus contains nearly 50 species, most of which occur in North America and a few in Eurasia. One species is recorded from China, distributed in the Northwest, North and Northeast regions. Species of the genus are usually small with largely blue body marked with black spots. Wings with arculus just below the second antenodal,

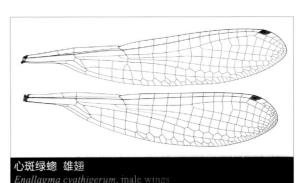

心斑绿螅 雄翅
Enallagma cyathigerum, male wings

postnodals 10-12, the upper side of quadrangle about two thirds length of the lower. Females with a vulvar spine on the underside of S8.

Enallagma species inhabit standing water with plenty of emergent plants. Males often perch on floaters or grass nearby water. They oviposit in tandem on aquatic plants.

心斑绿螅 雄 ｜莫善濂 摄

Enallagma cyathigerum, male ｜ Photo by Shanlian Mo

心斑绿螅 *Enallagma cyathigerum* (Charpentier, 1840)

【形态特征】雄性面部黑色具蓝色斑纹；胸部和腹部大面积蓝色具黑色斑纹。雌性与雄性相似但色彩稍淡。【长度】体长 29~36 mm，腹长 22~28 mm，后翅 15~21 mm。【栖息环境】海拔 500 m以下的静水环境。【分布】黑龙江、吉林、内蒙古、新疆、西藏、宁夏、河北；整个欧洲、亚洲的大部分温带地区和俄罗斯远东。【飞行期】5—9月。

[Identification] Male face black with blue markings. Thorax and abdomen largely blue with black markings. Female similar to male but body color paler. [Measurements] Total length 29-36 mm, abdomen 22-28 mm, hind wing 15-21 mm. [Habitat] Standing water below 500 m elevation. [Distribution] Heilongjiang, Jilin, Inner Mongolia, Xinjiang, Tibet, Ningxia, Hebei; Throughout the whole Europe and much of the temperate Asia to the Russian Far East. [Flight Season] May to September.

心斑绿螅 雄，黑龙江 | 莫善濂 摄
Enallagma cyathigerum, male from Heilongjiang | Photo by Shanlian Mo

心斑绿螅 雄，黑龙江 | 莫善濂 摄
Enallagma cyathigerum, male from Heilongjiang | Photo by Shanlian Mo

红眼蟌属 Genus *Erythromma* Charpentier, 1840

　　本属全球已知4种，分布于欧亚大陆的温带地区。中国仅知1种，主要分布于东北地区。本属豆娘体型较小，雄性复眼和面部红色，是本属区分于其他较近似蓝色豆娘的显著特征。翅的弓脉位于第2条结前横脉下方，结后横脉10～12条，四边室的前边大约为后边长度的1/2。

　　本属豆娘栖息于水草茂盛的静水环境。雄性经常停落在水面的漂浮物上或者挺水植物丛中。雌雄在水草上连结产卵。

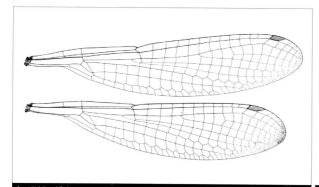

红眼蟌
Erythromma najas

红眼蟌 雄翅
Erythromma najas, male wings

红眼蟌属 雄性肛附器
Genus *Erythromma*, male anal appendages

The genus contains four species confined to temperate Eurasia. Only one species is recorded from China, mainly found in the Northeast region. Species of the genus are small-sized, males are remarkable for their red eyes and face, a character that separates them from similar blue damselflies. Wings with arculus just below the second antenodal, postnodals 10-12, the upper side of quadrangle about half length of the lower.

Erythromma species inhabit standing water with plenty of emergent plants. Males often perch on floating or emergent plants. They oviposit in tandem on aquatic plants.

红眼蟌 雌雄连结 | 金洪光 摄
Erythromma najas, pair in tandem | Photo by Hongguang Jin

红眼蟌 *Erythromma najas* (Hansemann, 1823)

【形态特征】雄性复眼和面部红色；胸部背面黑色具红褐色肩前条纹，侧面蓝色；腹部黑色，第1节、第9～10节蓝色。雌性黄褐色具黑色斑纹。【长度】体长 30～36 mm，腹长 25～30 mm，后翅 19～24 mm。【栖息环境】海拔 500 m以下水草茂盛的静水环境。【分布】黑龙江、吉林、新疆；从欧洲西部至日本广布。【飞行期】5—8月。

[Identification] Male eyes and face red. Thorax black dorsally with reddish brown antehumeral stripes, sides blue. Abdomen black, S1 and S9-S10 blue. Female yellowish brown with black markings. [Measurements] Total length 30-36 mm, abdomen 25-30 mm, hind wing 19-24 mm. [Habitat] Standing water with emergent plants below 500 m elevation. [Distribution] Heilongjiang, Jilin, Xinjiang; Widespread from western Europe to Japan. [Flight Season] May to August.

红眼蟌 雄, 吉林 | 金洪光 摄
Erythromma najas, male from Jilin | Photo by Hongguang Jin

红眼蟌 雌, 吉林 | 金洪光 摄
Erythromma najas, female from Jilin | Photo by Hongguang Jin

红眼蟌 雄, 黑龙江 | 莫善濂 摄
Erythromma najas, male from Heilongjiang | Photo by Shanlian Mo

火螅属 Genus *Huosoma* Guan, Dumont, Yu, Han & Vierstraete, 2013

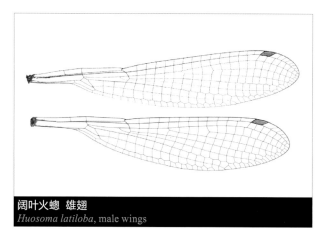

阔叶火螅 雄翅
Huosoma latiloba, male wings

本属全球已知2种,中国特有,分布于西南地区的高海拔山区。本属与欧洲分布的红螅属近似,依靠分子鉴定的结果建立。本属是一类体型较小的豆娘,由于雄性大面积鲜红色,因而得名"火螅"。

本属豆娘栖息于高海拔湿地,甚至可以在海拔3000 m以上的高山上生活。雄性通常停落在水面的植物上或者水边的草丛中。雌雄连结产卵。

The genus is endemic to China and contains two species confined to the high mountains in the Southwest region. The genus is similar to the European *Pyrrhosoma* Charpentier, 1840 and was erected largely based on molecular profile. Species of the genus are small damselflies, since the males are largely reddish, they are called "Huo damselfly" ("Huo" in Chinese means fire).

Huosoma species frequent well vegetated wetlands in the high mountains, some species can survive in the mountains above 3000 m elevation. Males usually perch on plants above water or grass nearby. They oviposit in tandem.

阔叶火螅 雌雄连结
Huosoma latiloba, pair in tandem

阔叶火螅 *Huosoma latiloba* (Yu, Yang & Bu, 2008)

【形态特征】雄性面部黑色具淡黄色条纹；胸部背面黑绿色具金属光泽，具红色肩条纹，侧面黄色；腹部红色，第7~10节具黑斑。雌性黄色具黑褐色条纹。【长度】体长 29~33 mm，腹长 23~26 mm，后翅 19~21 mm。【栖息环境】海拔 2000~3500 m水草茂盛的静水环境。【分布】中国云南（丽江、大理、保山）特有。【飞行期】4—7月。

阔叶火螅 雄,云南(保山)
Huosoma latiloba, male from Yunnan (Baoshan)

[Identification] Male face black with pale yellow stripes. Thorax metallic blackish green dorsally with red humeral stripes, sides yellow. Abdomen red, S7-S10 with black spots. Female yellow with blackish brown markings. [Measurements] Total length 29-33 mm, abdomen 23-26 mm, hind wing 19-21 mm. [Habitat] Standing water with emergent plants at 2000-3500 m elevation. [Distribution] Endemic to Yunnan (Lijiang, Dali, Baoshan) of China. [Flight Season] April to July.

阔叶火螅 雄,云南(保山)
Huosoma latiloba, male from Yunnan (Baoshan)

阔叶火螅 雌,云南(保山)
Huosoma latiloba, female from Yunnan (Baoshan)

异痣蟌属 Genus *Ischnura* Charpentier, 1840

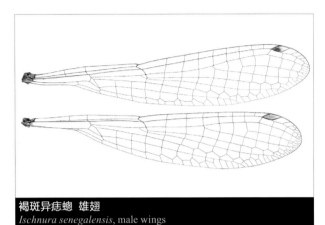

褐斑异痣蟌 雄翅
Ischnura senegalensis, male wings

本属全球已知约70种, 世界性分布。中国已知8种, 全国广布。本属豆娘体型较小, 身体色彩多变; 翅的弓脉位于第2条结前横脉下方, 结后横脉6~8条, 四边室的前边明显短于后边, 翅痣菱形, 雄性前翅翅痣的色彩通常为双色, 与后翅不同。

本属豆娘栖息于水草茂盛的静水环境和流速缓慢的溪流。雄性经常停落在挺水植物上或水面的漂浮物上。雌性单独在水草上产卵。

The genus contains about 70 species of worldwide distribution. Eight species are recorded from China, widespread throughout the country. Species of the genus are small-sized with body coloration variable. Wings with arculus just below the second antenodal, postnodals 6-8, the upper side of quadrangle much shorter than the lower, pterostigma rhomboid, males with bicolored pterostigma in fore wings and different from hind wings.

Ischnura species inhabit standing water with plenty of emergent plants and slow flowing streams. Males often perch on emergent plants or floaters above water. Females lay eggs alone on aquatic plants.

东亚异痣蟌 交尾
Ischnura asiatica, mating pair

东亚异痣蟌 *Ischnura asiatica* (Brauer, 1865)

【形态特征】雄性面部黑色具蓝色斑点；胸部背面黑色，具黄绿色的肩前条纹，侧面黄绿色；腹部黑色，侧面具黄色条纹，第8~10节具蓝斑。雌性未熟时红色，成熟后黄绿色或褐色具黑色条纹。【长度】体长 27~29 mm，腹长 22~24 mm，后翅 10~11 mm。【栖息环境】海拔 2000 m以下水草茂盛的静水环境，包括水稻田。【分布】东北、华北、华中、西南地区广布；朝鲜半岛、日本、俄罗斯远东。【飞行期】2—10月。

东亚异痣蟌 雄,吉林 |金洪光 摄
Ischnura asiatica, male from Jilin | Photo by Hongguang Jin

[Identification] Male face black with blue spots. Thorax black dorsally with yellowish green antehumeral stripes, sides yellowish green. Abdomen black with yellow stripes laterally, S8-S10 with blue spots. Female red when immature, yellowish green or brown with black markings when mature. [Measurements] Total length 27-29 mm, abdomen 22-

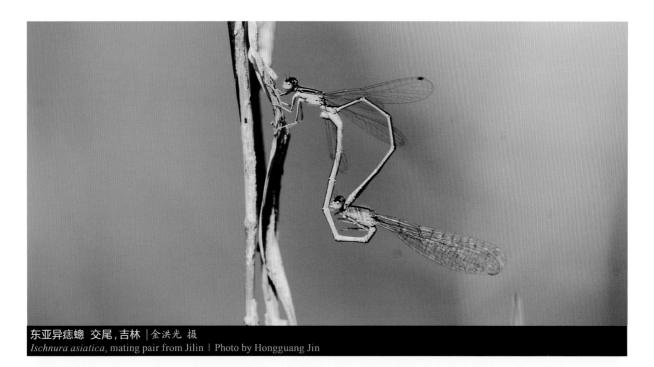

东亚异痣蟌 交尾,吉林 |金洪光 摄
Ischnura asiatica, mating pair from Jilin | Photo by Hongguang Jin

东亚异痣蟌 雌，湖北
Ischnura asiatica, female from Hubei

东亚异痣蟌 未熟雌，湖北
Ischnura asiatica, immature female from Hubei

24 mm, hind wing 10-11 mm. [Habitat] Standing water with emergent plants including paddy fields below 2000 m elevation. [Distribution] Widespread in the Northeast, North, Central and Southwest regions; Korean peninsula, Japan, Russian Far East. [Flight Season] February to October.

黄腹异痣蟌 *Ischnura aurora* (Brauer, 1865)

黄腹异痣蟌 雌雄连结，云南（红河）｜莫善濂 摄
Ischnura aurora, pair in tandem from Yunnan (Honghe) ｜ Photo by Shanlian Mo

黄腹异痣蟌 雄，云南（西双版纳）
Ischnura aurora, male from Yunnan (Xishuangbanna)

【形态特征】雄性面部黑色具蓝色斑点；胸部背面黑色，具黄绿色肩前条纹，侧面黄绿色；腹部第1~7节橙黄色，第8~9节蓝色，第6~7节和第10节背面具黑斑。雌性身体黄色具黑色条纹。【长度】体长 21~24 mm，腹长17~19 mm，后翅 10~11 mm。【栖息环境】海拔 2000 m以下水草茂盛的静水环境，包括水稻田。【分布】云南、台湾；日本、南亚、东南亚、大洋洲。【飞行期】全年可见。

[Identification] Male face black with blue spots. Thorax black dorsally with yellowish green antehumeral stripes, sides yellowish green. S1-S7 orange yellow, S8-S9 blue, S6-S7 and S10 with dorsal black spots. Female yellow with black markings. [Measurements] Total length 21-24 mm, abdomen 17-19 mm, hind wing 10-11 mm. [Habitat] Standing water with emergent plants including paddy fields below 2000 m elevation. [Distribution] Yunnan, Taiwan; Japan, South and Southeast Asia, Oceania. [Flight Season] Throughout the year.

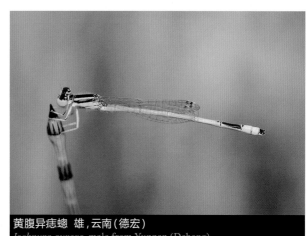

黄腹异痣蟌 雄，云南（德宏）
Ischnura aurora, male from Yunnan (Dehong)

长叶异痣蟌 *Ischnura elegans* (Vander Linden, 1820)

　　【形态特征】雄性面部黑色具蓝色和绿色斑纹；胸部背面黑色，具蓝色肩前条纹，侧面蓝色；腹部黑色，第1~3节、第7~10节具蓝斑。雌性多型，未熟时身体蓝紫色、橙色或黄色，成熟后身体大面积蓝色或黄色。【长度】体长30~35 mm，腹长 22~30 mm，后翅 14~23 mm。【栖息环境】海拔 1000 m以下水草茂盛的池塘、水稻田和流速缓慢的溪流。【分布】东北、华北地区广布；从欧洲西部至朝鲜半岛和日本广布。【飞行期】5—9月。

长叶异痣蟌 交尾,吉林 | 金洪光 摄
Ischnura elegans, mating pair from Jilin | Photo by Hongguang Jin

长叶异痣蟌 交尾,吉林 | 金洪光 摄
Ischnura elegans, mating pair from Jilin | Photo by Hongguang Jin

长叶异痣蟌 雌,北京 | 陈炜 摄
Ischnura elegans, female from Beijing | Photo by Wei Chen

长叶异痣蟌 雌,黑龙江 | 莫善濂 摄
Ischnura elegans, female from Heilongjiang | Photo by Shanlian Mo

[Identification] Male face black with blue and green spots. Thorax black dorsally with blue antehumeral stripes, sides blue. Abdomen black, S1-S3 and S7-S10 with blue spots. Female polymorphic, body bluish violet, orange or yellow when immature, body largely sky-blue or yellow when mature. [Measurements] Total length 30-35 mm, abdomen 22-30 mm, hind wing 14-23 mm. [Habitat] Ponds with emergent plants, paddy fields and slow flowing streams below 1000 m elevation. [Distribution] Widespread in the Northeast and North regions; Widespread from western Europe to Korean peninsula and Japan. [Flight Season] May to September.

长叶异痣蟌 雄,黑龙江 | 莫善濂 摄
Ischnura elegans, male from Heilongjiang | Photo by Shanlian Mo

蓝壮异痣蟌 *Ischnura pumilio* (Charpentier, 1825)

蓝壮异痣蟌 雌, 芬兰 | Sami Karjalainen 摄
Ischnura pumilio, female from Finland | Photo by Sami Karjalainen

【形态特征】雄性面部黑色具蓝色和绿色斑纹；胸部背面黑色，具蓝色肩前条纹，侧面蓝色；腹部主要黑色，第1~3节和第6~10节具蓝斑。雌性身体主要橙黄色具黑色斑纹。【长度】体长 26~31 mm，腹长 22~25 mm，后翅 14~18 mm。【栖息环境】海拔 1000 m以下水草茂盛的静水环境，包括水稻田。【分布】新疆、内蒙古；从欧洲西部至蒙古和中国西部。【飞行期】5—9月。

[Identification] Male face black with blue and green spots. Thorax black dorsally with blue antehumeral stripes, sides blue. Abdomen mainly black, S1-S3 and S6-S10 with blue markings. Female body mainly orange-yellow with black markings. [Measurements] Total length 26-31 mm, abdomen 22-25 mm, hind wing 14-18 mm. [Habitat] Standing water with plenty of emergent plants including paddy fields below 1000 m elevation. [Distribution] Xinjiang, Inner Mongolia; Widespread from western Europe to Mongolia and western China. [Flight Season] May to September.

蓝壮异痣蟌 交尾, 芬兰 | Sami Karjalainen 摄
Ischnura pumilio, mating pair from Finland | Photo by Sami Karjalainen

蓝壮异痣蟌 雄，芬兰 | Sami Karjalainen 摄
Ischnura pumilio, male from Finland | Photo by Sami Karjalainen

赤斑异痣蟌 *Ischnura rufostigma* Selys, 1876

【形态特征】雄性面部黑色具蓝色和绿色斑纹；胸部背面黑色，具黄绿色肩前条纹，侧面黄绿色；腹部第2~6节橙色，第7~10节黑色，第8节背面有时具1个蓝色斑。雌性多型，全身黄褐色具黑色条纹或与雄性相似。【长度】体长29~33 mm，腹长 23~26 mm，后翅 10~12 mm。【栖息环境】海拔 2500 m以下水草茂盛的池塘、水稻田和流速缓慢的溪流。【分布】四川、云南、贵州、福建、广西、广东；南亚、东南亚。【飞行期】全年可见。

赤斑异痣蟌 雄，云南（西双版纳）
Ischnura rufostigma, male from Yunnan (Xishuangbanna)

赤斑异痣蟌 雌，云南（西双版纳）｜莫善濂 摄
Ischnura rufostigma, female from Yunnan
(Xishuangbanna) | Photo by Shanlian Mo

赤斑异痣蟌，交尾，云南（大理）
Ischnura rufostigma, mating pair from Yunnan (Dali)

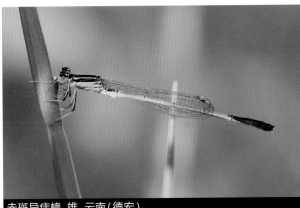

赤斑异痣蟌 雄,云南(德宏)
Ischnura rufostigma, male from Yunnan (Dehong)

赤斑异痣蟌 雌,贵州
Ischnura rufostigma, female from Guizhou

[Identification] Male face black with blue and green spots. Thorax black dorsally with yellowish green antehumeral stripes, sides yellowish green. S2-S6 orange, S7-S10 black, S8 sometimes with a dorsal blue spot. Female polymorphic, body yellowish brown with black markings or similar to male. [Measurements] Total length 29-33 mm, abdomen 23-26 mm, hind wing 10-12 mm. [Habitat] Ponds with emergent plants, paddy fields and slow flowing streams below 2500 m elevation. [Distribution] Sichuan, Yunnan, Guizhou, Fujian, Guangxi, Guangdong; South and Southeast Asia. [Flight Season] Throughout the year.

褐斑异痣蟌 *Ischnura senegalensis* (Rambur, 1842)

褐斑异痣蟌 雄,云南(德宏)
Ischnura senegalensis, male from Yunnan (Dehong)

【形态特征】雄性面部黑色具蓝色和绿色斑纹；胸部背面黑色，具黄绿色肩前条纹，侧面黄绿色；腹部主要黑色，第8～9节具蓝斑。雌性多型，身体黄绿色或淡蓝色具黑色条纹，未熟时胸部有时橙黄色。【长度】体长28～30 mm，腹长 21～24 mm，后翅 13～16 mm。【栖息环境】海拔 2500 m以下水草茂盛的池塘、水稻田和流速缓慢的溪流周边。【分布】华中、华南、西南地区广布；日本、南亚、东南亚、新几内亚、非洲。【飞行期】全年可见。

褐斑异痣蟌 未熟雌，湖北
Ischnura senegalensis, immature female from Hubei

褐斑异痣蟌 交尾，云南（德宏）
Ischnura senegalensis, mating pair from Yunnan (Dehong)

[Identification] Male face black with blue and green spots. Thorax black dorsally with yellowish green antehumeral stripes, sides yellowish green. Abdomen mainly black, S8-S9 with blue spots. Female polymorphic, body yellowish green or pale blue with black stripes. Immature female sometimes with orange-yellow thorax. [Measurements] Total length 28-30 mm, abdomen 21-24 mm, hind wing 13-16 mm. [Habitat] Ponds with emergent plants, paddy fields and slow flowing streams below elevation 2500 m. [Distribution] Widespread in the Central, South and Southwest regions; Japan, South and Southeast Asia, New Guinea, Africa. [Flight Season] Throughout the year.

褐斑异痣蟌 交尾，云南（大理）
Ischnura senegalensis, mating pair from Yunnan (Dali)

褐斑异痣蟌 交尾, 云南（大理）
Ischnura senegalensis, mating pair from Yunnan (Dali)

异痣蟌属待定种 *Ischnura* sp.

【形态特征】雄性面部黑色具黄绿色斑点；胸部背面黑色，具黄绿色肩前条纹，侧面黄绿色；腹部第1~7节橙黄色具黑色斑纹，第8~9节具蓝斑。雌性身体绿色具黑色条纹。【长度】体长 26~27 mm，腹长 20~22 mm，后翅13~15 mm。【栖息环境】海拔 2000~3000 m水草茂盛的静水环境。【分布】云南（丽江、大理）。【飞行期】5—8月。

[Identification] Male face black with yellowish green spots. Thorax black dorsally with yellowish green antehumeral stripes, sides yellowish green. S1-S7 orange yellow with black markings, S8-S9 with blue spots. Female body green with black markings. [Measurements] Total length 26-27 mm, abdomen 20-22 mm, hind wing 13-15 mm. [Habitat] Standing water with emergent plants from 2000-3000 m elevation. [Distribution] Yunnan (Lijiang, Dali). [Flight Season] May to August.

异痣蟌属待定种 雌，云南（丽江）
Ischnura sp., female from Yunnan (Lijiang)

异痣蟌属待定种 交尾，云南（丽江）
Ischnura sp., mating pair from Yunnan (Lijiang)

异痣蟌属待定种 雄，云南（丽江）
Ischnura sp., male from Yunnan (Lijiang)

妹蟌属 Genus *Mortonagrion* Fraser, 1920

本属全世界已知16种，分布于亚洲。中国已知3种，分布于华南和西南地区。本属豆娘体型较小；翅的弓脉位于第2条结前横脉后方，结后横脉6~8条，四边室的前边稍短于后边；雄性的下肛附器长于上肛附器或近等长。

本属豆娘栖息于水草茂盛的静水环境。雄性经常停落在挺水植物上或者水边的草丛中。雌性单独在水草上产卵。

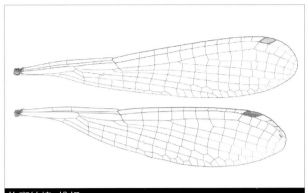

钩斑妹蟌 雄翅
Mortonagrion selenion, male wings

The genus contains 16 species distributed in Asia. Three species are recorded from China, found in the South and Southwest regions. Species of the genus are small-sized damselflies. Wings with arculus beginning after the second antenodal and 6-8 postnodals, the upper side of quadrangle slightly shorter than the lower. Male inferior appendages longer or the same length as the superiors.

Mortonagrion species inhabit standing water with plenty of emergent plants. Males often perch on emergent plants or grass nearby. Females lay eggs alone on aquatic plants.

蓝尾妹蟌 雄
Mortonagrion aborense, male

蓝尾妹蟌 *Mortonagrion aborense* (Laidlaw, 1914)

蓝尾妹蟌 雄，云南（西双版纳）
Mortonagrion aborense, male from Yunnan (Xishuangbanna)

【形态特征】雄性面部黑色具黄绿色和蓝色斑点；胸部背面黑色具绿色肩前条纹，侧面蓝色；腹部黑色具蓝斑。【长度】雄性体长 31 mm，腹长 26 mm，后翅 15 mm。【栖息环境】海拔 1000 m 以下的阴暗池塘和沟渠。【分布】云南（西双版纳）；印度、泰国、老挝、柬埔寨、越南、马来半岛、印度尼西亚。【飞行期】全年可见。

[Identification] Male face black with yellowish green and blue spots. Thorax black dorsally with green antehumeral stripes, sides blue. Abdomen black with blue markings. [Measurements] Male total length 31 mm, abdomen 26 mm, hind wing 15 mm. [Habitat] Shady ponds and ditches below 1000 m elevation. [Distribution] Yunnan (Xishuangbanna); India, Thailand, Laos, Cambodia, Vietnam, Peninsular Malaysia, Indonesia. [Flight Season] Throughout the year.

广濑妹蟌 *Mortonagrion hirosei* Asahina, 1972

广濑妹蟌 雄，香港 | 祁麟峰 摄
Mortonagrion hirosei, male from Hong Kong | Photo by Mahler Ka

【形态特征】雄性面部黑色具黄绿色斑点；胸部背面黑色，具4个黄绿色斑点，侧面黄绿色；腹部主要黑色，第1~8节具淡黄色斑纹。雌性多型，身体橙黄色或与雄性相似。【长度】腹长23~25 mm，后翅13~16 mm。【栖息环境】海岸的沼泽和红树林。【分布】广东、香港、台湾；日本。【飞行期】4—9月。

[Identification] Male face black with yellowish green spots. Thorax black dorsally with four yellowish green spots, sides yellowish green. Abdomen mainly black, S1-S8 with pale yellow markings. Female polymorphic, body orange yellw or similar to male.

[Measurements] Abdomen 23-25 mm, hind wing 13-16 mm. [Habitat] Marshes and mangroves along the coast.
[Distribution] Guangdong, Hong Kong, Taiwan; Japan. [Flight Season] April to September.

广濑妹螅, 雌, 香港 | 梁嘉景 摄
Mortonagrion hirosei, female from Hong Kong | Photo
by Kenneth Leung

广濑妹螅 未熟雌, 香港 | 梁嘉景 摄
Mortonagrion hirosei, immature female from Hong Kong | Photo
by Kenneth Leung

广濑妹螅 雄, 香港 | 祁麟峰 摄
Mortonagrion hirosei, male from Hong Kong | Photo by Mahler Ka

钩斑妹蟌 *Mortonagrion selenion* (Ris, 1916)

【形态特征】雄性面部黑色具黄绿色斑点；胸部背面黑色具蓝色肩前条纹，侧面绿色；腹部第1~6节主要褐色，第7~10节橙黄色。雌性未熟时通体白色；成熟后主要绿色，腹部第7~8节橙黄色。【长度】体长 27~30 mm，腹长 22~25 mm，后翅 15~16 mm。【栖息环境】海拔 1000~2500 m水草茂盛的静水环境。【分布】云南、贵州、台湾；朝鲜半岛、日本、俄罗斯远东。【飞行期】5—9月。

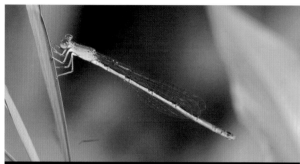

钩斑妹蟌 雌,云南(大理) | 莫善濂 摄
Mortonagrion selenion, female from Yunnan (Dali) | Photo by Shanlian Mo

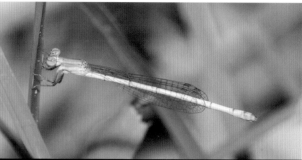

钩斑妹蟌 未熟雌,贵州 | 莫善濂 摄
Mortonagrion selenion, immature female from Guizhou | Photo by Shanlian Mo

[Identification] Male face black with yellowish green spots. Thorax black dorsally with blue antehumeral stripes, sides green. S1-S6 mainly brown, S7-S10 orange yellow. Female body white throughout when immature. Body mainly green, S7-S8 orange yellow when fully mature. [Measurements] Total length 27-30 mm, abdomen 22-25 mm, hind wing 15-16 mm. [Habitat] Standing water with emergent plants at 1000-2500 m elevation. [Distribution] Yunnan, Guizhou, Taiwan; Korean peninsula, Japan, Russian Far East. [Flight Season] May to September.

钩斑妹蟌 雄,云南(大理) | 莫善濂 摄
Mortonagrion selenion, male from Yunnan (Dali) | Photo by Shanlian Mo

绿背蟌属 Genus *Nehalennia* Selys, 1850

　　本属全球已知6种，主要分布于美洲。仅有1种分布于欧亚大陆的温带地区，在中国分布于西北和东北地区。本属是体型较小的豆娘，身体具绿色金属光泽；翅的弓脉位于第2条结前横脉下方，结后横脉6～8条，四边室的前边约为后边的1/2。

　　本属豆娘栖息于水草茂盛的静水环境。雄性经常停落在挺水植物上或者水边的草丛中。

　　This genus contains six species, mainly distributed in Americas. One species is widespread in temperate Eurasia, in China it is found in the Northwest and Northeast regions. Species of the genus are small-sized with a metallic green body. Wings with arculus just below the second antenodal, postnodals 6-8, the upper side of quadrangle about half length of the lower.

　　Nehalennia species inhabit standing water with plenty of emergent plants. Males often perch on emergent plants or grass nearby.

黑面绿背蟌 雄 | Sami Karjalainen 摄
Nehalennia speciosa, male | Photo by Sami Karjalainen

黑面绿背蟌 *Nehalennia speciosa* (Charpentier, 1840)

　　【形态特征】雄性面部墨绿色具蓝色和黄绿色斑点；胸部背面墨绿色具金属光泽，侧面淡蓝色；腹部墨绿色具金属光泽，第8～10节具蓝斑。雌性与雄性相似。【长度】体长 24～26 mm，腹长 19～25 mm，后翅 11～16 mm。【栖息环境】海拔 1000 m以下水草茂盛的静水环境。【分布】黑龙江、新疆、内蒙古；朝鲜半岛、日本、欧洲。【飞行期】5—9月。

[Identification] Male face metallic dark green with blue and yellowish green spots. Thorax metallic dark green dorsally, sides pale blue. Abdomen metallic dark green, S8-S10 with blue spots. Female similar to male. [Measurements] Total length 24-26 mm, abdomen 19-25 mm, hind wing 11-16 mm. [Habitat] Standing water with plenty of emergent plants below 1000 m elevation. [Distribution] Heilongjiang, Xinjiang, Inner Mongolia; Korean peninsula, Japan, Europe. [Flight Season] May to September.

黑面绿背螅 交尾, 芬兰 | Sami Karjalainen 摄
Nehalennia speciosa, mating pair from Finland | Photo by Sami Karjalainen

尾螅属 Genus *Paracercion* Weekers & Dumont, 2004

本属全球已知12种，分布于亚洲。中国已知7种，全国广布。本属豆娘体型较小，身体通常蓝色具黑色斑纹；翅的弓脉位于第2条结前横脉下方，结后横脉10～12条，四边室的前边为后边的1/3～1/2；雄性肛附器非常短小。本属豆娘的雄性可以通过腹部第2节和腹部末端黑斑的形状以及肛附器的构造区分。

本属豆娘栖息于水草茂盛的静水环境和流速缓慢的溪流。雄性经常停落在挺水植物上或者水面的漂浮物上。雌雄连结产卵。一些种类可以雌雄连结潜水产卵。

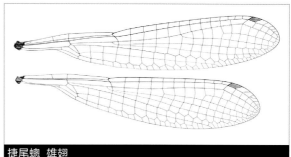

捷尾螅 雄翅
Paracercion v-nigrum, male wings

This genus contains 12 species, distributed in Asia. Seven species are recorded from China where they are widespread throughout the country. Species of the genus are small-sized damselflies, body usually blue with black markings. Wings with arculus just below the second antenodal, postnodals 10-12, the upper side of quadrangle about one third to half length of the lower. Male anal appendages are very short. Males of the genus can be distinguished by the shape of black spots on S2 and the end of abdomen as well as the shape of the anal appendages.

Paracercion species inhabit standing water with plenty of emergent plants and slow flowing streams. Males often perch on emergent plants or floaters. They oviposit in tandem. Some species lay eggs in tandem under water.

钳尾螅 雄
Paracercion dorothea, male

挫齿尾蟌 *Paracercion barbatum* (Needham,1930)

【形态特征】雄性身体大面积蓝色具黑色斑纹；腹部第2节背面具U形黑斑。雌性身体主要淡蓝色具黑褐色条纹。【长度】体长 32~35 mm，腹长 26~28 mm，后翅 20~23 mm。【栖息环境】海拔 2000~3000 m水草茂盛的静水环境。【分布】中国云南（丽江）特有。【飞行期】5—7月。

[Identification] Male body largely blue with black markings. S2 with a U-shaped black marking dorsally. Female body mainly pale blue with blackish brown markings. [Measurements] Total length 32-35 mm, abdomen 26-28 mm, hind wing 20-23 mm. [Habitat] Standing water with emergent plants from 2000-3000 m elevation. [Distribution] Endemic to Yunnan (Lijiang) of China. [Flight Season] May to July.

挫齿尾蟌 雄，云南（丽江）
Paracercion barbatum, male from Yunnan (Lijiang)

蓝纹尾蟌 *Paracercion calamorum* (Ris, 1916)

【形态特征】雄性面部黑色具蓝色斑点；胸部背面黑色，侧面蓝色，随年纪增长逐渐覆盖粉霜；腹部主要黑色，第8~10节具蓝斑。雌性身体主要黄绿色具黑色条纹。【长度】体长 26~32 mm，腹长 22~25 mm，后翅 15~17 mm。【栖息环境】海拔 1500 m以下水草茂盛的静水环境和流速缓慢的溪流。【分布】全国广布；朝鲜半岛、日本、俄罗斯远东、南亚、东南亚。【飞行期】2—11月。

[Identification] Male face black with blue spots. Thorax black dorsally, sides blue, gradually pruinosed with age. Abdomen mainly black, S8-S10 with blue markings. Female body mainly yellowish green with black markings. [Measurements] Total length 26-32 mm, abdomen 22-25 mm, hind wing 15-17 mm. [Habitat] Standing water with

蓝纹尾螅 雌，湖北
Paracercion calamorum, female from Hubei

蓝纹尾螅 连结产卵，湖北
Paracercion calamorum, laying eggs in tandem from Hubei

emergent plants and slow flowing streams below 1500 m elevation. **[Distribution]** Widespread throughout China; Korean peninsula, Japan, Russian Far East, South and Southeast Asia. **[Flight Season]** February to November.

蓝纹尾螅 雄，安徽
Paracercion calamorum, male from Anhui

钳尾螅 *Paracercion dorothea* (Fraser, 1924)

钳尾螅 雄,云南(丽江)
Paracercion dorothea, male from Yunnan (Lijiang)

【形态特征】雄性身体大面积蓝色具黑色斑纹；腹部第2节背面具U形黑斑，第8节侧面具1对三角形黑色斑。雌性身体主要黄色具黑褐色条纹；腹部末端具淡蓝色斑。本种与挫齿尾螅相似，但雄性腹部第8节具黑斑。【长度】体长 36~37 mm，腹长 28~30 mm，后翅 21~23 mm。【栖息环境】海拔 2000~3000 m水草茂盛的静水环境。【分布】中国云南(迪庆、丽江)特有。【飞行期】5—7月。

钳尾螅 雌,云南(丽江)
Paracercion dorothea, female from Yunnan (Lijiang)

钳尾螅 连结产卵,云南(丽江)
Paracercion dorothea, laying eggs in tandem from Yunnan (Lijiang)

[Identification] Male body largely blue with black markings. S2 with a U-shaped black marking dorsally, S8 with a pair of triangle-shaped black spots. Female body mainly yellow with blackish brown markings. Abdomen with apical pale blue spots. Similar to *P. barbatum*, but male S8 with black spots. [Measurements] Total length 36-37 mm, abdomen 28-30 mm, hind wing 21-23 mm. [Habitat] Standing water with emergent plants from 2000-3000 m elevation. [Distribution] Endemic to Yunnan (Diqing, Lijiang) of China. [Flight Season] May to July.

隼尾螅 *Paracercion hieroglyphicum* (Brauer, 1865)

【形态特征】雄性身体大面积蓝绿色具黑色条纹。雌性头部和胸部主要绿色；腹部橙黄色具褐色条纹。【长度】体长 25～28 mm，腹长 20～22 mm，后翅 12～15 mm。【栖息环境】海拔 500 m以下水草茂盛的静水环境。【分布】东北、华北、华中地区广布；朝鲜半岛、日本、俄罗斯远东。【飞行期】4—10月。

[Identification] Male body largely bluish green with black markings. Female head and thorax mainly green. Abdomen orange-yellow with brown markings. [Measurements] Total length 25-28 mm, abdomen 20-22 mm, hind

隼尾螅 雌，湖北
Paracercion hieroglyphicum, female from Hubei

隼尾螅 雌雄连结，湖北
Paracercion hieroglyphicum, pair in tandem from Hubei

隼尾螅 雄，湖北
Paracercion hieroglyphicum, male from Hubei

隼尾螅 交尾，湖北
Paracercion hieroglyphicum, mating pair from Hubei

wing 12-15 mm. [Habitat] Standing water with emergent plants below 500 m elevation. [Distribution] Widespread in the Northeast, North and Central regions; Korean peninsula, Japan, Russian Far East. [Flight Season] April to October.

黑背尾螅 *Paracercion melanotum* (Selys, 1876)

【形态特征】雄性身体主要蓝色具黑色条纹。雌性身体主要褐黄色具黑色条纹。本种雄性与挫齿尾螅腹部第2节黑斑的形状不同，与捷尾螅的差异在于腹部第3~7节背面的黑色条纹更长。【长度】体长 28~30 mm，腹长 21~25 mm，后翅 14~17 mm。【栖息环境】海拔 2500 m 以下水草茂盛的静水环境。【分布】全国广布；朝鲜半岛、日本、越南。【飞行期】2—11月。

黑背尾螅 雄，广东 ｜宋黎明 摄
Paracercion melanotum, male from Guangdong ｜ Photo by Liming Song

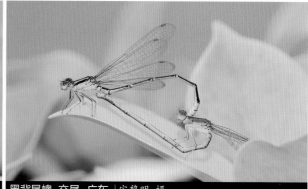

黑背尾螅 交尾，广东 ｜宋黎明 摄
Paracercion melanotum, mating pair from Guangdong ｜ Photo by Liming Song

[Identification] Male body mainly blue with black markings. Female body mainly brownish yellow with black markings. Male of the species is different from *P. barbatum* by the shape of the black spot on S2, differs from *P. v-nigrum* by the longer black stripes on dorsum of S3-S7. [Measurements] Total length 28-30 mm, abdomen 21-25 mm, hind wing 14-17 mm. [Habitat] Standing water with emergent plants below 2500 m elevation. [Distribution] Widespread throughout China; Korean peninsula, Japan, Vietnam. [Flight Season] February to November.

黑背尾蟌 雄,北京 | 陈炜 摄
Paracercion melanotum, male from Beijing | Photo by Wei Chen

黑背尾蟌 雄,云南(大理)
Paracercion melanotum, male from Yunnan (Dali)

七条尾螅 *Paracercion plagiosum* (Needham, 1930)

【形态特征】雄性身体蓝色具黑色条纹。雌性多型，身体黄色或淡蓝色具黑色条纹。【长度】体长 39～49 mm，腹长 29～37 mm，后翅 21～26 mm。【栖息环境】海拔 1000 m以下水草茂盛的静水环境。【分布】东北、华北地区广布；朝鲜半岛、日本、俄罗斯远东。【飞行期】5—9月。

[Identification] Male body blue with black markings. Female polymorphic, body yellow or pale blue with black markings. [Measurements] Total length 39-49 mm, abdomen 29-37 mm, hind wing 21-26 mm. [Habitat] Standing water with emergent plants below 1000 m elevation. [Distribution] Widespread in the Northeast and North regions; Korean peninsula, Japan, Russian Far East. [Flight Season] May to September.

七条尾螅 雄，北京 | 陈炜 摄
Paracercion plagiosum, male from Beijing | Photo by Wei Chen

七条尾螅 雌，北京 | 陈炜 摄
Paracercion plagiosum, female from Beijing | Photo by Wei Chen

七条尾螅 雌雄连结，北京 ｜陈炜 摄
Paracercion plagiosum, pair in tandem from Beijing ｜ Photo by Wei Chen

捷尾螈 *Paracercion v-nigrum* (Needham, 1930)

【形态特征】雄性身体蓝色具黑色条纹。雌性多型,身体黄色或淡蓝色具黑色条纹。【长度】体长 34~38 mm,腹长 27~30 mm,后翅 20~23 mm。【栖息环境】海拔 2500 m 以下水草茂盛的池塘、流速缓慢的溪流和水库。【分布】华北、华中、华南、西南地区广布;朝鲜半岛、俄罗斯远东、越南。【飞行期】5—10月。

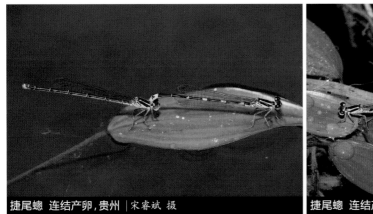

捷尾螈 连结产卵,贵州 | 宋睿斌 摄
Paracercion v-nigrum, laying eggs in tandem from Guizhou | Photo by Ruibin Song

捷尾螈 连结产卵,贵州 | 莫善濂 摄
Paracercion v-nigrum, laying eggs in tandem from Guizhou | Photo by Shanlian Mo

[Identification] Male body blue with black markings. Female polymorphic, body yellow or pale blue with black markings. [Measurements] Total length 34-38 mm, abdomen 27-30 mm, hind wing 20-23 mm. [Habitat] Ponds with emergent plants, slow flowing streams and reservoirs below 2500 m elevation. [Distribution] Widespread in the North, Central, South and Southwest regions; Korean peninsula, Russian Far East, Vietnam. [Flight Season] May to October.

捷尾螈 雄,贵州
Paracercion v-nigrum, male from Guizhou

尾螋属待定种1 *Paracercion* sp. 1

　　【形态特征】本种与钳尾螋相似但腹部的斑纹不同。【长度】雄性体长 36~38 mm，腹长 29~31 mm，后翅 20~21 mm。【栖息环境】海拔 1200 m左右的静水环境。【分布】贵州。【飞行期】6—8月。

[Identification] The species is similar to *P. dorothea* but differs in the color pattern of abdomen. [Measurements] Male total length 36-38 mm, abdomen 29-31 mm, hind wing 20-21 mm. [Habitat] Standing water at about 1200 m elevation. [Distribution] Guizhou. [Flight Season] June to August.

尾螋属待定种1 雄，贵州
Paracercion sp. 1, male from Guizhou

尾蟌属待定种2 *Paracercion* sp. 2

尾蟌属待定种2 雄,吉林 | 金洪光 摄
Paracercion sp. 2, male from Jilin | Photo by Hongguang Jin

尾蟌属待定种2 雌,湖北
Paracercion sp. 2, female from Hubei

尾蟌属待定种2 连结产卵,吉林 | 金洪光 摄
Paracercion sp. 2, laying eggs in tandem from Jilin | Photo by Hongguang Jin

【形态特征】本种与捷尾蟌相似,但雄性面部黄色,雌性具黄色肩条纹。【长度】体长 29～32 mm,腹长 23～25 mm,后翅 16～19 mm。【栖息环境】海拔 500 m以下的静水环境。【分布】吉林、湖北、安徽。【飞行期】4—7月。

[Identification] Similar to *P. v-nigrum* but male face yellow, female with yellow humeral stripes. [Measurements] Total length 29-32 mm, abdomen 23-25 mm, hind wing 16-19 mm. [Habitat] Standing water below 500 m elevation. [Distribution] Jilin, Hubei, Anhui. [Flight Season] April to July.

斑螅属 Genus *Pseudagrion* Selys, 1876

本属全世界已知约150种，分布于亚洲、非洲和大洋洲。中国已知7种，分布于华中、华南和西南地区。本属豆娘体型小至中型；翅的弓脉位于第2条结前横脉下方，结后横脉10～12条。

本属豆娘栖息于水草茂盛的静水环境和流速缓慢的溪流。雄性经常停落在挺水植物上。它们在水草上雌雄连结产卵。

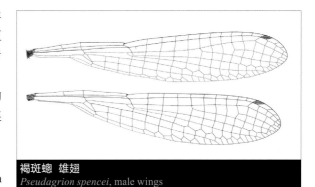

褐斑螅 雄翅
Pseudagrion spencei, male wings

The genus contains almost 150 species, distributed in Asia, Africa and Oceania. Seven species are known from China, found in the Central, South and Southwest regions. Species of the genus are small to medium sized damselflies. Wings with arculus just below the second antenodal, postnodals 10-12.

Pseudagrion species inhabit standing water with plenty of emergent plants and slow flowing streams. Males often perch on emergent plants. They oviposit in tandem in aquatic plants.

亚澳斑螅
Pseudagrion australasiae

绿斑螅
Pseudagrion microcephalum

斑螅属 雄性肛附器
Genus *Pseudagrion,* male anal appendages

赤斑螅 雄
Pseudagrion pruinosum, male

亚澳斑螈 *Pseudagrion australasiae* Selys, 1876

【形态特征】雄性身体主要蓝色具黑色斑纹。雌性身体主要黄色具黑色斑纹。【长度】雄性体长 41 mm，腹长 33 mm，后翅 22 mm。【栖息环境】海拔 500 m以下的池塘。【分布】海南、广东；南亚、东南亚。【飞行期】4—10月。

[Identification] Male body mainly blue with black markings. Female body mainly yellow with black markings. [Measurements] Male total length 41 mm, abdomen 33 mm, hind wing 22 mm. [Habitat] Ponds below 500 m elevation. [Distribution] Hainan, Guangdong; South and Southeast Asia. [Flight Season] April to October.

亚澳斑螈 雄，广东｜吴宏道 摄
Pseudagrion australasiae, male from Guangdong ｜ Photo by Hongdao Wu

亚澳斑螈 雌，广东｜吴宏道 摄
Pseudagrion australasiae, female from Guangdong ｜ Photo by Hongdao Wu

绿斑蟌 *Pseudagrion microcephalum* (Rambur, 1842)

【形态特征】本种与亚澳斑蟌相似，但本种上肛附器几乎与腹部第10节等长，而亚澳斑蟌上肛附器仅为第10节长度的1/2。此外两种雄性腹部第8~9节黑色条纹的宽度不同。【长度】体长 38~42 mm，腹长 27~29 mm，后翅17~20 mm。【栖息环境】海拔 1000 m以下的静水环境和流速缓慢的溪流。【分布】云南、福建、广东、香港、台湾；日本、南亚、东南亚、澳新界。【飞行期】3—12月。

绿斑蟌 交尾, 广东 | 宋黎明 摄
Pseudagrion microcephalum, mating pair from Guangdong | Photo by Liming Song

[Identification] The species is similar to *P. australasiae*, but male superior appendages almost as long as S10, only half length as S10 in male *P. australasiae*. The width of black stripes on S8 -S9 also different. [Measurements] Total length 38-42 mm, abdomen 27-29 mm, hind wing 17-20 mm. [Habitat] Standing water and slow flowing streams below 1000 m elevation. [Distribution] Yunnan, Fujian, Guangdong, Hong Kong, Taiwan; Japan, South and Southeast Asia, Australasia. [Flight Season] March to December.

绿斑蟌 雄, 云南 | 莫善濂 摄
Pseudagrion microcephalum, male from Yunnan | Photo by Shanlian Mo

红玉斑蟌 *Pseudagrion pilidorsum* (Brauer, 1868)

【形态特征】雄性身体主要红色；腹部第4~8节具黑斑。雌性黄褐色具黑色条纹。【长度】体长 42~47 mm。【栖息环境】海拔 1000 m 以下流速缓慢的溪流。【分布】中国台湾；日本、菲律宾、印度尼西亚。【飞行期】全年可见。

[Identification] Male body mainly red. S4-S8 with black markings. Female yellowish brown with black markings. [Measurements] Total length 42-47 mm. [Habitat] Slow flowing streams below 1000 m elevation. [Distribution] Taiwan of China; Japan, Philippines, Indonesia. [Flight Season] Throughout the year.

红玉斑蟌 雄，台湾｜嘎嘎 摄
Pseudagrion pilidorsum, male from Taiwan｜Photo by Gaga

红玉斑蟌 雌雄连结，台湾｜嘎嘎 摄
Pseudagrion pilidorsum, pair in tandem from Taiwan｜Photo by Gaga

赤斑蟌 *Pseudagrion pruinosum* (Burmeister, 1839)

【形态特征】雄性面部红褐色；身体主要黑色，胸部侧面和腹部第8~10节覆盖蓝灰色粉霜。雌性身体黄褐色具黑色条纹。【长度】腹长 34~37 mm，后翅 24~25 mm。【栖息环境】海拔 1500 m 以下流速缓慢的溪流。【分布】云南、贵州、福建、海南、广东、广西、香港；东南亚。【飞行期】全年可见。

赤斑蟌 雌，海南｜莫善濂 摄
Pseudagrion pruinosum, female from Hainan｜Photo by Shanlian Mo

赤斑蟌 交尾，海南
Pseudagrion pruinosum, mating pair from Hainan

赤斑螅 雄, 云南 (德宏)
Pseudagrion pruinosum, male from Yunnan (Dehong)

[Identification] Male face reddish brown. Body mainly black, sides of thorax and S8-S10 with bluish grey pruinescence. Female body yellowish brown with black markings. [Measurements] Abdomen 34-37 mm, hind wing 24-25 mm. [Habitat] Slow flowing streams below 1500 m elevation. [Distribution] Yunnan, Guizhou, Fujian, Hainan, Guangdong, Guangxi, Hong Kong; Southeast Asia. [Flight Season] Throughout the year.

丹顶斑螅 *Pseudagrion rubriceps* Selys, 1876

【形态特征】雄性面部大面积橙黄色；身体大面积蓝色具黑色斑纹。雌性身体黄褐色和黄绿色具黑色条纹。【长度】腹长 29~31 mm，后翅 18~21 mm。【栖息环境】海拔 1500 m 以下的池塘、水库和流速缓慢的溪流。【分布】云南、贵州、广西、广东、海南、香港；南亚、东南亚。【飞行期】全年可见。

[Identification] Male face largely orange yellow. Body largely blue with black markings. Female body yellowish brown and yellowish green with black stripes. [Measurements] Abdomen 29-31 mm,

丹顶斑螅 雄, 云南 (德宏)
Pseudagrion rubriceps, male from Yunnan (Dehong)

hind wing 18-21 mm. [Habitat] Ponds, reservoirs and slow flowing streams below 1500 m elevation. [Distribution] Yunnan, Guizhou, Guangxi, Guangdong, Hainan, Hong Kong; South and Southeast Asia. [Flight Season] Throughout the year.

丹顶斑螅 雄, 云南（德宏）
Pseudagrion rubriceps, male from Yunnan (Dehong)

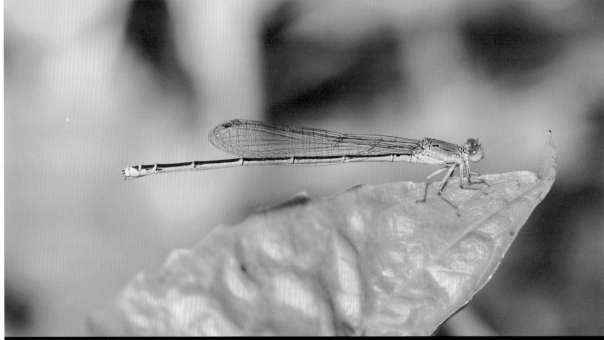

丹顶斑螅 雌, 云南（德宏）
Pseudagrion rubriceps, female from Yunnan (Dehong)

褐斑蟌 *Pseudagrion spencei* Fraser, 1922

褐斑蟌 雄,广东
Pseudagrion spencei, male from Guangdong

褐斑蟌 雌,广东
Pseudagrion spencei, female from Guangdong

褐斑蟌 雄,云南(德宏)
Pseudagrion spencei, male from Yunnan (Dehong)

【形态特征】雄性身体大面积蓝色具黑色条纹。雌性金褐色具黑色条纹。【长度】体长 30~32 mm,腹长 22~24 mm,后翅 15~16 mm。【栖息环境】海拔 1000 m以下的静水环境和流速缓慢的溪流。【分布】华中、华南、西南地区广布;南亚、越南。【飞行期】3—11月。

[Identification] Male body largely blue with black markings. Female orange brown with black stripes. [Measurements] Total length 30-32 mm, abdomen 22-24 mm, hind wing 15-16 mm. [Habitat] Standing water and slow flowing streams below 1000 m elevation. [Distribution] Widespread in the Central, South and Southwest regions; South Asia, Vietnam. [Flight Season] March to November.

13 扁蟌科 Family Platystictidae

除了非洲和大洋洲，扁蟌科豆娘在热带地区极为繁盛，全球已知10属约270种。中国已知4属19种。中国的扁蟌分布于华南和西南地区的热带和亚热带森林中，华南地区研究较多但西部地区研究较少，本书所包含的许多待定种多是近期在云南和广西地区所发现，多数可能是新种。本科是小至中型豆娘，体色较暗，腹部极为细长；翅窄而短，具较长的翅柄，翅末端卷曲。大多数种类翅透明。

本科豆娘栖息于茂盛森林中的林荫小溪和渗流地。它们经常停落在极其阴暗处的枝条和植物上。有些雄性具领域行为。雌性在植物茎干上产卵。

The family is rich in the tropics apart from Africa and Oceania, about 270 species in ten genera are known worldwide. 19 species in four genera are recorded from China. In China the family is confined to the tropical and subtropical forests in the South and Southwest. Species from the South are relatively well known but knowledge of the species from the Southwest is inadequate. Many species not yet identified to species level are included here, most of them are recently discovered in Yunnan and Guangxi, they are likely to be undescribed species. The platystictids are small to medium sized damselflies, body is usually dark with an extremely long and very slender abdomen. Wings are narrow and relatively short with a long stalk at base and falcate tip. Most species have hyaline wings.

Species of the family inhabit shady streams and seepages in dense forest. They usually perch on the branches or plants in the shade. Males exhibit territorial behavior. Females lay eggs into the stalks of plants.

巨镰扁螅 雄
Drepanosticta magna, male

周氏镰扁螅 雄
Drepanosticta zhoui, male

镰扁蟌属 Genus *Drepanosticta* Laidlaw, 1917

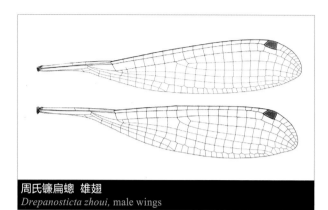

本属全球已知超过120种，主要分布于亚洲的热带地区。中国已知5种，主要分布于华南地区。本属豆娘拥有极长的腹部；翅透明，基臀区通常仅有1条横脉。

本属豆娘栖息于森林中的狭窄小溪和渗流地，喜欢潮湿阴暗的环境。雄性通常停落在小溪边缘较低处的斜坡或树枝上。雌性产卵于植物的茎干中。

周氏镰扁蟌 雄翅
Drepanosticta zhoui, male wings

The genus contains over 120 species, mainly distributed in tropical Asia. Five species are recorded from China, mainly in the South region. Species of the genus have a very long abdomen. Wings hyaline, the cubital space usually with only one crossvein.

Drepanosticta species inhabit narrow streams and seepages in forest, favouring damp and shady habitats. Males usually perch on the slopes or low vegetation at the edge of streams. Females lay eggs into the stalks of plants.

周氏镰扁蟌 雄
Drepanosticta zhoui, male

周氏镰扁螈 雌，产卵
Drepanosticta zhoui, female laying eggs

白尾镰扁螈 *Drepanosticta brownelli* (Tinkham, 1938)

白尾镰扁螈 雄，广东 | 莫善濂 摄
Drepanosticta brownelli, male from Guangdong | Photo by Shanlian Mo

【形态特征】雄性身体主要黑色；胸部侧面具1个白色斑；腹部第8~10节背面具白斑。雌性与雄性相似，腹部具更多白色斑纹。【长度】体长 41~45 mm，腹长 35~38 mm，后翅 23~26 mm。【栖息环境】海拔 1000 m以下森林中的渗流地、狭窄小溪和沟渠。【分布】中国特有，分布于广东、广西。【飞行期】4—7月。

白尾镰扁螅 雌，广东 ｜莫善濂 摄
Drepanosticta brownelli, female from Guangdong ｜ Photo by Shanlian Mo

[Identification] Male body mainly black. Thorax with a lateral white spot. S8-S10 with white spots dorsally. Female similar to male, abdomen with more white markings. [Measurements] Total length 41-45 mm, abdomen 35-38 mm, hind wing 23-26 mm. [Habitat] Seepages, narrow streams and ditches in forest below 1000 m elevation. [Distribution] Endemic to China, recorded from Guangdong, Guangxi. [Flight Season] April to July.

修长镰扁螅 *Drepanosticta elongata* Wilson & Reels, 2001

【形态特征】雄性面部蓝黑色具金属光泽，上唇和前唇基淡蓝色；前胸白色，合胸黑色，侧面具2条蓝白色条纹；腹部甚长，深褐色，第1~7节具白斑，第8~10节淡蓝色。雌性腹部较短，末端无淡蓝色斑。【长度】体长 48~62 mm，腹长 40~54 mm，后翅 25~28 mm。【栖息环境】海拔 1000 m以下森林中的渗流地、狭窄小溪和沟渠。【分布】中国海南特有。【飞行期】4—7月。

[Identification] Male face metallic bluish black, labrum and anteclypeus pale blue. Prothorax white, synthorax black with two bluish white stripes laterally. Abdomen very long, dark brown with white spots on S1-S7, S8-S10 pale blue. Female abdomen shorter without apical pale blue spots. [Measurements] Total length 48-62 mm, abdomen 40-54 mm, hind wing 25-28 mm. [Habitat] Seepages, narrow streams and ditches in forest below 1000 m elevation. [Distribution] Endemic to Hainan of China. [Flight Season] April to July.

修长镰扁螈 雄，海南
Drepanosticta elongata, male from Hainan

修长镰扁螈 雌，海南
Drepanosticta elongata, female from Hainan

香港镰扁蟌 *Drepanosticta hongkongensis* Wilson, 1996

香港镰扁蟌 雄,广东 | 宋睿斌 摄
Drepanosticta hongkongensis, male from Guangdong | Photo by Ruibin Song

香港镰扁蟌 雌,香港 | 祁麟峰 摄
Drepanosticta hongkongensis, female from Hong Kong | Photo by Mahler Ka

【形态特征】雄性与白尾镰扁蟌相似,但胸部侧面无白斑;腹部第8~10节具蓝斑。雌性则非常相似,较难区分。【长度】体长 33~45 mm,腹长 26~37 mm,后翅 20~25 mm。【栖息环境】海拔 1000 m以下森林中的渗流地、狭窄小溪和沟渠。【分布】福建、广东、广西、香港;越南。【飞行期】5—8月。

[Identification] Male similar to *D. brownelli*, but thorax without white spot laterally. Abdomen S8-S10 with blue spots. Female very similar to *D. brownelli* and difficult to be distinguished. [Measurements] Total length 33-45 mm, abdomen 26-37 mm, hind wing 20-25 mm. [Habitat] Seepages, narrow streams and ditches in forest below 1000 m elevation. [Distribution] Fujian, Guangdong, Guangxi, Hong Kong; Vietnam. [Flight Season] May to August.

巨镰扁蟌 *Drepanosticta magna* Wilson & Reels, 2003

【形态特征】雄性面部黑色，上唇白色；前胸背面黄色，合胸黑色，侧面具2条黄色条纹；腹部甚长，深褐色，第1~8节具黄白色条纹，第8~10节具蓝斑。雌性腹部较短，末端无蓝斑。【长度】体长 53~67 mm，腹长 45~58 mm，后翅 27~28 mm。【栖息环境】海拔 1000 m以下森林中的渗流地、狭窄小溪和沟渠。【分布】中国广西特有。【飞行期】5—7月。

巨镰扁蟌 雄，广西
Drepanosticta magna, male from Guangxi

[Identification] Male face black, labrum white. Prothorax yellow dorsally, synthorax black with two yellow stripes laterally. Abdomen very long, dark brown with yellowish white strips on S1-S8, S8-S10 with blue spots. Female abdomen shorter without apical blue spots. [Measurements] Total length 53-67 mm, abdomen 45-58 mm, hind wing 27-28 mm. [Habitat] Seepages, narrow streams and ditches in forest below 1000 m elevation. [Distribution] Endemic to Guangxi of China. [Flight Season] May to July.

巨镰扁蟌 雌，广西
Drepanosticta magna, female from Guangxi

周氏镰扁蟌 *Drepanosticta zhoui* Wilson & Reels, 2001

【形态特征】雄性面部蓝黑色具金属光泽，上唇白色；胸部黑色，侧面具2条蓝白色条纹，翅透明，有时端部具褐斑；腹部黑色，第1~7节具白斑，第8~10节淡蓝色。雌性腹部较短，具丰富的白斑。【长度】体长 35~46 mm，腹长 28~39 mm，后翅 19~24 mm。【栖息环境】海拔 1000 m以下森林中的渗流地、狭窄小溪和沟渠。【分布】中国海南特有。【飞行期】4—7月。

周氏镰扁蟌 雄，海南
Drepanosticta zhoui, male from Hainan

周氏镰扁蟌 雌，海南
Drepanosticta zhoui, female from Hainan

[Identification] Male face metallic bluish black, labrum white. Thorax black with two bluish white stripes laterally, wings hyaline, sometimes with brown tips. Abdomen black, S1-S7 with white spots, S8-S10 pale blue. Female abdomen shorter with abundant white spots. [Measurements] Total length 35-46 mm, abdomen 28-39 mm, hind wing 19-24 mm. [Habitat] Seepages, narrow streams and ditches in forest below 1000 m elevation. [Distribution] Endemic to Hainan of China. [Flight Season] April to July.

周氏镰扁蟌 交尾，海南
Drepanosticta zhoui, mating pair from Hainan

镰扁蟌属待定种1 *Drepanosticta* sp. 1

镰扁蟌属待定种1　雄，云南（红河）
Drepanosticta sp. 1, male from Yunnan (Honghe)

【形态特征】本种与香港镰扁蟌相似，但雄性第8节蓝色斑的形状不同，肛附器构造也稍有差异。【长度】雄性体长 45~47 mm，腹长 38~39 mm，后翅 26~27 mm。【栖息环境】海拔 500 m以下森林中的狭窄小溪。【分布】云南（红河）。【飞行期】4—7月。

[Identification] The species is similar to *D. hongkongensis*, but the shape of blue spot on male S8 different, anal appendages also slightly different. [Measurements] Male total length 45-47 mm, abdomen 38-39 mm, hind wing 26-27 mm. [Habitat] Narrow streams in forest below 500 m elevation. [Distribution] Yunnan (Honghe). [Flight Season] April to July.

镰扁蟌属待定种2 *Drepanosticta* sp. 2

【形态特征】雄性面部黑色；胸部黑色，侧面具2条黄色条纹；腹部主要黑褐色，第8～10节膨大并具淡蓝色斑。【长度】雄性体长 45～46 mm，腹长 38～39 mm，后翅 24 mm。【栖息环境】海拔 1000 m以下森林中的狭窄小溪和渗流地。【分布】云南（普洱）。【飞行期】5—7月。

[Identification] Male face black. Thorax black with two yellow stripes laterally. Abdomen mainly blackish brown, S8-S10 expanded with pale blue spots. [Measurements] Male total length 45-46 mm, abdomen 38-39 mm, hind wing 24 mm. [Habitat] Narrow streams and seepages in forest below 1000 m elevation. [Distribution] Yunnan (Pu'er). [Flight Season] May to July.

镰扁蟌属待定种2 雄，云南（普洱）
Drepanosticta sp. 2, male from Yunnan (Pu'er)

镰扁蟌属待定种3 *Drepanosticta* sp. 3

【形态特征】雄性面部黑色；胸部黑色，脊具1条黄色细条纹，肩条纹淡黄色，侧面具2条甚阔的黄色条纹；腹部大面积黑褐色具白斑，第8～10节膨大具淡蓝色斑，第8节腹面具1簇綮。【长度】雄性体长 47 mm，腹长 40 mm，后翅23 mm。【栖息环境】海拔 500 m以下森林中的狭窄小溪。【分布】云南（德宏）。【飞行期】4—6月。

[Identification] Male face black. Thorax black, dorsal carina with a slender yellow stripe, humeral stripes pale yellow, sides with two broad yellow stripes. Abdomen largely blackish brown with white spots, S8-S10 expanded with pale blue spots, underside of S8 with a tuft of setae. [Measurements] Male total length 47 mm, abdomen 40 mm, hind wing 23 mm. [Habitat] Narrow streams in forest below 500 m elevation. [Distribution] Yunnan (Dehong). [Flight Season] April to June.

镰扁蟌属待定种3 雄，云南（德宏）
Drepanosticta sp. 3, male from Yunnan (Dehong)

镰扁蟌属待定种4 *Drepanosticta* sp. 4

【形态特征】雄性面部黑色,上唇和前唇基白色;胸部黑色,侧面具2条黄色条纹;腹部黑色具白斑,第8~9节稍微膨大,第8节腹面具1簇鬃。雌性与雄性相似,但腹部较短。【长度】体长 38~43 mm,腹长 31~36 mm,后翅20~21 mm。【栖息环境】海拔 1000~1500 m森林中的狭窄小溪。【分布】云南(西双版纳、普洱)。【飞行期】5—7月。

镰扁蟌属待定种4 雌,云南(普洱)
Drepanosticta sp. 4, female from Yunnan (Pu'er)

镰扁蟌属待定种4 雄,云南(普洱)
Drepanosticta sp. 4, male from Yunnan (Pu'er)

[Identification] Male face black, labrum and anteclypeus white. Thorax black with two yellow stripes laterally. Abdomen black with white spots, S8-S9 slightly expanded, underside of S8 with a tuft of setae. Female similar to male with shorter abdomen. [Measurements] Total length 38-43 mm, abdomen 31-36 mm, hind wing 20-21 mm. [Habitat] Narrow streams in forest at 1000-1500 m elevation. [Distribution] Yunnan (Xishuangbanna, Pu'er). [Flight Season] May to July.

镰扁蟌属待定种4 雄,云南(普洱)
Drepanosticta sp. 4, male from Yunnan (Pu'er)

镰扁蟌属待定种5 *Drepanosticta* sp. 5

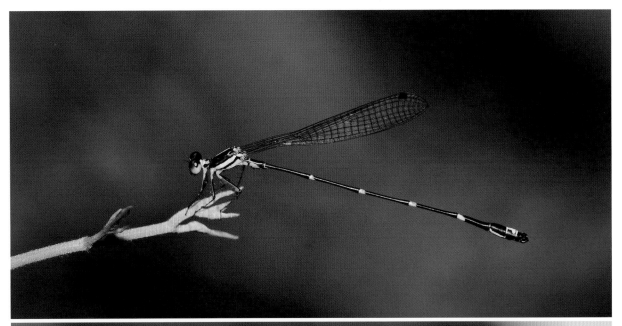

镰扁蟌属待定种5 雄，云南（德宏）
Drepanosticta sp. 5, male from Yunnan (Dehong)

　　【形态特征】雄性面部黑色；前胸黄色，合胸脊具1条黄色条纹，侧面具2条黄色条纹；腹部主要黑褐色，第8～10节膨大具淡蓝色斑。【长度】雄性体长 46 mm，腹长 39 mm，后翅 25 mm。【栖息环境】海拔 1000 m左右森林中的狭窄小溪和渗流地。【分布】云南（德宏）。【飞行期】5—7月。

　　[Identification] Male face black. Prothorax yellow, synthorax with a yellow stripe along dorsal carina, sides with two yellow stripes. Abdomen mainly blackish brown, S8-S10 expanded with pale blue spots. [Measurements] Male total length 46 mm, abdomen 39 mm, hind wing 25 mm. [Habitat] Narrow streams and seepages in forest at about 1000 m elevation. [Distribution] Yunnan (Dehong). [Flight Season] May to July.

原扁蟌属 Genus *Protosticta* Selys, 1885

本属全球已知约50种，分布于亚洲的热带和亚热带地区。中国已知8种，分布于华南和西南地区。本属是一类具有极为细长腹部的豆娘；翅透明，基臀区通常仅有1条横脉。本属很多种类十分相似，雄性可以依靠肛附器的构造区分。本属翅上无残留的"V"形臀脉，可以与镰扁蟌属区分。

本属豆娘栖息于森林中的狭窄小溪和渗流地。雄性通常停落在小溪边缘的阴暗斜坡和植物上。雌性产卵于植物的茎干中。

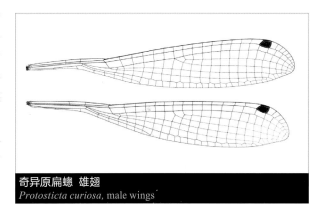

奇异原扁蟌 雄翅
Protosticta curiosa, male wings

The genus contains about 50 species, distributed in tropical and subtropical Asia. Eight species are recorded from China, found in the South and Southwest regions. Species of the genus have a very long and slender abdomen. Wings hyaline, usually only one crossvein in cubital space. Many species of the genus are similar in appearance, males can be separated by the shape of the anal appendages. The genus can be distinguished from *Drepanosticta* by the absence of the V-shaped vestigial anal vein.

Protosticta species inhabit narrow streams and seepages in forest. Males usually perch on the shady slopes and vegetation at the edge of streams. Females lay eggs into the stalks of plants.

卡罗原扁蟌 雄
Protosticta caroli, male

黄颈原扁蟌
Protosticta beaumonti

卡罗原扁蟌
Protosticta caroli

暗色原扁蟌
Protosticta grandis

泰国原扁蟌
Protosticta khaosoidaoensis

奇异原扁蟌
Protosticta curiosa

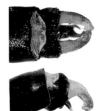

白瑞原扁蟌
Protosticta taipokauensis

原扁蟌属 雄性肛附器
Genus *Protosticta*, male anal appendages

黄颈原扁蟌 *Protosticta beaumonti* Wilson, 1997

黄颈原扁蟌 雄，广西
Protosticta beaumonti, male from Guangxi

黄颈原扁蟌 雄，广西
Protosticta beaumonti, male from Guangxi

黄颈原扁蟌 雌，香港｜祁麟峰 摄
Protosticta beaumonti, female from Hong Kong｜Photo by Mahler Ka

　　【形态特征】雄性面部黑色，上唇和前唇基白色；前胸淡黄色，合胸黑褐色，侧面具2条淡黄色条纹；腹部甚长，黑褐色具白斑。雌性与雄性相似，腹部较短。【长度】体长 38～50 mm，腹长 32～44 mm，后翅 20～22 mm。【栖息环境】海拔 1000 m 以下森林中的渗流地、狭窄小溪和沟渠。【分布】广西、广东、香港；越南。【飞行期】5—7月。

　　[Identification] Male face black, labrum and anteclypeus white. Prothorax pale yellow, synthorax blackish brown with two pale yellow laterally. Abdomen very long, blackish brown with white spots. Female similar to male with shorter abdomen. [Measurements] Total length 38-50 mm, abdomen 32-44 mm, hind wing 20-22 mm. [Habitat] Seepages, narrow streams and ditches in forest below 1000 m elevation. [Distribution] Guangxi, Guangdong, Hong Kong; Vietnam. [Flight Season] May to July.

卡罗原扁蟌 *Protosticta caroli* Van Tol, 2008

卡罗原扁蟌 雄，云南（红河）
Protosticta caroli, male from Yunnan (Honghe)

卡罗原扁蟌 雌，云南（西双版纳）
Protosticta caroli, female from Yunnan (Xishuangbanna)

【形态特征】本种与黄颈原扁蟌相似，但腹部白斑更少，雄性肛附器构造和雌性前胸构造不同。【长度】体长 37~47 mm，腹长 31~41 mm，后翅 19~22 mm。【栖息环境】海拔 1000 m 以下森林中的渗流地、狭窄小溪和沟渠。【分布】云南（西双版纳、红河）；柬埔寨、老挝、越南。【飞行期】4—6月。

[Identification] The species is similar to *P. beaumonti*, but abdomen with fewer white spots, male anal appendages and female prothorax different. [Measurements] Total length 37-47 mm, abdomen 31-41 mm, hind wing 19-22 mm. [Habitat] Seepages, narrow streams and ditches in forest below 1000 m elevation. [Distribution] Yunnan (Xishuangbanna, Honghe); Cambodia, Laos, Vietnam. [Flight Season] April to June.

暗色原扁蟌 *Protosticta grandis* Asahina, 1985

【形态特征】雄性面部黑色，上唇白色；胸部黑色，侧面具2条白色条纹，翅端稍染褐色；腹部黑色具白色斑纹，肛附器白色。雌性与雄性相似，腹部稍短。【长度】体长 53～60 mm，腹长 43～51 mm，后翅 31～32 mm。【栖息环境】海拔 1000 m 以下森林中的渗流地、狭窄小溪和沟渠。【分布】云南（红河）；泰国、柬埔寨、老挝、越南。【飞行期】4—7月。

暗色原扁蟌 雄，云南（红河）
Protosticta grandis, male from Yunnan (Honghe)

暗色原扁蟌 雌，云南（红河）
Protosticta grandis, female from Yunnan (Honghe)

[Identification] Male face black, labrum white. Thorax black with two white stripes laterally, wings with brown tips. Abdomen black with white spots, anal appendages white. Female similar to male with shorter abdomen. [Measurements] Total length 53-60 mm, abdomen 43-51 mm, hind wing 31-32 mm. [Habitat] Seepages, narrow streams and ditches in forest below 1000 m elevation. [Distribution] Yunnan (Honghe); Thailand, Cambodia, Laos, Vietnam. [Flight Season] April to July.

暗色原扁蟌 雄，云南（红河）
Protosticta grandis, male from Yunnan (Honghe)

泰国原扁蟌 *Protosticta khaosoidaoensis* Asahina, 1984

【形态特征】本种与黄颈原扁蟌相似，但雄性肛附器和雌性前胸构造不同。【长度】体长 39～52 mm，腹长 33～46 mm，后翅 19～21 mm。【栖息环境】海拔 1000 m 以下森林中的渗流地、狭窄小溪和沟渠。【分布】云南（红河）；泰国、老挝、越南。【飞行期】4—7月。

泰国原扁蟌 雄，云南（红河）
Protosticta khaosoidaoensis, male from Yunnan (Honghe)

[Identification] The species is similar to *P. beaumonti*, but male anal appendages and female prothorax are different. [Measurements] Total length 39-52 mm, abdomen 33-46 mm, hind wing 19-21 mm. [Habitat] Seepages, narrow streams and ditches in forest below 1000 m elevation. [Distribution] Yunnan (Honghe); Thailand, Laos, Vietnam. [Flight Season] April to July.

泰国原扁蟌 雌，云南（红河）
Protosticta khaosoidaoensis, female from Yunnan (Honghe)

黑胸原扁蟌 *Protosticta nigra* Kompier, 2016

【形态特征】雄性面部黑色，上唇和前唇基白色；前胸淡黄色，合胸黑褐色，后胸后侧板具1个较大的白斑；腹部甚长，黑褐色具白斑。雌性与雄性相似，腹部较短。【长度】体长 40～47 mm，腹长 34～41 mm，后翅 22～23 mm。【栖息环境】海拔 1000 m 以下森林中的渗流地、狭窄小溪和沟渠。【分布】广西；越南。【飞行期】5—7月。

[Identification] Male face black, labrum and anteclypeus white. Prothorax pale yellow, synthorax blackish brown with a white spot on metepimeron. Abdomen very long, blackish brown with white spots. Female similar to male with shorter abdomen. [Measurements] Total length 40-47 mm, abdomen 34-41 mm, hind wing 22-23 mm. [Habitat] Seepages, narrow streams and ditches in forest below 1000 m elevation. [Distribution] Guangxi; Vietnam. [Flight Season] May to July.

黑胸原扁蟌 雄，广西
Protosticta nigra, male from Guangxi

黑胸原扁蟌 雌，广西
Protosticta nigra, female from Guangxi

白瑞原扁蟌 *Protosticta taipokauensis* Asahina & Dudgeon, 1987

【形态特征】本种与暗色原扁蟌相似，但翅端无褐斑，雄性肛附器构造明显不同。【长度】体长 49～58 mm，腹长 41～50 mm，后翅 26～32 mm。【栖息环境】海拔 1000 m以下森林中的渗流地、狭窄小溪和沟渠。【分布】福建、广西、广东、香港；老挝。【飞行期】4—8月。

[Identification] The species is similar to *P. grandis*, but wings without brown tips, male anal appendages clearly different. [Measurements] Total length 49-58 mm, abdomen 41-50 mm, hind wing 26-32 mm. [Habitat] Seepages, narrow streams and ditches in forest below 1000 m elevation. [Distribution] Fujian, Guangxi, Guangdong, Hong Kong; Laos. [Flight Season] April to August.

白瑞原扁蟌 雌，广东 | 宋黎明 摄
Protosticta taipokauensis, female from Guangdong | Photo by Liming Song

白瑞原扁蟌 交尾，广东 | 宋黎明 摄
Protosticta taipokauensis, mating pair from Guangdong | Photo by Liming Song

白瑞原扁蟌 雄，广东 | 宋睿斌 摄
Protosticta taipokauensis, male from Guangdong | Photo by Ruibin Song

奇异原扁蟌 *Protosticta curiosa* Fraser, 1934

奇异原扁蟌 雄，云南（西双版纳）
Protosticta curiosa, male from Yunnan (Xishuangbanna)

奇异原扁蟌 雌，云南（西双版纳）
Protosticta curiosa, female from Yunnan (Xishuangbanna)

【形态特征】本种与黄颈原扁蟌和泰国原扁蟌相似，但前胸背面具显著的黑斑，雄性肛附器和雌性前胸构造不同。【长度】体长 38~51 mm，腹长 31~44 mm，后翅 19~21 mm。【栖息环境】海拔 1000 m 以下森林中的渗流地、狭窄小溪和沟渠。【分布】中国云南（西双版纳、临沧）特有。【飞行期】4—6月。

[Identification] The species is similar to *P. beaumonti* and *P. khaosoidaoensis*, but prothorax with clear black spots, different in male anal appendages and female prothorax. [Measurements] Total length 38-51 mm, abdomen 31-44 mm, hind wing 19-21 mm. [Habitat] Seepages, narrow streams and ditches in forest below 1000 m elevation. [Distribution] Endemic to Yunnan (Xishuangbanna, Lincang) of China. [Flight Season] April to June.

原扁蟌属待定种1 *Protosticta* sp. 1

【形态特征】本种与暗色原扁蟌和白瑞原扁蟌相似，但胸部侧面仅有1条黄白色条纹，雄性肛附器构造不同。【长度】体长 51~60 mm，腹长 43~51 mm，后翅 29~32 mm。【栖息环境】海拔 1000~1500 m茂密森林中的狭窄小溪。【分布】云南（西双版纳、普洱）。【飞行期】4—6月。

[Identification] The species is similar to *P. grandis* and *P. taipokauensis*, but thorax with only one lateral yellowish white stripe, male anal appendages different. [Measurements] Total length 51-60 mm, abdomen 43-51 mm, hind wing 29-32 mm. [Habitat] Narrow streams in forest at 1000~1500 m elevation. [Distribution] Yunnan (Xishuangbanna, Pu'er). [Flight Season] April to June.

原扁蟌属待定种1 雄，云南（普洱）
Protosticta sp. 1, male from Yunnan (Pu'er)

原扁蟌属待定种1 雌，云南（普洱）
Protosticta sp. 1, female from Yunnan (Pu'er)

原扁蟌属待定种2 *Protosticta* sp. 2

　　【形态特征】本种与暗色原扁蟌和白瑞原扁蟌相似，但胸部具褐色肩前条纹，雄性肛附器构造不同。【长度】体长 47~58 mm，腹长 39~50 mm，后翅 27~30 mm。【栖息环境】海拔 1000 m以下茂密森林中的狭窄小溪和渗流地。【分布】云南（西双版纳）。【飞行期】4—6月。

[Identification] The species is similar to *P. grandis* and *P. taipokauensis*, but thorax with brown antehumeral stripes, male anal appendages different. [Measurements] Total length 47-58 mm, abdomen 39-50 mm, hind wing 27-30 mm. [Habitat] Narrow streams and seepages in forest below 1000 m elevation. [Distribution] Yunnan (Xishuangbanna). [Flight Season] April to June.

原扁蟌属待定种2 雄，云南（西双版纳）
Protosticta sp. 2, male from Yunnan (Xishuangbanna)

原扁螅属待定种3 *Protosticta* sp. 3

【形态特征】本种与黄颈原扁螅和泰国原扁螅相似，但前胸背面具显著的黑斑，合胸侧面的黄色条纹中央间断。【长度】雄性体长 48 mm，腹长 42 mm，后翅 21 mm。【栖息环境】海拔 800 m处森林中的渗流地。【分布】云南（红河）。【飞行期】5—6月。

原扁螅属待定种3 雄，云南（红河）
Protosticta sp. 3, male from Yunnan (Honghe)

[Identification] The species is similar to *P. beaumonti* and *P. khaosoidaoensis*, but prothorax with clear black spots, sides of synthorax with yellow stripes interrupted at mid point. [Measurements] Male total length 48 mm, abdomen 42 mm, hind wing 21 mm. [Habitat] Seepages in forest at 800 m elevation. [Distribution] Yunnan (Honghe). [Flight Season] May to June.

原扁螅属待定种4 *Protosticta* sp. 4

【形态特征】本种与黄颈原扁螅和泰国原扁螅相似，但身体色彩较淡，合胸脊具1条白色细条纹，有时具较短的肩条纹。【长度】体长 37~50 mm，腹长 32~44 mm，后翅 19~20 mm。【栖息环境】海拔 500 m以下森林中的渗流地和具有渗流的石壁。【分布】云南（德宏）。【飞行期】5—7月。

[Identification] The species is similar to *P. beaumonti* and *P. khaosoidaoensis*, but body color paler, thorax with a slender white stripe along dorsal carina, sometimes short antehumeral stripes present. [Measurements] Total length 37-50 mm, abdomen 32-44 mm, hind wing 19-20 mm. [Habitat] Seepages and precipices with trickles in forest below 500 m elevation. [Distribution] Yunnan (Dehong). [Flight Season] May to July.

原扁蟌属待定种4 雄, 云南 (德宏)
Protosticta sp. 4, male from Yunnan (Dehong)

原扁蟌属待定种4 雌, 云南 (德宏)
Protosticta sp. 4, female from Yunnan (Dehong)

原扁蟌属待定种5 *Protosticta* sp. 5

【形态特征】本种与黑胸原扁蟌相似，但体型稍大，下唇色彩较深，为黑褐色，前胸背面黑色，雌性前胸构造不同。【长度】体长 41～48 mm，腹长 35～42 mm，后翅 23～24 mm。【栖息环境】海拔 500～1000 m森林中的渗流地。【分布】广西。【飞行期】5—7月。

原扁蟌属待定种5 雄，广西
Protosticta sp. 5, male from Guangxi

[Identification] The species is similar to *P. nigra*, but size slightly larger, labium darker, blackish brown coloring, prothorax black dorsally, female prothorax structure different. [Measurements] Total length 41-48 mm, abdomen 35-42 mm, hind wing 23-24 mm. [Habitat] Seepages in forest at 500-1000 m elevation. [Distribution] Guangxi. [Flight Season] May to July.

原扁蟌属待定种5 雌，广西
Protosticta sp. 5, female from Guangxi

原扁蟌属待定种6 *Protosticta* sp. 6

原扁蟌属待定种6 雄，云南（西双版纳）
Protosticta sp. 6, male from Yunnan (Xishuangbanna)

【形态特征】本种雄性肛附器的形状与越南分布的林奈原扁蟌相似，但本种体型稍大，腹部第10节具1对小白斑。【长度】雄性体长 50～52 mm，腹长 44～46 mm，后翅 22～23 mm。【栖息环境】海拔 1000 m处森林中的狭窄小溪。【分布】云南（西双版纳）。【飞行期】5—7月。

[Identification] Male anal appendages of this species are similar to *P. linnaei* Van Tol, 2008 known from Vietnam, but size slightly larger, S10 with a pair of white spots. [Measurements] Male total length 50-52 mm, abdomen 44-46 mm, hind wing 22-23 mm. [Habitat] Narrow streams in forest at 1000 m elevation. [Distribution] Yunnan (Xishuangbanna). [Flight Season] May to July.

华扁蟌属 Genus *Sinosticta* Wilson, 1997

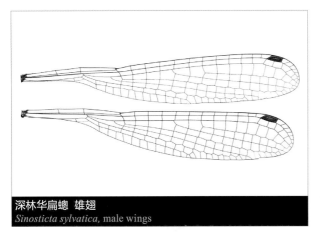

深林华扁蟌 雄翅
Sinosticta sylvatica, male wings

本属全球已知4种，分布于中国和越南。它们相比镰扁蟌和原扁蟌属略显粗壮，身体色彩更鲜艳。雄性肛附器的构造相对简单。

本属豆娘栖息于森林中的狭窄小溪，喜欢阴暗潮湿的环境。雌雄都会在小溪边缘的草丛中活动，通常停落在叶片上。

The genus contains four species, distributed in China and Vietnam. Compared with species of genera *Drepanosticta* and *Protosticta* they are stouter, body color is brighter. The male anal appendages are relatively simple.

Sinosticta species inhabit narrow streams in forest, preferring damp and dark places. Both sexes can be seen in grass near streams, usually they perch on the leaves.

深林华扁蟌 雄
Sinosticta sylvatica, male

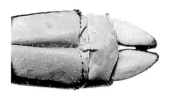

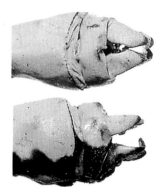

戴波华扁蟌
Sinosticta debra

深林华扁蟌
Sinosticta sylvatica

华扁蟌属 雄性肛附器
Genus *Sinosticta*, male anal appendages

戴波华扁蟌 *Sinosticta debra* Wilson & Xu, 2007

【形态特征】雄性面部黑色，上唇和前唇基黄色；胸部背面黑色，具细长肩前条纹，侧面大面积黄色；腹部黑色，第1~6节具黄斑，第9~10节及上肛附器淡蓝色。【长度】雄性体长 46 mm，腹长 37 mm，后翅 30 mm。【栖息环境】海拔 1000 m 以下森林中的狭窄小溪和沟渠。【分布】广东；越南。【飞行期】5—7月。

[Identification] Male face black, labrum and anteclypeus yellow. Thorax black dorsally with slender antehumeral stripes, sides largely yellow. Abdomen black, S1-S6 with yellow spots, S9-S10 and superior appendages pale blue. [Measurements] Male total length 46 mm, abdomen 37 mm, hind wing 30 mm. [Habitat] Narrow streams and ditches in forest below 1000 m elevation. [Distribution] Guangdong; Vietnam. [Flight Season] May to July.

戴波华扁蟌 雄，广东
Sinosticta debra, male from Guangdong

海南华扁蟌 *Sinosticta hainanense* Wilson & Reels, 2001

海南华扁蟌 雄，海南 | Graham Reels 摄
Sinosticta hainanense, male from Hainan | Photo by Graham Reels

【形态特征】雄性面部黑色，上唇和前唇基淡黄色；胸部黑色，前胸具黄色斑点，合胸具甚细的黄色肩前条纹，侧面具2条黄色条纹；腹部黑褐色，第1~7节侧面具黄斑，第8~10节蓝色，肛附器黑色。雌性与雄性相似，腹部黑色具黄斑。【长度】腹长 38~48 mm，后翅 29~34 mm。【栖息环境】海拔 1000 m以下森林中的渗流地、狭窄小溪和沟渠。【分布】中国海南特有。【飞行期】4—7月。

[Identification] Male face black, labrum and anteclypeus pale yellow. Thorax black, prothorax with yellow spots, synthorax with slender yellow antehumeral stripes, sides with two yellow stripes. Abdomen blackish brown, S1-S7 with yellow spots laterally, S8-S10 blue, anal appendages black. Female similar to male, abdomen black with yellow spots.

海南华扁蟌 雌, 海南 | Graham Reels 摄
Sinosticta hainanense, female from Hainan | Photo by Graham Reels

[Measurements] Abdomen 38-48 mm, hind wing 29-34 mm. [Habitat] Seepages, narrow streams and ditches in forest below 1000 m elevation. [Distribution] Endemic to Hainan of China. [Flight Season] April to July.

绪方华扁蟌 *Sinosticta ogatai* (Matsuki & Saito, 1996)

绪方华扁蟌 雄, 香港 | 梁嘉景 摄
Sinosticta ogatai, male from Hong Kong | Photo by Kenneth Leung

【形态特征】雄性面部黑色，上唇淡黄色；胸部黑色，前胸具蓝斑，合胸具淡黄色肩前条纹，侧面具1个淡黄色斑；腹部黑褐色，第3~7节侧面具白斑，第8~10节及上肛附器淡蓝色。雌性与雄性相似。【长度】体长 45~49 mm，腹长 37~40 mm，后翅 26~29 mm。【栖息环境】海拔 1000 m以下森林中的渗流地、狭窄小溪和沟渠。【分布】中国特有，分布于广东、香港。【飞行期】4—6月。

[Identification] Male face black, labrum pale yellow. Thorax black, prothorax with blue spots, synthorax with pale yellow antehumeral stripes, sides with a pale yellow spot. Abdomen blackish brown, S3-S7 with white spots laterally, S8-S10 and superior appendages pale blue. Female similar to male. [Measurements] Total length 45-49 mm, abdomen 37-40 mm, hind wing 26-29 mm. [Habitat] Seepages, narrow streams and ditches in forest below 1000 m elevation. [Distribution] Endemic to China, recorded from Guangdong, Hong Kong. [Flight Season] April to June.

绪方华扁螅 雄，香港｜祁麟峰 摄
Sinosticta ogatai, male from Hong Kong｜Photo by Mahler Ka

绪方华扁螈 雌，香港 | 祁麟峰 摄
Sinosticta ogatai, female from Hong Kong | Photo by Mahler Ka

深林华扁螅 *Sinosticta sylvatica* **Yu & Bu, 2009**

【形态特征】本种与戴波华扁螅相似，但胸部的黄色条纹更发达，雄性肛附器的构造不同。【长度】体长 42~48 mm，腹长 34~38 mm，后翅 21~30 mm。【栖息环境】海拔 1000 m以下森林中的渗流地、狭窄小溪和沟 渠。【分布】中国海南特有。【飞行期】4—7月。

深林华扁螅 雄，海南
Sinosticta sylvatica, male from Hainan

深林华扁螅 雌，海南
Sinosticta sylvatica, female from Hainan

深林华扁螅 交尾, 海南 | 莫善濂 摄
Sinosticta sylvatica, mating pair from Hainan | Photo by Shanlian Mo

深林华扁螅 雌, 海南 | 莫善濂 摄
Sinosticta sylvatica, female from Hainan | Photo by Shanlian Mo

[Identification] This species is similar to *S. debra* but the thoracic yellow stripes more developed, male anal appendages different. [Measurements] Total length 42-48 mm, abdomen 34-38 mm, hind wing 21-30 mm. [Habitat] Seepages, narrow streams and ditches in forest below 1000 m elevation. [Distribution] Endemic to Hainan of China. [Flight Season] April to July.

华扁螅属待定种 *Sinosticta* sp.

华扁螅属待定种 雌，云南（红河）｜莫善濂 摄
Sinosticta sp., female from Yunnan (Honghe)｜Photo by Shanlian Mo

【形态特征】雌性面部黑褐色；胸部黑色，具黄色肩前条纹，侧面具2条黄色条纹；腹部黑色，第1~7节侧面具黄色条纹。【长度】雌性体长 44 mm，腹长 35 mm，后翅 29 mm。【栖息环境】海拔 500 m左右森林中的狭窄小溪。【分布】云南（红河）。【飞行期】5—6月。

[Identification] Female face dark brown. Thorax black with yellow antehumeral stripes, sides with two yellow stripes. Abdomen black, S1-S7 with yellow stripes laterally. [Measurements] Female total length 44 mm, abdomen 35 mm, hind wing 29 mm. [Habitat] Narrow streams in forest at about 500 m elevation. [Distribution] Yunnan (Honghe). [Flight Season] May to June.

云扁螅属 Genus *Yunnanosticta* Dow & Zhang, 2018

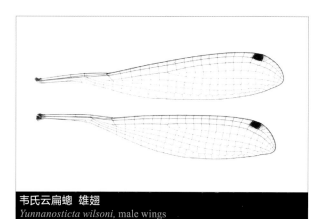

韦氏云扁螅 雄翅
Yunnanosticta wilsoni, male wings

本属目前已知2种，仅发现于云南西部。本属是一类体型细小的豆娘，合胸大面积黑褐色，缺乏浅色斑纹。翅透明，翅端略钩曲；肛附器较粗壮。

本属豆娘栖息于森林中阴暗的渗流地。雄性通常停落在具有渗流的斜坡或水边的植物上。

The genus contains two species and currently only found in the west of Yunnan. Species of the genus are small and slender damselflies, synthorax largely dark brown and lacking pale markings. Wings hyaline with tips slightly falcated. Anal appendages stout.

Yunnanosticta species inhabit shady seepages in forest. Males usually perch on the slopes with seepages or plants near water.

韦氏云扁螅 雄
Yunnanosticta wilsoni, male

蓝颈云扁螅 *Yunnanosticta cyaneocollaris* Dow & Zhang, 2018

【形态特征】雄性面部黑色，上唇和前唇基白色；合胸黑色，前胸背面天蓝色；腹部黑色，第8～9节背面具蓝白色斑。【长度】雄性腹长 25～26 mm，后翅 16～17 mm。【栖息环境】海拔 700～1300 m森林中阴暗的狭窄溪流和渗流地。【分布】中国云南（德宏）特有。【飞行期】5—7月。

[Identification] Male face black, labrum and anteclypeus white. Synthorax black, prothorax sky-blue dorsally; Abdomen black, S8-S9 with bluish white spots. [Measurements] Male abdomen 25-26 mm, hind wing 16-17 mm. [Habitat] Shady narrow streams and seepages in forest at 700-1300 m elevation. [Distribution] Endemic to Yunnan (Dehong) of China. [Flight Season] May to July.

蓝颈云扁螅 雄，云南（德宏）
Yunnanosticta cyaneocollaris, male from Yunnan (Dehong)

韦氏云扁蟌 *Yunnanosticta wilsoni* Dow & Zhang, 2018

【形态特征】本种与蓝颈云扁蟌相似，但前胸主要黑色，有时具蓝斑，合胸腹面黄色，雄性肛附器构造不同。**【长度】**雄性腹长 24～29 mm，后翅 17～20 mm。**【栖息环境】**海拔 700～1300 m森林中阴暗的狭窄溪流和渗流地。**【分布】**中国云南（德宏）特有。**【飞行期】**5—7月。

[Identification] The species is similar to *Y. cyaneocollaris*, but prothorax mainly black, sometimes with blue spots, synthorax yellow ventrally, male anal appendages different. [Measurements] Male abdomen 24-29 mm, hind wing 17-20 mm. [Habitat] Shady narrow streams and seepages in forest at 700-1300 m elevation. [Distribution] Endemic to Yunnan (Dehong) of China. [Flight Season] May to July.

韦氏云扁蟌 雄，云南（德宏）
Yunnanosticta wilsoni, male from Yunnan (Dehong)

韦氏云扁蟌 雄，云南（德宏）
Yunnanosticta wilsoni, male from Yunnan (Dehong)

云扁螅属待定种 *Yunnanosticta* sp.

【形态特征】雌性面部黑色，上唇和前唇基白色；胸部黑色，前胸背面具1对甚小的蓝斑，合胸侧面具1条黄色条纹；腹部黑色，第9节背面具1对淡蓝色斑。目前云扁螅属2个已知种的雌性未知，但本种雌性体型、翅的形状与本属特征较吻合。【长度】雌性体长 35 mm，腹长 28 mm，后翅 22 mm。【栖息环境】海拔 1500 m森林中的狭窄小溪。【分布】云南（德宏）。【飞行期】5—6月。

[Identification] Female face black, labrum and anteclypeus white. Thorax black, prothorax with a pair of small blue spots dorsally, sides of synthorax with a yellow stripe. Abdomen black, S9 with a pair of pale blue spots. Females of the two named *Yunnanosticta* species are still unknown, but the size and wing shape of this female match fairly well the character of the genus. [Measurements] Female total length 35 mm, abdomen 28 mm, hind wing 22 mm. [Habitat] Narrow streams in forest at 1500 m elevation. [Distribution] Yunnan (Dehong). [Flight Season] May to June.

云扁螅属待定种 雌，云南（德宏）
Yunnanosticta sp., female from Yunnan (Dehong)

褐狭扇蟌
Copera vittata

黄狭扇蟌（左）
Copera marginipes (left)

丹顶斑蟌（右）
Pseudagrion rubriceps (right)

褐斑蟌
Pseudagrion spencei

褐斑异痣蟌 | 莫善濂 摄
Ischnura senegalensis | Photo by Shanlian Mo

中国蜻蜓稚虫

DRAGONFLY LARVAE OF CHINA

图鉴系列......

中国昆虫生态大图鉴	张巍巍	李元胜	
中国蜘蛛生态大图鉴	张志升	王露雨	
中国鸟类生态大图鉴	郭冬生	张正旺	
中国蜻蜓大图鉴	张浩淼		
青藏高原野花大图鉴	牛洋	王辰	彭建生
中国蝴蝶生活史图鉴	朱建青	谷宇	陈志兵
	陈嘉霖		

常见园林植物识别图鉴（第2版）	吴棣飞	尤志勉
药用植物生态图鉴	赵素云	
凝固的时空	张巍巍	
琥珀中的昆虫及其他无脊椎动物		
常见兰花400种识别图鉴	吴棣飞	
中国湿地植物图鉴	王辰	王英伟

自然观察手册系列......

云与大气现象	张超	王燕平	王辰
天体与天象	朱江		
中国常见古生物化石	唐永刚	邢立达	
矿物与宝石	朱江		
岩石与地貌	朱江		

好奇心单本......

昆虫之美1精灵物语			李元胜
昆虫之美2雨林秘境			李元胜
与万物同行			李元胜
昆虫家谱			张巍巍
蜜蜂邮花	王荫长	张巍巍	缪晓青

野外识别手册系列......

常见昆虫野外识别手册	张巍巍	
常见鸟类野外识别手册	郭冬生	
常见植物野外识别手册	刘全儒	王辰
常见蝴蝶野外识别手册	黄灏	张巍巍
常见蘑菇野外识别手册	肖波	范宇光
常见蜘蛛野外识别手册	张志升	
常见南方野花识别手册	江珊	
常见天牛野外识别手册	林美英	
常见蜗牛野外识别手册	吴岷	
常见海滨动物野外识别手册	刘文亮	严莹
常见爬行动物野外识别手册	齐硕	

凝固的时空

琥珀中的昆虫及其他无脊椎动物

FROZEN DIMENSIONS

一部疯狂 缜密 伟大的工具书

人类一直在透过琥珀看远古，但这是看得最清楚的一次

著　　者：张巍巍
定　　价：498.00 元
出版单位：重庆大学出版社

- 本书精选了产自缅甸、波罗的海和多米尼加的虫珀 800 件，向广大读者全面系统地介绍了琥珀中出现的无脊椎动物 6 门 12 纲 67 目的 600 余种，并简要介绍了其他琥珀内含物（脊椎动物、植物、菌类等）的基本情况和世界各国的主要琥珀产地。

- 全书照片多达 2 000 余幅，是关于虫珀收藏和研究的重要文献资料。